Laser-Tissue Interactions

Markolf H. Niemz

Laser-Tissue Interactions

Fundamentals and Applications

Fourth Edition

 Springer

Markolf H. Niemz
MABEL Institute
Heidelberg University
Mannheim, Baden-Württemberg, Germany

ISBN 978-3-030-11919-5   ISBN 978-3-030-11917-1   (eBook)
https://doi.org/10.1007/978-3-030-11917-1

This Springer imprint is published by the registered company Springer Nature Switzerland AG
The registered company address is: Gewerbestrasse 11, 6330 Cham, Switzerland

Foreword to the First Edition

Dr. Markolf Niemz has undertaken the formidable task of writing a monograph on virtually all aspects of the current use of lasers in medicine, using laser–tissue interaction mechanisms as a guide throughout this book. The professional background of the author is in physics, in bioengineering, and in biomedical optics. In 1995, he was awarded the Karl–Freudenberg Prize by the Heidelberg Academy of Sciences, Germany, for his basic studies on laser–tissue interactions. Such a background is excellently suited to achieving the goals of this book, which are to offer an interdisciplinary approach to the basics of laser–tissue interactions and to use this knowledge for a review of clinical laser applications including laser safety.

His own research applying ultrashort laser pulses has enabled the author to provide profound discussions on photoablation, plasma-induced ablation, and photodisruption. Several aspects of related effects were first described by himself. Moreover, photodynamic therapy, photothermal applications, and laser-induced interstitial thermotherapy are extensively addressed in this book. The reader thus obtains a comprehensive survey of the present state of the art.

This book is intended mainly for scientists and engineers in this field, but medical staff will also find many important aspects of interest. There is no doubt that this book will fulfil a need for all of us working in the field of lasers in medicine, and I expect that it will be received very well.

Academic Medical Center   Martin J. C. van Gemert
Amsterdam, 1996   Director of the Laser Center

Dedicated to my wife Alexandra

Preface

Welcome to a fascinating field of medical research! Welcome to this fourth edition of "Laser–Tissue Interactions" which has meanwhile turned into the standard reference book at universities all around the world.

There are two basic kinds of being in this universe: light and matter. Light always travels at the speed of light and seems to be everywhere around us. I sometimes wonder what light really is, but nature won't tell us. Matter is the substance we are all made of. Within our bodies this substance is either hard tissue or soft tissue or fluid or gas. But light and matter aren't two different worlds. Light is permanently acting on matter, and matter is permanently acting on light. So we always have a light–matter **inter**action. The very same holds true for laser light and the various tissues of our bodies.

It really is amazing how much progress has been made in medical laser applications during the past decade. That's why Springer asked me to prepare a fourth edition of this book with all the new insights and techniques being added. Only a few things had to be modified in Chapters 1 and 2, since the basic mechanisms described here won't change. Light will always move at the speed of light. Some major modifications have been made in Chapter 3:

- the central map of laser–tissue interactions (Fig. 3.1) is improved,
- three basic questions on how to select a suitable laser are added,
- photochemical interactions are now classified into three groups,
- the new term "photobiomodulation" is introduced,
- the effects of wound healing and pain relief are explained,
- a stunning photograph (Fig. 3.13a) is added which very clearly illustrates the various effects of photothermal interaction.

And, of course, Chapter 4 has been updated with all the newest medical laser applications: femto-LASIK, smile-LASIK, photo-activated disinfection of teeth, lasers in endodontics and implantology, various holmium laser applications in urology, transmyocardial laser revascularization, and cosmetic laser surgery—just to mention a few of them. All of these very exciting techniques are discussed in great detail and always with the newest references from the best journals available.

Yet there is one more highlight brought to you by this fourth edition, and again it's got to do with light and matter: **color** has been added to many illustrations. Colors are a powerful tool. They help us better understand how things relate to each other. It's variety that makes colors and life so wonderful. In my new spiritual book "How science can help us live in peace – Darwin, Einstein, Whitehead" (Universal Publishers, Irvine, 2018) I put it this way: "Variety gives reality warmth and charm. How unbearably monotonous the cosmos would be without it!"

Heidelberg University,
November 2018   MARKOLF H. NIEMZ

Preface to the Third Edition

Do you like the idea of scrabble? Well, let's just give it a try:

```
                                    I
     T                     A C T I O N S
     I                             I     T
L  A  S  E  R       or             S     E
     S                     L  A  S  E  R
     U                             U
     E                             E
```

I like playing around with words and letters. You probably know that LASER is an artificial word derived from "**L**ight **A**mplification by **S**timulated **E**mission of **R**adiation". When starting my lecture on "Laser–Tissue Interactions" I tend to write this derivation on the board. I continue with "LIGHT: **L**asers **I**rradiate **G**erminated and **H**ealthy **T**issues".

Why? Lasers cut everything, if appropriate laser parameters are selected. There is no shield around healthy tissue. And there is no laser that fits all sizes as some clothes do. Lasers never have been some kind of wonder instruments. A wrong selection of laser parameters easily induces more damage than cure.

Congratulations! You are just reading the third edition of the textbook "Laser–Tissue Interactions". Its main improvement is that a total of 40 comprehensive questions and solutions have been added to Chaps. 2 through 5. With these questions you can immediately test your acquired knowledge or prepare yourself for related exams.

Compared to the second edition, minor changes have been made throughout the book and a few figures have been modified. The new soft cover design helps to reduce costs. Thus, the third edition is now affordable by students looking for a textbook to lighten up their lectures. Let there be light—laser light. Or what about: "LIGHT: **L**ove **I**s **G**od's **H**int to **T**rust" ...

Heidelberg University,
September 2007

MARKOLF H. NIEMZ

Preface to the Second Edition

Since the publication of the first edition of this book six years ago both research and applications in laser medicine have undergone substantial growth. The demand for novel techniques based on minimally invasive surgery has increased tremendously, and there is no end to it yet. Therefore, as the first edition ran out of stock, the publisher has asked me to prepare a second edition taking all these new developments into account.

Well, here it is. Although minor changes and corrections have been made throughout the book, major changes have been limited to Chap. 4. The reason is that the theory presented in Chaps. 2 and 3 is basically complete and does not need any further modifications, except that the discussion on *laser-induced interstitial thermotherapy (LITT)* in Sect. 3.2 has been extended by the technique of a multi-fiber treatment. On the other hand, the contents of Chap. 4 – the chapter on applications – strongly depend on the current state of the art. The second edition of this book covers all applications addressed in the first edition plus novel techniques for refractive corneal surgery and the treatment of caries.

The success in refractive corneal surgery has significantly increased since the introduction of *laser in situ keratomileusis (LASIK)* described in Sect. 4.1. The quality of caries removal can be improved with the application of ultrashort laser pulses with durations in the femtosecond range as discussed in Sect. 4.2. Furthermore, descriptive graphics have been added as in Sects. 3.2 and 4.10, and the reference section has been updated with the newest citations available on each topic.

Enjoy reading your second edition ...

Heidelberg University,
January 2002

MARKOLF H. NIEMZ

Preface to the First Edition

This book has emerged from the need for a comprehensive presentation of the recently established field of laser–tissue interactions. So far, only publications dealing with specific issues and conference proceedings with contributions by several authors have been available for this subject. From these multi-author presentations, it is quite difficult for the reader to get to the bottom line of such a novel discipline. A textbook written by a single author is probably better suited for this purpose, although it might not provide the reader with all the details of a specific application.

The basic scope of the book was outlined during several lectures on biomedical optics which I held at Heidelberg University in the years 1992–1995. I have tried to include the most significant studies which are related to the field of laser–tissue interactions and which have been published during the past three decades. This comprises the description of experiments and techniques as well as their results and the theoretical background. Some parts of this book, especially the detailed discussion of ultrashort laser pulses, are naturally influenced by my own interests.

Due to the rapidly increasing number of medical laser applications, it is almost impossible to present a complete survey of all publications. Thus, this book will mainly serve as a starting guide for the newcomer and as a quick reference guide for the insider. For discussion of the newest techniques and results, the reader should consult the latest issues of scientific journals rather than a textbook. Regular coverage is provided by the journals *Lasers in Surgery and Medicine*, *Lasers in Medical Science*, *Biomedical Optics*, and the *SPIE Proceedings on Biomedical Optics*. Apart from these, related articles frequently appear in special issues of other journals, e.g. *Applied Physics B* and the *IEEE Journal of Quantum Electronics*, as well.

I wish to thank all authors and publishers who permitted me to reproduce their figures in this book. Some of the figures needed to be redrawn to improve readability and to obtain a uniform presentation. My special thanks are addressed to the participants of the seminar on Biomedical Optics of the *Studienstiftung des Deutschen Volkes* (German National Fellowship Foundation) which was held in Kranjska Gora, Slovenia, in September 1995. Furthermore, I acknowledge Prof. Dr. J. Bille and his students for their valuable advice concerning the manuscript, Dr. T. Pioch for providing several of the

XVI

pictures taken with scanning electron microscopy, the editorial and production staff of Springer-Verlag for their care and cooperation in producing this book, and last but definitely not least all friends who spent some of their precious time in reading the manuscript.

In spite of great care and effort on my part, I am fairly sure that some errors still remain in the book. I hope you will bring these to my attention for further improvements.

Heidelberg University,
February 1996   MARKOLF H. NIEMZ

Table of Contents

1. **Introduction** 1
   1.1 Historic Review 1
   1.2 Goal of the Book 6
   1.3 Outlook 7

2. **Matter Acting on Light** 9
   2.1 Reflection and Refraction 10
   2.2 Absorption 15
   2.3 Scattering 19
   2.4 Turbid Media 25
   2.5 Photon Transport Theory 27
   2.6 Measurement of Optical Tissue Properties 37
   2.7 Questions to Chapter 2 43

3. **Light Acting on Matter** 45
   3.1 Photochemical Interaction 47
       3.1.1 Photoactivated Disinfection (PAD) 47
       3.1.2 Photodynamic Therapy (PDT) 49
       3.1.3 Photobiomodulation (PBM) 56
       3.1.4 Summary of Photochemical Interaction 60
   3.2 Photothermal Interaction 60
       3.2.1 Heat Generation 70
       3.2.2 Heat Transport 70
       3.2.3 Heat Effects 79
       3.2.4 Laser-Induced Interstitial Thermotherapy (LITT) ... 83
       3.2.5 Summary of Photothermal Interaction 89
   3.3 Photoablation 90
       3.3.1 Model of Photoablation 98
       3.3.2 Cytotoxicity of UV Radiation 102
       3.3.3 Summary of Photoablation 104
   3.4 Plasma-Induced Ablation 105
       3.4.1 Model of Plasma-Induced Ablation 110
       3.4.2 Analysis of Plasma Parameters 123
       3.4.3 Summary of Plasma-Induced Ablation 127
   3.5 Photodisruption 128

      3.5.1   Plasma Formation 133

      3.5.2   Shock Wave Generation 137

      3.5.3   Cavitation 145

      3.5.4   Jet Formation 149

      3.5.5   Summary of Photodisruption 151

  3.6   Questions to Chapter 3 151

4.   Medical Applications of Lasers 153

  4.1   Lasers in Ophthalmology 154

  4.2   Lasers in Dentistry 181

  4.3   Lasers in Gynecology 204

  4.4   Lasers in Urology 210

  4.5   Lasers in Neurosurgery 216

  4.6   Lasers in Angioplasty and Cardiology 224

  4.7   Lasers in Orthopedics 230

  4.8   Lasers in Dermatology and Cosmetics 235

  4.9   Lasers in Gastroenterology 240

  4.10 Lasers in Otorhinolaryngology and Pulmology 244

  4.11 Questions to Chapter 4 249

5.   Laser Safety 251

  5.1   Introduction 251

  5.2   Laser Hazards 251

  5.3   Eye Hazards 252

  5.4   Skin Hazards 253

  5.5   Associated Hazards from High Power Lasers 255

  5.6   Laser Safety Standards and Hazard Classification 255

  5.7   Viewing Laser Radiation 260

  5.8   Eye Protection 262

  5.9   Laser Beam Calculations 264

  5.10 Questions to Chapter 5 265

A. Appendix 267

  A.1 Medical Neodymium Laser System 267

  A.2 Physical Constants and Parameters 271

B. Solutions 275

References 277

Index 305

About the Author 313

1. Introduction

1.1 Historic Review

Since the first report on laser radiation by Maiman (1960), many potential fields for its application have been investigated. Among these, medical laser surgery certainly is one of the most significant advances in modern technology. Actually, various kinds of lasers have already become irreplaceable tools of modern medicine. Although clinical applications were first limited to ophthalmology – a very spectacular and today well-established laser surgery being argon ion laser coagulations in the case of retinal detachment – the fields of medical laser treatment have meanwhile considerably widened. Due to the variety of existing laser systems, the diversity of their physical parameters, and last but not least the enthusiasm of several research groups almost every branch of surgical medicine has been involved. This should not be interpreted as criticism, although much damage has been done in some cases – especially in the field of biostimulation – when researchers have lost orientation due to striving for new publications and success, and industries have praised laser systems that later turned out to be completely useless. In general, though, many really useful laser techniques have been developed and clinically established with the help of all kinds of scientists. These methods of treatment have been reconfirmed by other researchers and properly documented in a variety of well-accepted scientific journals. And, even with early laser applications primarily aimed at therapeutic results, several interesting diagnostic techniques have meanwhile been added. Only some of them will be addressed in this book wherever appropriate, for instance diagnosis of tumors by fluorescence dyes and diagnosis of caries by spectroscopical analysis of laser-induced plasma sparks. However, the discussion of these diagnostic applications is not the main goal of the author, and the interested reader is referred to detailed descriptions found elsewhere.

From the historic point of view, lasers were first applied in ophthalmology. This was obvious, since the eye and its interior belong to the easiest accessible organs because of their high transparency. And it was only a few years earlier that Meyer-Schwickerath (1956) had successfully investigated the coagulative effects of xenon flash lamps on retinal tissue. In 1961, just one year after the invention of the laser, first experimental studies were published by Zaret et al. (1961). Shortly afterwards, patients with retinal de-

© Springer Nature Switzerland AG 2019
M. H. Niemz, *Laser-Tissue Interactions*,
https://doi.org/10.1007/978-3-030-11917-1_1

tachment were already being treated as reported by Campbell et al. (1963) and Zweng et al. (1964). At the same time, investigations were first carried out in dentistry by Goldman et al. (1964) and Stern and Sognnaes (1964). In the beginning, laser treatment was limited to the application of ruby lasers. Later on, other types of lasers followed. And, accordingly, clinical research extended within the disciplines of ophthalmology and dentistry.

Starting in the late 1960s, lasers were introduced to other medical disciplines, as well. And today, a large variety of laser procedures is performed all over the world. Most of them belong to the family of *minimally invasive surgery (MIS)*, a special term reserved for low-trauma and bloodless surgical procedures. It's basically these two characteristics that have promoted lasers to being a universal scalpel and tool for treatment. Many patients, and also surgeons as sketched in Fig. 1.1, believed in lasers as if they were some kind of magical instruments. This attitude evoked misleading statements and unjustified hopes. Careful judgment of new developments is always appropriate, and not every reported laser-induced cure can be taken for granted until it is reconfirmed by independent studies. Laser-induced effects are manifold as will be shown in this book. Most of them can be scientifically explained. However, the same effect which might be good for a certain treatment can be disastrous for another. For instance, heating of cancerous tissue by means of laser radiation might lead to desired tumor necrosis. On the other hand, using the same laser parameters for retinal coagulation can burn the retina, resulting in irreversible blindness. Thermal effects, in particular, tend to be irreversible if temperatures $> 60°C$ are achieved.

"Wow, this magic laser burns it all away ..."

Fig. 1.1. Cartoon

Laser systems are classified as *continuous wave (CW) lasers* and *pulsed lasers*. Whereas most gas lasers and to some extent also solid-state lasers belong to the first group, the family of pulsed lasers mainly includes solid-state lasers, excimer lasers, and some diode lasers. In Table 1.1, a list of medical laser types and two of their characteristic parameters is given: *wavelength* and *pulse duration*. The list is arranged with respect to the latter one, since the duration of exposure primarily characterizes the type of interaction with biological tissue, as we will evaluate in Chap. 3. The wavelength is a second important laser parameter. It determines how deep laser radiation penetrates into a tissue, that is how effectively it is absorbed and scattered. Frequently, a third parameter – the applied *energy density* – is also considered as being significant. However, its value only serves as a necessary condition for the occurrence of a certain effect and then determines its extent. Actually, we will see in Chap. 3 that all medically relevant effects take place at energy densities between $1\,\mathrm{J/cm}^2$ and $1000\,\mathrm{J/cm}^2$. This is a rather narrow range compared to the 15 orders of magnitude of potential pulse durations. A fourth parameter – the applied *intensity* – is given as the ratio of energy density and pulse duration. For a detailed discussion of all these dependences, the reader is referred to Chap. 3. Each laser type listed in Table 1.1 is used for particular clinical applications as described in Chap. 4.

Table 1.1. List of some medical laser systems

Laser type	Wavelength	Typical pulse duration
Argon ion	488/514 nm	CW
Krypton ion	531/568/647 nm	CW
He-Ne	633 nm	CW
CO_2	9.6/10.6 μm	CW or pulsed
Dye laser	450–900 nm	CW or pulsed
Diode laser	405–3330 nm	CW or pulsed
Ruby	694 nm	1–250 μs
Nd:YLF	1053 nm	100 ns – 250 μs
Nd:YAG	1064 nm	100 ns – 250 μs
Ho:YAG	2120 nm	100 ns – 250 μs
Er:YSGG	2780 nm	100 ns – 250 μs
Er:YAG	2940 nm	100 ns – 250 μs
Alexandrite	720–800 nm	50 ns – 100 μs
XeCl	308 nm	20–300 ns
XeF	351 nm	10–20 ns
KrF	248 nm	10–20 ns
ArF	193 nm	10–20 ns
Nd:YLF	1053 nm	30–100 ps
Nd:YAG	1064 nm	30–100 ps
Free electron laser	800–6000 nm	2–10 ps
Ti:Sapphire	700–1000 nm	10 fs – 100 ps

Two recent laser developments have become more and more important for medical research: diode lasers and free electron lasers. Diode lasers can emit either CW or pulsed radiation and are extremely compact and cheap. Free electron lasers provide very short laser pulses but are huge machines which are driven by powerful electron accelerators and are available at a few selected locations only.

The progress in laser surgery can be primarily attributed to the rapid development of pulsed laser systems. As already mentioned above, it is the pulse duration which finally determines the effect on biological tissue. In particular, *thermal* and *non-thermal effects* may be distinguished. A rough approximation is the "1 µs rule" stating that pulse durations $> 1\,\mu s$ are often associated with measurable thermal effects. At pulse durations $< 1\,\mu s$, on the other hand, thermal effects usually become negligible if a moderate repetition rate is chosen (see Sect. 3.2 for further details). Without the implementation of additional features, lasers will either emit CW radiation or pulses with durations $> 1\,\mu s$. Investigations are thus limited to the study of potential thermal effects. Only when generating shorter laser pulses do other types of interactions become accessible. Among these are very efficient ablation mechanisms such as photoablation, plasma-induced ablation, and photodisruption which take place on the nanosecond or picosecond scale. Today, even shorter pulses in the femtosecond range can be realized. They revolutionized refractive corneal surgery (femto-LASIK) as we will see in Sect. 4.1. Both plasma-induced ablation and photodisruption originate from a physical phenomenon called *optical breakdown* which creates a physical plasma consisting of free electrons and ions. However, as will be shown in a theoretical analysis in Sect. 3.4, the threshold parameters of optical breakdown do not decrease any further when going below 100 femtoseconds. So it's very unlikely that surgery will benefit from pulse durations shorter than 100 femtoseconds.

In Fig. 1.2, the progress in the development of pulsed laser systems is illustrated. In the case of *solid-state lasers*, two milestones were reached when discovering the technique of mode locking and when developing novel laser media with extremely large bandwidths like Ti:Sapphire (Zhou et al. (1994)). It's these two events that are responsible for the two steep declines in Fig. 1.2. The other group of lasers capable of providing ultrashort pulses consists of *dye lasers*. They were invented after the first solid-state lasers. Their progress in shorter pulse durations was not as steep but proceeded rather smoothly. Several new techniques such as colliding pulse mode locking were developed which also lead to very short pulse durations comparable to those of solid-state lasers. However, medical applications of dye lasers will be rather limited because of their inconvenience and complicated maintenance. In contrast to long-living solid-state crystals, dyes need to be recirculated and exchanged on a regular basis which often disables a push-button operation.

The very first laser was a ruby laser pumped by a xenon flash lamp. The output of such a laser is characterized by several spikes. Their overall

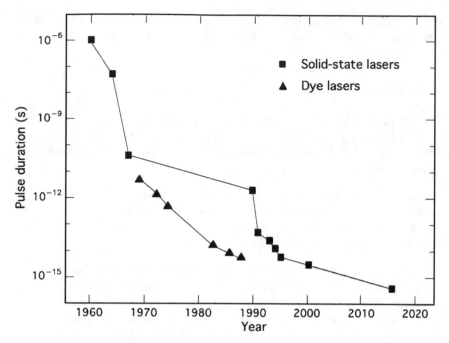

Fig. 1.2. Shortest pulse durations achieved with solid-state lasers and dye lasers

duration is determined by the flash itself which is matched to the lifetime of the upper state of the laser transition (in ruby approximately 1 ms). With the invention of *Q-switching* ("quality switching"), pulses as short as 50 ns could be obtained. Either mechanical devices (rotating apertures or laser mirrors) or optical devices (electrooptic or acoustooptic Pockels crystals) may serve as a Q-switch. In both cases, losses inside the resonator are kept artificially high until an extremely large inversion of the energy levels is achieved. Then, when removing the artificial losses, all energy stored in the laser medium is suddenly converted by means of stimulated emission. Even shorter pulses were obtained when initiating *mode locking* inside the laser cavity. During mode locking, a modulation of the electromagnetic field is induced by using either fast modulating crystals (active mode locking) or saturable absorbers (passive mode locking). By this means, the phases of all oscillating axial laser modes are forced to coincide, resulting in picosecond pulses. A typical representative is the Nd:YAG laser with an optical bandwidth of the order of 1 nm. This bandwidth limits the shortest achievable pulse duration to a few picoseconds. Thus, the realization of femtosecond lasers mainly depended on the discovery of novel laser media with larger optical bandwidths. These were found in crystals such as Ti:Sapphire or Cr:LiSAF. The shortest visible laser pulse to date was measured by Hassan et al. (2016) and has a duration of only 0.38 fs. This duration is equivalent to a spatial pulse extent of less than a single wavelength. The most significant techniques of ultrashort pulse generation are described in detail in the excellent book written by Siegman (1986).

1.2 Goal of the Book

The main goal of this book is to offer an interdisciplinary approach to the basics of laser–tissue interactions. It thus addresses all kinds of scientists, engineers, medical doctors, and graduate students involved in this field. Special emphasis is put on

- giving a detailed description of the physical background of potential interaction mechanisms between laser light and biological tissue,
- providing an updated review of clinical laser applications,
- including a chapter on laser safety.

In Chap. 2, all the relevant information is provided which is essential for understanding the five interaction mechanisms discussed in Chap. 3. Basic phenomena dealing with light and matter such as *reflection*, *refraction*, *absorption*, and *scattering* are explained with basic physics. In each case, special attention is paid to their indispensable mathematical handling. The informed reader may well skip these sections and directly proceed with Sect. 2.5. In that section, when discussing *photon transport theory*, important tools will be derived which are of considerable importance in modern theoretical research. In order to solve the governing equation of energy transfer, either the Kubelka–Munk theory, the method of diffusion approximation, or Monte Carlo simulations are most frequently used. All of them will be comprehensively reviewed in Sect. 2.5 and compared to each other along with their advantages and disadvantages. The interested reader should also consult the original works by Kubelka (1948), Metropolis and Ulam (1949), and the profound theory developed by Ishimaru (1978).

The main chapter of this book is Chap. 3. While Chap. 2 focuses on how matter is acting on light, here we are going to consider the opposite effect: How is light acting on matter? Starting with some general remarks and definitions, a classification scheme is developed with the exposure duration being the main physical parameter. In total, five different types of interaction mechanisms are presented: *photochemical interaction* including photobiomodulation, *photothermal interaction*, *photoablation*, *plasma-induced ablation*, and *photodisruption*. Each of them is thoroughly discussed including selected photographs and manifold illustrations. At the end of each section, a comprehensive summary is given highlighting significant features of each interaction mechanism. Special techniques such as photodynamic therapy (PDT) or laser-induced interstitial thermotherapy (LITT) are explained according to the latest references. Both techniques are concerned with the laser treatment of cancer, either photochemically or photothermally, as a good alternative to conventional methods being unsatisfactory for a large group of patients. When discussing photoablation, potential risks originating from UV radiation will be surveyed. The difference between plasma-induced ablation and photodisruption is explained and properly visualized. Novel theoretical models are introduced, describing the basic mechanism of plasma-induced ablation. They help us better understand the physical phenomena associated with optical breakdown and its threshold parameters.

In Chap. 4, the most important clinical applications are reviewed based on the latest results and references. Due to their historic sequence and present significance, applications in ophthalmology, dentistry, gynecology, and urology are considered first. In ophthalmology, various standard techniques are discussed such as coagulation of the retina, laser treatments of glaucoma, and fragmentation of the lens. The newest methods and results concerning refractive corneal surgery are presented, as well. In dentistry, special emphasis is put on various laser treatments of caries and on new applications of diode lasers. In gynecology and urology, various thermal effects of laser radiation are discussed with major tasks and clinical results being surveyed. Other disciplines of clinical importance follow as mentioned in the table of contents. In each case, experimental procedures and clinical results are reviewed along with any complications arising during treatment. Original photographs and custom-made artwork shall pass on professional insight to you.

Finally, Chap. 5 comprises the latest standard of laser safety. It outlines a careful selection of essential guidelines published by the Laser Institute of America, Orlando, Florida. Meanwhile, most of them have been adapted by other governments, as well. A laser classification scheme is introduced which is commonly used all over the world. Moreover, important exposure limits are listed to be taken into account when treating any patients. This chapter is meant to serve as a quick reference guide when operating lasers.

1.3 Outlook

It is interesting to observe that almost every new technique initially evokes a euphoric reaction among surgeons and patients. This period is often followed by indifference and rejection when long-term effects and limitations become obvious. Eventually, researchers agree on certain indications for applying the new technique which then leads to the final approval. One typical example for the occurrence of this sequence was the introduction of photodisruptive lasers to ophthalmology by Aron-Rosa et al. (1980).

At present, lasers have already contributed to the treatment of a wide variety of maladies. However, today's clinical lasers and their applications most probably represent only the infancy of laser medicine. In the near future, other lasers will evolve and take their places in hospitals and medical centers. Miniaturization will enhance their usefulness and applicability, and highly specialized delivery optics will expand the surgeon's ability to establish new and very precise therapies. Moreover, combinations of different wavelengths distributed both spatially and temporally may provide tissue effects superior to those of single wavelengths.

Ultimately, all these endeavors will advance minimally invasive surgery beyond our present horizon. This progress, however, will rely on our creativity and cooperation. Further scientific research is as essential as the promotion of its results to clinical applications. The future of medical lasers cannot be created by physicists, engineers, or surgeons alone, but must be realized through collective human sources of science, medicine, industry, and government.

2. Matter Acting on Light

In this and the following chapter, we will discuss basic phenomena occurring when matter is exposed to light. While here we will be concerned with various actions of matter on light, the opposite effect (actions of light on matter) will be addressed in Chap. 3. Matter can act on electromagnetic radiation in manifold ways. In Fig. 2.1, a typical situation is shown, where a light beam is incident on a slice of matter, for example a piece of skin. In principle, there are four phenomena that have an effect on the beam's propagation:

- reflection and refraction,
- absorption,
- scattering.

Reflection and refraction are strongly related to each other by *Fresnel's laws*. Therefore, these two effects will be addressed in the same section. In Fig. 2.1, refraction is visualized by a tilted beam inside the piece of matter. In medical laser applications, however, refraction plays a significant role only when irradiating transparent media like corneal tissue. In opaque media, usually, the effect of refraction is difficult to measure due to absorption and scattering.

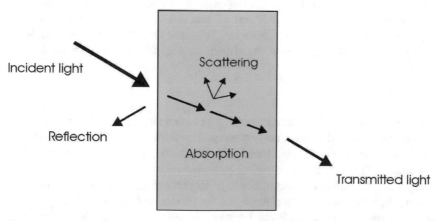

Fig. 2.1. Geometry of reflection, refraction, absorption, and scattering

M. H. Niemz, *Laser-Tissue Interactions*,
https://doi.org/10.1007/978-3-030-11917-1_2

Only non-reflected and non-absorbed or forward scattered photons are transmitted by the slice and contribute to the intensity detected behind the slice. The ratio of transmitted and incident intensities is called *transmittance*. Which of the losses – reflection, absorption, or scattering – is dominant primarily depends on the type of material and the incident wavelength. As we will encounter in the following sections, the wavelength is a very important parameter indeed. It determines the index of refraction as well as the absorption and scattering coefficients. The index of refraction measures the optical density of a given material. This index strongly depends on wavelength in regions of high absorption only. The scattering coefficient, on the other hand, can scale inversely with the fourth power of wavelength as we will evaluate in Sect. 2.3 when discussing Rayleigh scattering.

In laser surgery, knowledge of reflection, refraction, absorption, and scattering is essential for the purpose of performing a successful treatment. The index of refraction, for example, is of considerable interest when applying laser radiation to the eye as in refractive corneal surgery or in the treatment of retinal detachment. In this book, no specific kind of target or biological tissue will be assumed unless otherwise stated in some figures or tables. Instead, we will focus on general physical relations which apply for hard and soft tissues. In reality, of course, limitations are given by the inhomogeneity of biological tissue which is the main reason for our inability to provide other than mean tissue parameters.

2.1 Reflection and Refraction

Reflection is defined as the returning of electromagnetic radiation by surfaces upon which it is incident. In general, a reflecting surface is the physical boundary between two materials of different indices of refraction such as air and tissue. The simple law of reflection requires the wave normals of the incident and reflected beams and the normal of the reflecting surface to lie within one plane, called the *plane of incidence*. It states that the reflection angle θ' equals the angle of incidence θ, as shown in Fig. 2.2 and expressed by the simple relation

$$\theta = \theta' \ . \tag{2.1}$$

The angles θ and θ' are always measured between the surface normal and the incident and reflected beams, respectively. The surface itself is assumed to be smooth, with surface irregularities being small compared to the wavelength of radiation. This results in so-called *specular reflection*.

In contrast, i.e. when the roughness of the reflecting surface is comparable or even larger than the wavelength of radiation, *diffuse reflection* occurs. Then, several beams are reflected which do not necessarily lie within the plane of incidence, and (2.1) no longer applies. Diffuse reflection is a common

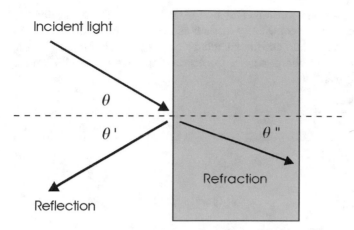

Fig. 2.2. Geometry of specular reflection and refraction

phenomenon of all tissues, since none of them is provided with highly polished surfaces such as optical mirrors. Only in special cases such as wet tissue surfaces might specular reflection surpass diffuse reflection.

Refraction always occurs at the surface of two media with different indices of refraction. It basically originates from a change in speed of the light wave. The simple mathematical relation governing refraction is known as *Snell's law*. It is given by

$$\frac{\sin \theta}{\sin \theta''} = \frac{v}{v'} \ , \tag{2.2}$$

where θ'' is the angle of refraction, and v and v' are the speeds of light in the two media in front of and behind the reflecting surface, respectively. Since the corresponding indices of refraction are defined by

$$n = \frac{c}{v} \ , \tag{2.3}$$

$$n' = \frac{c}{v'} \ ,$$

where c denotes the speed of light in vacuum, (2.2) turns into

$$n \sin \theta = n' \sin \theta'' \ . \tag{2.4}$$

Only for $\sin \theta > n'/n$ can (2.4) not be fulfilled, meaning that refraction will not occur. This event is also referred to as *total reflection*.

The *reflectivity* of a surface is a measure of the amount of reflected radiation. It is defined as the ratio of reflected and incident electric field amplitudes. The *reflectance* is the ratio of the correponding intensities and is thus equal to the square of the reflectivity. Reflectivity and reflectance depend on the angle of incidence, the polarization of radiation, and the indices of refraction of the materials forming the boundary surface. Relations for reflectivity and refraction are commonly known as *Fresnel's laws*. In this book, we will merely state them and consider their principal physical impact. Exact derivations are found elsewhere, e.g. in books dealing with electrodynamics. Fresnel's laws are given by

$$\frac{E_s{}'}{E_s} = -\frac{\sin(\theta - \theta'')}{\sin(\theta + \theta'')} , \tag{2.5}$$

$$\frac{E_p{}'}{E_p} = \frac{\tan(\theta - \theta'')}{\tan(\theta + \theta'')} , \tag{2.6}$$

$$\frac{E_s{}''}{E_s} = \frac{2\sin\theta'' \cos\theta}{\sin(\theta + \theta'')} , \tag{2.7}$$

$$\frac{E_p{}''}{E_p} = \frac{2\sin\theta'' \cos\theta}{\sin(\theta + \theta'') \cos(\theta - \theta'')} , \tag{2.8}$$

where E, E', and E'' are the amplitudes of the electric field vectors of the incident, reflected, and refracted light, respectively. The subscripts "s" and "p" denote the two planes of oscillation with "s" being perpendicular to the plane of incidence – from the German *senkrecht* – and "p" being parallel to the plane of incidence.

Further interaction of incident light with the slice of matter is limited to the refracted beam. One might expect that the intensity of the refracted beam would be complementary to the reflected one so that the addition of both would give the incident intensity. However, this is not correct, because intensity is defined as the power per unit area, and the cross-section of the refracted beam is different from that of the incident and reflected beams except at normal incidence. It is only the total energy in these beams that is conserved. The reflectances in either plane are given by

$$R_s = \left(\frac{E_s{}'}{E_s}\right)^2 , \tag{2.9}$$

$$R_p = \left(\frac{E_p{}'}{E_p}\right)^2 . \tag{2.10}$$

In Fig. 2.3, the reflectances R_s and R_p are plotted as a function of the angle of incidence. Here we assume that $n = 1$ and $n' = 1.33$ which are the indices of refraction of air and water, respectively. In particular, Fig. 2.3 describes the specular reflectance on wet surfaces.

The angle at which $R_p = 0$ is called the *Brewster angle*. In the case of water, it is equal to $53°$. At normal incidence, i.e. $\theta = 0$, the reflectances in either plane are approximately 2%. This value is not directly evident from (2.5) and (2.6), since insertion of $\theta = \theta'' = 0$ gives an indeterminate result. However, we can evaluate it as follows: Since both θ and θ'' become very small when approaching normal incidence, we may set the tangents in (2.6) equal to the sines and obtain

$$R_p \simeq R_s = \frac{\sin^2(\theta - \theta'')}{\sin^2(\theta + \theta'')} = \left(\frac{\sin\theta \, \cos\theta'' - \cos\theta \, \sin\theta''}{\sin\theta \, \cos\theta'' + \cos\theta \, \sin\theta''} \right)^2 . \tag{2.11}$$

When dividing numerator and denominator of (2.11) by $\sin\theta''$ and replacing $\sin\theta / \sin\theta''$ by n', i.e. assuming $n = 1$, it reduces to

$$R_p \simeq R_s = \left(\frac{n' \cos\theta'' - \cos\theta}{n' \cos\theta'' + \cos\theta} \right)^2 \simeq \left(\frac{n' - 1}{n' + 1} \right)^2 . \tag{2.12}$$

The approximate equality becomes exact within the limit of normal incidence. Inserting $n' = 1.33$ yields

$$R_p \simeq R_s \simeq 2\% .$$

In several cases, this fraction of incident radiation is not negligible. Thus, regarding laser safety, it is one of the main reasons why proper eye protection is always required (see Chap. 5).

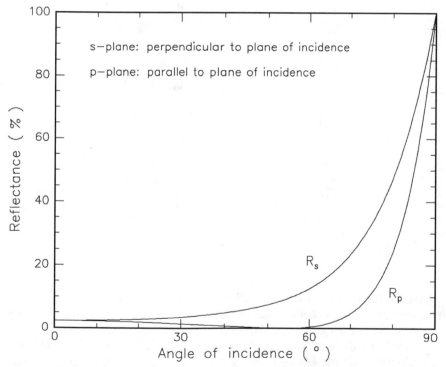

Fig. 2.3. Reflectances in s- and p-plane for water ($n = 1.33$)

For water, the indices of refraction and the corresponding reflectances at different wavelengths are listed in Table 2.1. Two strong absorption bands occur at about 2.9 μm and 6.0 μm. They result from vibrational and rotational oscillations of the water molecule. The major aspects of absorption will be addressed in the next section.

Table 2.1. Indices of refraction and reflectances of water. Data according to Hale and Querry (1973)

Wavelength λ (μm)	Index of refraction n	Reflectance R
0.2	1.396	0.027
0.3	1.349	0.022
0.4	1.339	0.021
0.5	1.335	0.021
0.6	1.332	0.020
0.7	1.331	0.020
0.8	1.329	0.020
0.9	1.328	0.020
1.0	1.327	0.020
1.6	1.317	0.019
2.0	1.306	0.018
2.6	1.242	0.012
2.7	1.188	0.007
2.8	1.142	0.004
2.9	1.201	0.008
3.0	1.371	0.024
3.1	1.467	0.036
3.2	1.478	0.037
3.3	1.450	0.034
3.4	1.420	0.030
3.5	1.400	0.028
4.0	1.351	0.022
5.0	1.325	0.020
6.0	1.265	0.014
7.0	1.317	0.019
8.0	1.291	0.016
9.0	1.262	0.013
10.0	1.218	0.010

Even if the dependence of the index of refraction on wavelength is rather weak in the visible spectrum, it should be taken into account when irradiating biological tissues. In general, indices of refraction for various kinds of tissue are difficult to measure due to absorption and scattering. Reflection from these tissues must be obtained empirically. In many cases, the indices of the refraction of water may be considered a rough estimate.

2.2 Absorption

The first law of photobiology says: Photons *must be* absorbed for biological effects to occur. So, we always have to think about *absorption* first. During absorption, the intensity of an incident electromagnetic wave is attenuated in passing through a medium. The *absorbance* of a medium is defined as the ratio of absorbed and incident intensities. A perfectly *transparent* medium permits the passage of light without any absorption, i.e. the total radiant energy entering into and emerging from such a medium is the same. Among biological tissues, cornea and lens can be considered as being highly transparent for visible light. In contrast, media in which incident radiation is reduced practically to zero are called *opaque*.

The terms "transparent" and "opaque" are relative, since they certainly are wavelength-dependent. Cornea and lens, for instance, mainly consist of water which shows a strong absorption at wavelengths in the infrared spectrum. Hence, these tissues appear opaque in this spectral region. Actually, no medium is known to be either transparent or opaque to all wavelengths of the electromagnetic spectrum.

A substance is said to show *general absorption* if it reduces the intensity of all wavelengths in the considered spectrum by a similar fraction. In the case of visible light, such substances will thus appear grey to our eye. *Selective absorption*, on the other hand, is the absorption of certain wavelengths in preference to others. The existence of colors actually originates from selective absorption. Usually, *body colors* and *surface colors* are distinguished. Body color is generated by light which penetrates a certain distance into the substance. By backscattering, it is then deviated and escapes backwards from the surface but only after being partially absorbed at selected wavelengths. In contrast, surface color originates from reflection at the surface itself. It mainly depends on the reflectances which are related to the wavelength of incident radiation by (2.12).

The ability of a medium to absorb electromagnetic radiation depends on a number of factors, mainly the electronic constitution of its atoms and molecules, the wavelength of radiation, the thickness of the absorbing layer, and internal parameters such as the temperature or concentration of absorbing agents. Two laws are frequently applied which describe the effect of either thickness or concentration on absorption, respectively. They are commonly called *Lambert's law* and *Beer's law*, and are expressed by

$$I(z) = I_0 \exp(-\alpha z) , \tag{2.13}$$

and

$$I(z) = I_0 \exp(-k'cz) , \tag{2.14}$$

where z denotes the optical axis, $I(z)$ is the intensity at a distance z, I_0 is the incident intensity, α is the absorption coefficient of the medium, c is the concentration of absorbing agents, and k' depends on internal parameters other than concentration. Since both laws describe the same behavior of absorption, they are also known as the *Lambert–Beer law*. From (2.13), we obtain

$$z = \frac{1}{\alpha} \ln \frac{I_0}{I(z)} \ . \tag{2.15}$$

The inverse of the absorption coefficient α is also referred to as absorption length or optical penetration depth L, i.e.

$$L = \frac{1}{\alpha} \ . \tag{2.16}$$

The absorption length measures the distance z in which the intensity $I(z)$ has dropped to $1/e$ of its incident value I_0.

In biological tissues, absorption is mainly caused by either water molecules or macromolecules such as proteins and pigments. Whereas absorption in the IR region of the spectrum can be primarily attributed to water molecules, proteins as well as pigments mainly absorb in the UV and visible range of the spectrum. Proteins, in particular, have an absorption peak at approximately 280 nm according to Boulnois (1986). The absorption spectrum of water – the main constituent of most tissues – will be discussed in detail in Sect. 3.2 when addressing photothermal interactions.

In Fig. 2.4, absorption coefficients of water, tooth enamel, and the two most important absorbers of human tissues are shown. They belong to melanin and hemoglobin (Hb), respectively. Melanin is the basic pigment of the skin and is by far our most important epidermal chromophore. Its absorption coefficient monotonically increases across the visible spectrum toward the UV. Hemoglobin is dominant in vascularized tissue and is the main absorber during the laser coagulation of blood. A general feature of most biomolecules is their complex band structure between 400 nm and 600 nm. Since neither macromolecules nor water strongly absorb in the near infrared, a *therapeutic window* is delineated between roughly 600 nm and 1200 nm. In this spectral range, radiation penetrates biological tissues at a lower loss, thus enabling treatment of deeper tissue structures.

The absorption spectra of three typical tissues are presented in Fig. 2.5. They are obtained from the skin, aortic wall, and cornea, respectively. Among these, skin is the highest absorber, whereas the cornea is almost perfectly transparent[1] in the visible region of the spectrum. Because of the uniqueness of the absorption spectra, each of them can be regarded as a fingerprint of the corresponding tissue. Of course, slight deviations from these spectra can occur due to the inhomogeneity of most tissues.

---

[1] Actually, it is amazing how nature was able to create a stable tissue with such a transparency. The latter is due to the extremely regular structure of collagen fibrils inside the cornea and its high water content.

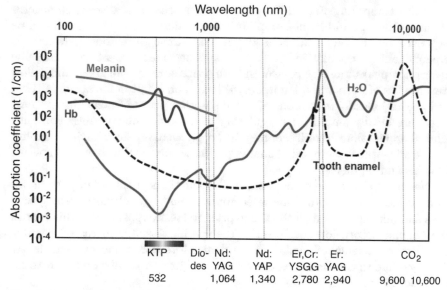

Fig. 2.4. Approximate absorption coefficients of water, tooth enamel, melanin, and hemoglobin (Hb). Data according to Coluzzi (2008)

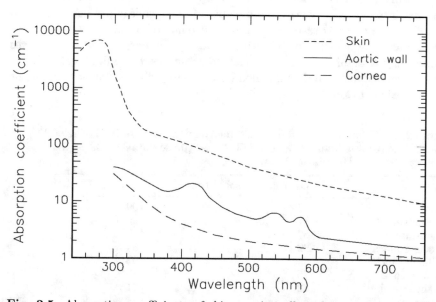

Fig. 2.5. Absorption coefficients of skin, aortic wall, and cornea. In the visible spectrum, the absorption of skin is 20–30 times higher than the absorption of corneal tissue. Data according to Parrish and Anderson (1983), Keijzer et al. (1989), and Eichler and Seiler (1991)

When comparing Figs. 2.4 and 2.5, we find that the absorption spectra of the aortic wall and hemoglobin are quite similar. This observation can be explained by the fact that hemoglobin – as already previously stated – is predominant in vascularized tissue. Thus, it becomes evident that the same absorption peaks must be present in both spectra. Since the green and yellow wavelengths of both argon ion lasers (514 nm) and krypton ion lasers (531 nm and 568 nm) closely match one of the absorption peaks of hemoglobin, these lasers are ideally suited for coagulating blood and blood vessels. For some clinical applications, dye lasers may be an alternative, since their tunability can be advantageously used to match particular absorption bands of specific proteins and pigments.

Not only the absorption of the biological tissue itself is important for medical laser surgery. In some laser applications, e.g. sclerostomies, special dyes and inks are added to the tissue prior to laser exposure. By this means, the original absorption coefficient of the specific tissue is increased, leading to a higher efficiency of the laser treatment. Moreover, adjacent tissue is less damaged due to the enhanced absorption. For further details on sclerostomy, the reader is referred to Sect. 4.1.

In Table 2.2, the effects of some selected dyes on the damage threshold are demonstrated in the case of scleral tissue. The dyes were applied to the sclera by means of electrophoresis, i.e. an electric current was used to direct the dye into the tissue. Afterwards, the samples were exposed to picosecond pulses from a Nd:YLF laser to achieve optical breakdown (see Sect. 3.4). The absolute and relative threshold values of pulse energy are listed for each dye. Obviously, the threshold for the occurrence of optical breakdown can be decreased by a factor of two when choosing the correct dye. Other dyes evoked only a slight decrease in threshold or no effect at all. In general, the application of dyes should be handled very carefully, since some of them might induce toxic side effects.

Table 2.2. Effect of selected dyes and inks on damage threshold of scleral tissue. Damage was induced by a Nd:YLF laser (pulse duration: 30 ps, focal spot size: 30 µm). Unpublished data

Dye	Damage threshold (µJ)	Relative threshold
None	87	100 %
Erythrosine	87	100 %
Nigrosine	87	100 %
Reactive black	82	94 %
Brilliant black	81	93 %
Amido black	75	86 %
Methylene blue	62	71 %
Tatrazine	62	71 %
Bismarck brown	56	64 %
India ink	48	55 %

2.3 Scattering

When elastically bound charged particles are exposed to electromagnetic waves, the particles are set into motion by the electric field. If the frequency of the wave equals the natural frequency of free vibrations of a particle, resonance occurs being accompanied by a considerable amount of absorption. *Scattering*, on the other hand, takes place at frequencies not corresponding to those natural frequencies of particles. The resulting oscillation is determined by forced vibration. In general, this vibration will have the same frequency and direction as that of the electric force in the incident wave. Its amplitude, however, is much smaller than in the case of resonance. Also, the phase of the forced vibration differs from the incident wave, causing photons to slow down when penetrating into a denser medium. Hence, scattering can be regarded as the basic origin of dispersion.

Inelastic and *elastic scattering* are distinguished, depending on whether part of the incident photon energy is converted during the process of scattering or not. In the following paragraphs, we will first consider elastic scattering, where incident and scattered photons have the same energy. A special kind of elastic scattering is *Rayleigh scattering*. Its only restriction is that the scattering particles be smaller than the wavelength of incident radiation. In particular, we will find a relationship between scattered intensity and index of refraction, and that scattering is inversely proportional to the fourth power of wavelength. The latter statement is also known as *Rayleigh's law* and shall be derived in the following paragraphs.

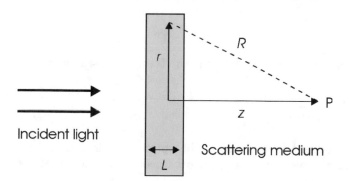

Fig. 2.6. Geometry of Rayleigh scattering

In Fig. 2.6, a simple geometry of Rayleigh scattering is shown. A plane electromagnetic wave is incident on a thin scattering medium with a total thickness L. At a particular time, the electric field of the incident wave can be expressed by

$$E(z) = E_0 \exp(ikz) \ ,$$

where E_0 is the amplitude of the incident electric field, k is the amount of the propagation vector, and z denotes the optical axis. In a first approximation, we assume that the wave reaching some point P on the optical axis will essentially be the original wave, plus a small contribution due to scattering. The loss in intensity due to scattering is described by a similar relation as absorption, i.e.

$$I(z) = I_0 \, \exp(-\alpha_s z) \,, \tag{2.17}$$

where α_s is the scattering coefficient. Differentiation of (2.17) with respect to z leads to

$$\mathrm{d}I = -\alpha_s I \, \mathrm{d}z \,.$$

The intensity scattered by a thin medium of a thickness L as shown in Fig. 2.6 is thus proportional to α_s and L:

$$I_s \sim \alpha_s L \,. \tag{2.18}$$

Let us now assume that there are NL atoms per unit area of the scattering medium. Herein, the parameter N shall denote the density of scattering atoms. Then, the intensity scattered by one of these atoms can be described by the relation

$$I_1 \sim \frac{\alpha_s L}{NL} = \frac{\alpha_s}{N} \,.$$

Thus, the amplitude of the corresponding electric field is

$$E_1 \sim \sqrt{\frac{\alpha_s}{N}} \,.$$

Due to the interference of all scattered waves, the total scattered amplitude can be expressed by

$$E_s \sim NL \sqrt{\frac{\alpha_s}{N}} = L\sqrt{\alpha_s N} \,.$$

The complex amplitude at a distance z on the optical axis is obtained by adding the amplitudes of all scattered spherical waves to the amplitude of the incident plane wave, i.e.

$$E(z) = E_0 \left(\mathrm{e}^{\mathrm{i}kz} + L\sqrt{\alpha_s N} \int_0^\infty \frac{\mathrm{e}^{\mathrm{i}kR}}{R} \, 2\pi r \, \mathrm{d}r \right) \,, \tag{2.19}$$

with $R^2 = z^2 + r^2 \,.$

At a given z, we thus obtain

$$r \, \mathrm{d}r = R \, \mathrm{d}R \,,$$

which reduces (2.19) to

$$E(z) = E_0 \left(e^{ikz} + L\sqrt{\alpha_s N}\, 2\pi \int_z^\infty e^{ikR}\, dR \right) .$$ (2.20)

Since wave trains always have a finite length, scattering from $R \to \infty$ can be neglected. Hence, (2.20) turns into

$$E(z) = E_0 \left(e^{ikz} - L\sqrt{\alpha_s N}\, \frac{2\pi}{ik}\, e^{ikz} \right) ,$$

and when inserting the wavelength $\lambda = 2\pi/k$

$$E(z) = E_0\, e^{ikz} \left(1 + i\lambda L\sqrt{\alpha_s N} \right) .$$ (2.21)

According to our assumption, the contribution of scattering – i.e. the second term in parentheses in (2.21) – is small compared to the first. Hence, they can be regarded as the first two terms of an expansion of

$$E(z) = E_0 \exp\left[i\left(kz + \lambda L\sqrt{\alpha_s N} \right) \right] .$$

Therefore, the phase of the incident wave is altered by the amount $\lambda L\sqrt{\alpha_s N}$ due to scattering. This value must be equal to the well-known expression of phase retardation given by

$$\Delta\phi = \frac{2\pi}{\lambda}\, (n-1)L ,$$

which occurs when light enters from free space into a medium with refractive index n. Hence,

$$\lambda L\sqrt{\alpha_s N} = \frac{2\pi}{\lambda}(n-1)L ,$$

$$n - 1 = \frac{\lambda^2}{2\pi}\sqrt{\alpha_s N} .$$ (2.22)

From (2.18) and (2.22), we finally obtain *Rayleigh's law* of scattering when neglecting the wavelength-dependence of n, i.e.

$$I_s \sim \frac{1}{\lambda^4} .$$ (2.23)

If the scattering angle θ is taken into account, a more detailed analysis yields

$$I_s(\theta) \sim \frac{1 + \cos^2(\theta)}{\lambda^4} ,$$ (2.24)

where $\theta = 0$ denotes forward scattering. Rayleigh's law is illustrated in Fig. 2.7. Within the visible spectrum, scattering is already significantly reduced when comparing blue and red light.

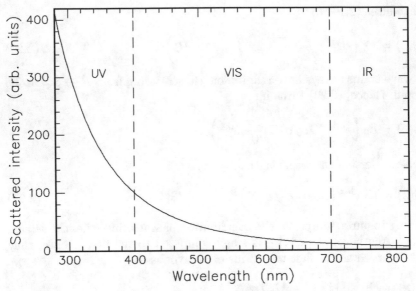

Fig. 2.7. Rayleigh's law of scattering for near UV, visible, and near IR light

Rayleigh scattering is elastic scattering, i.e. scattered light has the same values of k and λ as incident light. One important type of inelastic scattering is known as *Brillouin scattering*. It arises from acoustic waves propagating through a medium, thereby inducing inhomogeneities of the refractive index. Brillouin scattering of light to higher (or lower) frequencies occurs, because scattering particles are moving toward (or away from) the light source. It can thus be regarded as an optical *Doppler effect* in which the frequency of photons is shifted up or down. In laser–tissue interactions, Brillouin scattering becomes significant only during the generation of shock waves as will be discussed in Sect. 3.5.

In our derivation of Rayleigh's law, absorption has been neglected. Therefore, (2.22)–(2.24) are valid only for wavelengths far away from any absorption bands. Simultaneous absorption and scattering will be discussed in Sect. 2.4. Moreover, we did not take the spatial extent of scattering particles into account. If this extent becomes comparable to the wavelength of the incident radiation such as in blood cells, Rayleigh scattering no longer applies and another type of scattering called *Mie scattering* occurs. The theory of Mie scattering is rather complex and shall thus not be repeated here. However, it is emphasized that Mie scattering and Rayleigh scattering differ in two important respects. First, Mie scattering shows a weaker dependence on wavelength ($\sim \lambda^{-x}$ with $0.4 \leq x \leq 0.5$) compared to Rayleigh scattering ($\sim \lambda^{-4}$). Second, Mie scattering preferably takes place in the forward direction, whereas Rayleigh scattering is proportional to $1 + \cos^2(\theta)$ according to (2.24), i.e. forward and backward scattered intensities are the same.

In many biological tissues, it was found by Wilson and Adam (1983), Jacques et al. (1987b), and Parsa et al. (1989) that photons are preferably scattered in the forward direction. This phenomenon cannot be explained by Rayleigh scattering. On the other hand, the observed wavelength-dependence is somewhat stronger than predicted by Mie scattering. Thus, neither Rayleigh scattering nor Mie scattering precisely describe scattering in tissues. Therefore, it is very convenient to define a *probability function* $p(\theta)$ of a photon to be scattered by an angle θ which can be fitted to experimental data. If $p(\theta)$ does not depend on θ, we speak of *isotropic scattering*. Otherwise, *anisotropic scattering* occurs.

A measure of the anisotropy of scattering is given by the coefficient of anisotropy g, where $g = 1$ denotes purely forward scattering, $g = -1$ purely backward scattering, and $g = 0$ isotropic scattering. In polar coordinates, the coefficient g is defined by

$$g = \frac{\int_{4\pi} p(\theta) \, \cos\theta \, d\omega}{\int_{4\pi} p(\theta) \, d\omega} \,, \tag{2.25}$$

where $p(\theta)$ is a probability function and $d\omega = \sin\theta \, d\theta \, d\phi$ is the elementary solid angle. By definition, the coefficient of anisotropy g represents the average value of the cosine of the scattering angle θ. As a good approximation, it can be assumed that g ranges from 0.7 to 0.99 for most biological tissues. Hence, the corresponding scattering angles are most frequently between $8°$ and $45°$. The important term in (2.25) is the probability function $p(\theta)$. It is also called the *phase function* and is usually normalized by

$$\frac{1}{4\pi} \int_{4\pi} p(\theta) \, d\omega = 1 \,. \tag{2.26}$$

Several theoretical phase functions $p(\theta)$ have been proposed and are known as *Henyey–Greenstein*, *Rayleigh–Gans*, *δ–Eddington*, and *Reynolds* functions[2]. Among these, the first is best in accordance with experimental observations. It was introduced by Henyey and Greenstein (1941), and it is expressed as

$$p(\theta) = \frac{1 - g^2}{(1 + g^2 - 2g \cos\theta)^{3/2}} \,. \tag{2.27}$$

This phase function is mathematically very convenient to handle, since it is equivalent to the representation

$$p(\theta) = \sum_{i=0}^{\infty} (2i + 1) \, g^i \, P_i(\cos\theta) \,, \tag{2.28}$$

---

[2] Detailed information on these phase functions is provided in the reports by Henyey and Greenstein (1941), van de Hulst (1957), Joseph et al. (1976), and Reynolds and McCormick (1980).

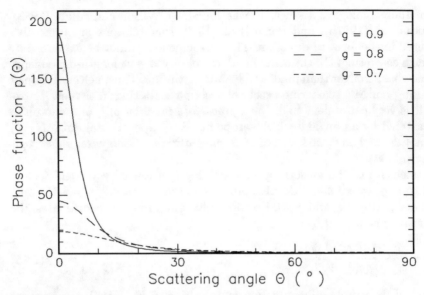

Fig. 2.8. Henyey–Greenstein function for different coefficients of anisotropy ranging from $g = 0.7$ to $g = 0.9$

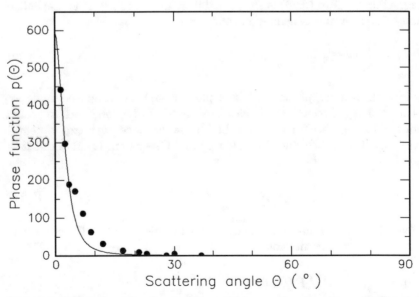

Fig. 2.9. Phase function for an 80 μm thick sample of aortic wall. The data are fitted to a composite function consisting of an isotropic term u and the anisotropic Henyey–Greenstein function (fit parameters: $g = 0.945$, $u = 0.071$). Data according to Yoon et al. (1987)

where P_i are the Legendre polynomials. In some cases, though, a composite function of an isotropic term u and the Henyey–Greenstein function does fit better to experimental data. According to Yoon et al. (1987), this modified function can be expressed by

$$p(\theta) = \frac{1}{4\pi} \frac{u + (1 - u)(1 - g^2)}{(1 + g^2 - 2g \cos \theta)^{3/2}} . \tag{2.29}$$

In Fig. 2.8, the Henyey–Greenstein phase function is graphically shown for different values of g. Obviously, it describes the dominant process of scattering in the forward direction very well. In Fig. 2.9, experimental data are plotted for an 80 μm sample of aortic tissue. The data are fitted to the modified Henyey–Greenstein function as determined by (2.29) with the parameters $g = 0.945$ and $u = 0.071$.

2.4 Turbid Media

In the previous two sections, we have considered the occurrences of either absorption or scattering. In most tissues, though, both of them will be present simultaneously. Such media are called *turbid media*. Their total attenuation coefficient can be expressed by

$$\alpha_t = \alpha + \alpha_s . \tag{2.30}$$

In turbid media, the mean free optical path of incident photons is thus determined by

$$L_t = \frac{1}{\alpha_t} = \frac{1}{\alpha + \alpha_s} . \tag{2.31}$$

Only in some cases, either α or α_s may be negligible with respect to each other, but it is important to realize the existence of both processes and the fact that usually both are operating. Also, it is very convenient to define an additional parameter, the *optical albedo a*, by

$$a = \frac{\alpha_s}{\alpha_t} = \frac{\alpha_s}{\alpha + \alpha_s} . \tag{2.32}$$

For $a = 0$, attenuation is exclusively due to absorption, whereas in the case of $a = 1$ only scattering occurs. For $a = 1/2$, (2.32) can be turned into the equality $\alpha = \alpha_s$, i.e. the coefficients of absorption and scattering are of the same magnitude. In general, both effects will take place but they will occur in variable ratios.

In Fig. 2.10, the albedo is shown as a function of the scattering coefficient. Three different absorption coefficients are assumed which are typical for biological tissue. In addition, the value $a = 1/2$ is indicated. For $\alpha_s \gg \alpha$, the albedo asymptotically approaches unity.

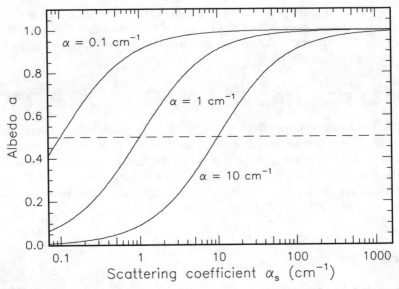

Fig. 2.10. Optical albedo as a function of scattering coefficient (absorption coefficient: as labeled). The albedo $a = 1/2$ is indicated

When dealing with turbid media, another useful parameter is the *optical depth d* which is defined by

$$d = \int_0^s \alpha_t \, ds' , \qquad (2.33)$$

where ds' denotes a segment of the optical path, and s is the total length of the optical path. In the case of homogeneous attenuation, i.e. a constant attenuation coefficient α_t, (2.33) turns into

$$d = \alpha_t s .$$

The advantage of using albedo a and optical depth d – instead of absorption coefficient α and scattering coefficient α_s – is that the former are dimensionless parameters. The information contained in the pair a and d, though, is the same as in the pair α and α_s.

When considering turbid media, the normalization of the phase function as given by (2.26) must be altered to

$$\frac{1}{4\pi} \int_{4\pi} p(\theta) \, d\omega = a ,$$

since the probability function should approach zero at negligible scattering. Hence, (2.27) and (2.28) are no longer valid and must be replaced by

$$p(\theta) = a \, \frac{1 - g^2}{(1 + g^2 - 2g \cos\theta)^{3/2}} ,$$

and

$$p(\theta) = a \sum_{i=0}^{\infty} (2i + 1)\, g^i\, P_i(\cos\theta) \ .$$

In the literature, reduced scattering and attenuation coefficients are often dealt with and are given by

$$\alpha_{\mathrm{s}}' = \alpha_{\mathrm{s}}\, (1 - g) \ , \tag{2.34}$$

and

$$\alpha_{\mathrm{t}}' = \alpha + \alpha_{\mathrm{s}}' \ , \tag{2.35}$$

since forward scattering alone, i.e. $g = 1$, would not lead to an attenuation of intensity. These definitions will turn out to be very useful, especially when discussing the diffusion approximation in the next section.

2.5 Photon Transport Theory

A mathematical description of the absorption and scattering characteristics of light can be performed in two different ways: analytical theory or transport theory. The first is based on the physics of *Maxwell's equations* and is, at least in principle, the most fundamental approach. However, its applicability is limited due to the complexities involved when deriving exact analytical solutions. Transport theory, on the other hand, directly addresses the transport of photons through absorbing and scattering media without taking Maxwell's equations into account. It is of a heuristic character and lacks the strictness of analytical theories. Nevertheless, transport theory has been used extensively when dealing with laser–tissue interactions, and experimental evidence is given that its predictions are satisfactory in many cases. That is why I decided to include this section on the transport theory of light.

In our presentation, we will follow the excellent review of transport theory given by Ishimaru (1989). The fundamental quantity in transport theory is called the *radiance*[3] $J(\boldsymbol{r}, \boldsymbol{s})$ and is expressed in units of $\mathrm{W\,cm^{-2}\,sr^{-1}}$. It denotes the power flux density in a specific direction $\boldsymbol{s}$ within a unit solid angle $\mathrm{d}\omega$. The governing differential equation for radiance is called the *radiative transport equation* and is given by

$$\frac{\mathrm{d}J(\boldsymbol{r}, \boldsymbol{s})}{\mathrm{d}s} = -\,\alpha_{\mathrm{t}} J(\boldsymbol{r}, \boldsymbol{s}) + \frac{\alpha_{\mathrm{s}}}{4\pi} \int_{4\pi} p(\boldsymbol{s}, \boldsymbol{s}')J(\boldsymbol{r}, \boldsymbol{s}')\, \mathrm{d}\omega' \ , \tag{2.36}$$

---

[3] Note that our definition of *radiance* differs from *intensity* by the extra unit $\mathrm{sr^{-1}}$. We are thus using two different symbols J and I, respectively. This mathematically inconvenient distinction becomes necessary, since scattered photons may proceed into any solid angle.

where $p(s, s')$ is the phase function of a photon to be scattered from direction s' into s, ds is an infinitesimal path length, and $d\omega'$ is the elementary solid angle about the direction s'. If scattering is symmetric about the optical axis, we may set $p(s, s') = p(\theta)$ with θ being the scattering angle as defined in Sect. 2.3.

When performing measurements of optical properties, the observable quantity is the *intensity* which is derived from radiance by integration over the solid angle

$$I(r) = \int_{4\pi} J(r, s) \, d\omega \; . \tag{2.37}$$

On the other hand, radiance may be expressed in terms of intensity by

$$J(r, s) = I(r) \, \delta(\omega - \omega_\text{s}) \; , \tag{2.38}$$

where $\delta(\omega - \omega_\text{s})$ is a solid angle delta function pointing into the direction given by s.

When a laser beam is incident on a turbid medium, the radiance inside the medium can be divided into a coherent and a diffuse term according to the relation

$$J = J_\text{c} + J_\text{d} \; .$$

The coherent radiance is reduced by attenuation due to absorption and scattering of the direct beam. It can thus be calculated from

$$\frac{dJ_\text{c}}{ds} = -\alpha_\text{t} J_\text{c} \; ,$$

with the solution

$$J_\text{c} = I_0 \, \delta(\omega - \omega_\text{s}) \, \exp(-d) \; ,$$

where I_0 is the incident intensity, and the dimensionless parameter d is the optical depth defined by (2.33). Hence, the coherent intensity in turbid media is characterized by an exponential decay.

The main problem transport theory has to deal with is the evaluation of the diffuse radiance, since scattered photons do not follow a determined path. Hence, adequate approximations and statistical approaches must be chosen, mainly depending on the value of the albedo, i.e. whether either absorption or scattering is the dominant process of attenuation. These methods are referred to as *first-order scattering*, *Kubelka–Munk theory*, *diffusion approximation*, *Monte Carlo simulations*, or *inverse adding–doubling*. In the following paragraphs, they will be discussed in this order. Each method is based on certain assumptions regarding initial and boundary conditions. In general, the complexity of either approach is closely related to its accuracy but also to the calculation time needed.

First-Order Scattering. Only if the diffuse radiance is considerably smaller than the coherent radiance, can an analytical solution be given by assuming that

$$J_c + J_d \simeq J_c .$$

This case is called *first-order scattering*, since scattered light can be treated in a similar manner as absorbed light. The intensity at a distance z from the tissue surface is thus given by Lambert's law

$$I(z) = I_0 \exp[-(\alpha + \alpha_s)z] ,$$

where z denotes the axis of the incident beam. Hence, first-order scattering is limited to plane incident waves, and multiple scattering is not taken into account. It is thus a very simple solution and might be useful for a few practical problems only, i.e. if $d \ll 1$ or $a \ll 0.5$. However, in the therapeutic window between 600 nm and 1200 nm, where most tissues are highly transparent, the albedo is close to unity. For these wavelengths, the first-order solution is often not applicable.

Kubelka–Munk Theory. A more useful model which is still restricted to linear geometries has been developed by Kubelka and Munk (1931). Since its parameters are often used in medical physics, the basic idea of this theory shall be described. In contrast to first-order scattering, the main assumption is that the radiance be diffuse, i.e.

$$J_c = 0 .$$

In Fig. 2.11, a geometry is shown which illustrates that two diffuse fluxes inside the tissue can be distinguished: a flux J_1 in the direction of the incident radiation and a backscattered flux J_2 in the opposite direction. Two Kubelka–Munk coefficients, A_{KM} and S_{KM}, are defined for the absorption and scattering of diffuse radiation, respectively[4]. With these parameters, two differential equations can be formed:

$$\frac{dJ_1}{dz} = -S_{KM}J_1 - A_{KM}J_1 + S_{KM}J_2 , \tag{2.39}$$

$$\frac{dJ_2}{dz} = -S_{KM}J_2 - A_{KM}J_2 + S_{KM}J_1 , \tag{2.40}$$

where z denotes the mean direction of incident radiation. Each of these equations states that radiance in either direction encounters two losses due to absorption and scattering and one gain due to scattering of photons from the opposite direction. The general solutions to (2.39) and (2.40) can be expressed by

---

[4] The coefficients A_{KM} and S_{KM} must be distinguished from α and α_s, since the latter are defined for coherent radiation only.

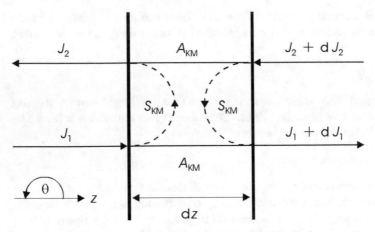

Fig. 2.11. Geometry of two fluxes in Kubelka–Munk theory

$$J_1(z) = c_{11} \exp(-\gamma z) + c_{12} \exp(+\gamma z) \,,$$

$$J_2(z) = c_{21} \exp(-\gamma z) + c_{22} \exp(+\gamma z) \,,$$

with

$$\gamma = \sqrt{A_{\mathrm{KM}}^2 + 2A_{\mathrm{KM}}S_{\mathrm{KM}}} \,.$$

The major problem of Kubelka–Munk theory is the description of A_{KM} and S_{KM} in terms of α and α_s. When considering $\mathrm{d}l$ as an infinitesimal path length of a scattered photon and $\mathrm{d}z$ as an infinitesimal path length of a coherent photon, we may write for the average values

$$\alpha <\mathrm{d}l> = \alpha \,(b <\mathrm{d}z>) = (\alpha \, b) <\mathrm{d}z> = A_{\mathrm{KM}} <\mathrm{d}z> \,, \qquad (2.41)$$

with some numerical factor $b > 1$. When also taking the geometry shown in Fig. 2.12 into account, we obtain

$$\frac{<\mathrm{d}l>}{<\mathrm{d}z>} = \frac{\int_{-1}^{+1} \frac{1}{|\cos\theta|} \, J(z) \, |\cos\theta| \, \mathrm{d}(\cos\theta)}{\int_{-1}^{+1} J(z) \, |\cos\theta| \, \mathrm{d}(\cos\theta)} \,,$$

and, since J does not depend on θ according to our assumption of purely diffuse scattering,

$$\frac{<\mathrm{d}l>}{<\mathrm{d}z>} = \frac{\int_{-1}^{+1} \mathrm{d}(\cos\theta)}{\int_{-1}^{+1} |\cos\theta| \, \mathrm{d}(\cos\theta)} = 2 \,. \qquad (2.42)$$

Combining (2.41) and (2.42) leads to

$$A_{\mathrm{KM}} = 2\alpha \,.$$

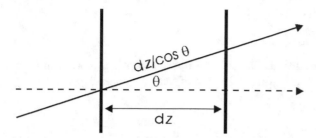

Fig. 2.12. Path length in a thin layer at scattering angle θ

Because only backscattering is assumed as pointed out in Fig. 2.11, the corresponding relation for S_{KM} is given by

$$S_{\text{KM}} = \alpha_{\text{s}} \ .$$

The Kubelka–Munk theory is a special case of the so-called *many flux theory*, where the transport equation is converted into a matrix differential equation by considering the radiance at many discrete angles. A detailed description of the many flux theory is found in the book by Ishimaru (1978). Beside the two fluxes proposed by Kubelka and Munk, other quantities of fluxes were also considered, as for instance seven fluxes by Yoon et al. (1987) or even twenty-two fluxes by Mudgett and Richards (1971). In general, though, all these many flux theories are restricted to a one-dimensional geometry and to the assumption that the incident light be already diffuse. Another disadvantage is the necessity of extensive computer calculations.

Diffusion Approximation. For albedos $a \gg 0.5$, i.e. if scattering overwhelms absorption, the diffuse part of (2.37) tends to be almost isotropic. According to Ishimaru (1989), we may then expand the diffuse radiance J_{d} in a series by

$$J_{\text{d}} = \frac{1}{4\pi} \left(I_{\text{d}} + 3 \boldsymbol{F}_{\text{d}} \boldsymbol{s} + \cdots \right) , \tag{2.43}$$

where I_{d} is the diffuse intensity, and the vector flux $\boldsymbol{F}_{\text{d}}$ is determined by

$$\boldsymbol{F}_{\text{d}}(\boldsymbol{r}) = \int_{4\pi} J_{\text{d}}(\boldsymbol{r}, \boldsymbol{s}) \, \boldsymbol{s} \, d\omega \ .$$

The first two terms of the expansion expressed by (2.43) constitute the diffusion approximation. The diffuse intensity I_{d} itself satisfies the following diffusion equation

$$\left(\nabla^2 - \kappa^2 \right) I_{\text{d}}(\boldsymbol{r}) = - Q(\boldsymbol{r}) , \tag{2.44}$$

where κ^2 is the diffusion parameter, and Q represents the source of scattered photons. It was shown by Ishimaru (1978) that

$$\kappa^2 = 3\alpha \left[\alpha + \alpha_s \left(1 - g \right) \right] \ ,$$

$$Q = 3\alpha_s (\alpha_t + g\alpha) \, F_0 \, \exp(-d) \ ,$$

with the incident flux amplitude F_0 and the optical depth d as determined by (2.33). According to our previous definition of reduced coefficients in (2.34) and (2.35), we thus have

$$\kappa^2 = 3\alpha\alpha_t{}' \ .$$

The diffusion equation (2.44) suggests the introduction of an effective diffusion length L_{eff} and an effective attenuation coefficient α_{eff} of diffuse light which can be defined by

$$L_{\mathrm{eff}} = \frac{1}{\kappa} = \frac{1}{\sqrt{3\alpha\alpha_t{}'}} \ ,$$

$$\alpha_{\mathrm{eff}} = \frac{1}{L_{\mathrm{eff}}} = \sqrt{3\alpha\alpha_t{}'} \ .$$

In general, the diffusion approximation thus states that

$$I = I_c + I_d = A \, \exp(-\alpha_t z) + B \, \exp(-\alpha_{\mathrm{eff}} z) \ , \tag{2.45}$$

with $A + B = I_0$. There exist different sets of values for α, α_s, and g which provide similar radiances in diffusion approximation calculations. They can be expressed in terms of each other by so-called *similarity relations* given by

$$\tilde{\alpha} = \alpha \ ,$$

$$\tilde{\alpha}_s \left(1 - \tilde{g} \right) = \alpha_s \left(1 - g \right) \ ,$$

where tildes indicate transformed parameters. One motive for applying similarity relations is the transformation of anisotropic scattering into isotropic scattering by using

$$\tilde{g} = 0 \ ,$$

$$\tilde{\alpha} = \alpha \ ,$$

$$\tilde{\alpha}_s = \alpha_s \left(1 - g \right) = \alpha_s{}' \ .$$

By this transformation, computer calculations are significantly facilitated, since only absorption coefficient and reduced scattering coefficients are needed for the characterization of optical tissue properties.

In Fig. 2.13, the dependence of the diffuse radiance on optical depth is illustrated in the case of isotropic scattering ($g = 0$) and different albedos ($0 < a < 1$). For $a = 0$, attenuation follows Lambert's law of absorption. For $a = 1$, the radiance obviously approaches an asymptotic value. It is also interesting to note that just beneath the surface of the scattering medium the diffuse intensity exceeds the incident intensity, since backscattered photons from deeper layers must be added to the incident intensity.

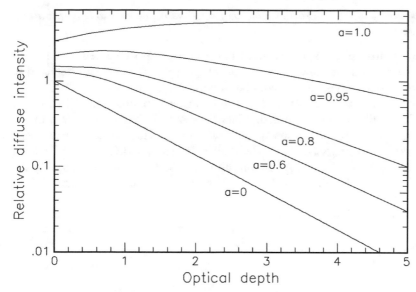

Fig. 2.13. Diffuse intensity as a function of optical depth, assuming the validity of the diffusion approximation and isotropic scattering. The ordinate expresses diffuse intensity in units of incident intensity. Data according to van Gemert et al. (1990)

Monte Carlo Simulations. A numerical approach to the transport equation (2.36) is based on Monte Carlo simulations. The Monte Carlo method essentially runs a computer simulation of the random walk of a number N of photons. It is thus a statistical approach. Since the accuracy of results based on statistics is proportional to $\sqrt{N}$, a large number of photons has to be taken into account to yield a valuable approximation. Therefore, the whole procedure is rather time-consuming and can be efficiently performed only on large computers. The Monte Carlo method was first proposed by Metropolis and Ulam (1949). It has meanwhile advanced to a powerful tool in many disciplines. Here, we will first point out the basic idea and then briefly discuss each step of the simulation.

The principal idea of Monte Carlo simulations applied to absorption and scattering phenomena is to follow the optical path of a photon through the turbid medium. The distance between two collisions is selected from a logarithmic distribution, using a random number generated by the computer. Absorption is accounted for by implementing a weight to each photon and permanently reducing this weight during propagation. If scattering is to occur, a new direction of propagation is chosen according to a given phase function and another random number. The whole procedure continues until the photon escapes from the considered volume or its weight reaches a given cutoff value. According to Meier et al. (1978) and Groenhuis et al. (1983), Monte Carlo simulations of absorption and scattering consist of five principal steps: *source photon generation, pathway generation, absorption, elimination,* and *detection.*

- Step 1: Source photon generation. Photons are generated at a surface of the considered medium. Their spatial and angular distribution can be fitted to a given light source, e.g. a Gaussian beam.
- Step 2: Pathway generation. After generating a photon, the distance to the first collision is determined. Absorbing and scattering particles in the turbid medium are supposed to be randomly distributed. Thus, the mean free path is $1/\varrho\sigma_s$, where ϱ is the density of the particles and σ_s is their scattering cross-section[5]. A random number $0 < \xi_1 < 1$ is generated by the computer, and the distance $L(\xi_1)$ to the next collision is calculated from

$$L(\xi_1) = - \frac{\ln \xi_1}{\varrho\sigma_s} \ .$$

Since

$$\int_0^1 \ln \xi_1 \ d\xi_1 = -1 \ ,$$

the average value of $L(\xi_1)$ is indeed $1/\varrho\sigma_s$. Hence, a scattering point has been obtained. The scattering angle is determined by a second random number ξ_2 in accordance with a certain phase function, e.g. the Henyey–Greenstein function. The corresponding azimuth angle Φ is chosen as

$$\Phi = 2\pi \, \xi_3 \ ,$$

where ξ_3 is a third random number between 0 and 1.
- Step 3: Absorption. To account for absorption, a weight is attributed to each photon. When entering the turbid medium, the weight of the photon is unity. Due to absorption – in a more accurate program also due to reflection – the weight is reduced by $\exp[-\alpha L(\xi_1)]$, where α is the absorption coefficient. As an alternative to implementing a weight, a fourth random number ξ_4 between 0 and 1 can be drawn. Then, instead of assuming only scattering events in Step 2, scattering takes place if $\xi_4 < a$, where a is the albedo. For $\xi_4 > a$, on the other hand, the photon is absorbed which then is equivalent to Step 4.
- Step 4: Elimination. This step only applies if a weight has been attributed to each photon (see Step 3.). When this weight reaches a certain cutoff value, the photon is eliminated. Then, a new photon is launched, and the program proceeds with Step 1.
- Step 5: Detection. After having repeated Steps 1–4 for a sufficient number of photons, a map of pathways is calculated and stored in the computer. Thus, statistical statements can be made about the fraction of incident photons being absorbed by the medium as well as the spatial and angular distribution of photons having escaped from it.

---

[5] Absorption will be taken into account in Step 3.

It has already been mentioned that the accuracy of Monte Carlo simulations strongly increases with the number of photons taken into account. Due to the necessity of extensive computer calculations, though, this is a rather time-consuming process. A very powerful approach has been introduced by Graaff et al. (1993a) in what they called *condensed Monte Carlo simulations*. The results of earlier calculations can be stored and used again if needed for the same phase function but for different values of the absorption coefficient and the albedo. When applying this alternative technique, a considerable amount of computing time can be saved.

Inverse Adding–Doubling Method. Another numerical approach to the transport equation is called *inverse adding–doubling* which was first introduced by Prahl et al. (1993). The term "inverse" implies a reversal of the usual process of calculating reflectance and transmittance from optical properties. The term "adding–doubling" refers to earlier techniques established by van de Hulst (1962) and Plass et al. (1973). The doubling method assumes that reflection and transmission of light incident at a certain angle is known for one layer of a tissue slab. The same properties for a layer twice as thick is found by dividing it into two equal slabs and adding the reflection and transmission contributions from either slab. Thus, reflection and transmission for an arbitrary slab of tissue can be calculated by starting with a thin slab with known properties, e.g. as obtained by absorption and single scattering measurements, and doubling it until the desired thickness is achieved. The adding method extends the doubling method to dissimilar slabs of tissue. With this supplement, layered tissues with different optical properties can be simulated, as well.

So far, we have always assumed that the radiance J is a scalar and polarization effects are negligible. In the 1980s, several extensive studies were done on transport theory pointing out the importance of additional polarizing effects. A good summary is given in the paper by Ishimaru and Yeh (1984). Herein, the radiance is replaced by a four-dimensional *Stokes vector*, and the phase function by a 4×4 *Müller matrix*. The Stokes vector accounts for all states of polarization. The Müller matrix describes the probability of a photon to be scattered into a certain direction at a given polarization. The transport equation then becomes a matrix integro-differential equation and is called a *vector transport equation*.

In this section, we have discussed different methods for solving the transport equation. Among them, the most important are the Kubelka–Munk theory, the diffusion approximation, and Monte Carlo simulations. In a short summary, we will now compare either method with each other. The Kubelka–Munk theory can deal with diffuse radiation only and is limited to those cases where scattering overwhelms absorption. Another disadvantage of the Kubelka–Munk theory is that it can be applied to one-dimensional geometries only. The diffusion approximation, on the other hand, is not restricted to diffuse radiation but is also limited to predominant scattering. In the latter case, though, it can be regarded as a powerful tool. Monte Carlo simulations, finally, provide the most accurate solutions, since a variety of input parameters may be considered in specially developed computer programs. Moreover, two-dimensional and three-dimensional geometries can be evaluated, even though they often require long computation times.

In Fig. 2.14, the intensity distributions inside a turbid medium calculated with either method are compared with each other. Because isotropic scattering is assumed, an analytical solution can also be considered which is labeled "transport theory". Two different albedos, $a = 0.9$ and $a = 0.99$, are taken into account. Hence, the coefficient of scattering surpasses the coefficient of absorption by a factor of 9 or 99, respectively. The accordance of all four approaches is fairly good. The Kubelka–Munk theory usually yields higher values, whereas diffusion approximation and Monte Carlo simulation frequently underestimate the intensity.

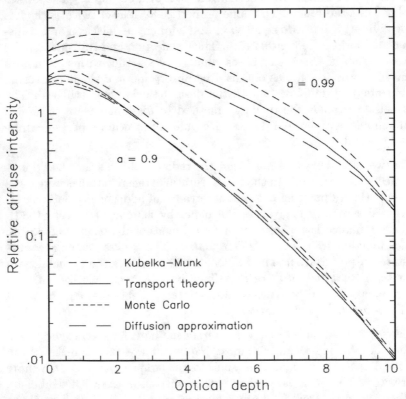

Fig. 2.14. Comparison of different methods for calculating the diffuse intensity as a function of optical depth. The ordinate expresses diffuse intensity in units of incident intensity. Isotropic scattering is assumed. Data according to Wilson and Patterson (1986)

2.6 Measurement of Optical Tissue Properties

In general, there exist several methods for the measurement of optical tissue properties. They focus on different quantities such as transmitted, reflected, and scattered intensities. The absorbance itself is difficult to determine, since photons absorbed by the tissue can no longer be used for detection. Therefore, the absorbed intensity is usually obtained when subtracting transmitted, reflected, and scattered intensities from the incident intensity. Depending on the experimental method, either only the total attenuation coefficient or the coefficients of both absorption and scattering can be evaluated. If the angular dependence of the scattered intensity is measured by rotating the corresponding detector, the coefficient of anisotropy can be obtained, as well.

In Fig. 2.15a–c, three commonly used arrangements are illustrated. The simplest setup for measuring total attenuation is shown in Fig. 2.15a. By means of a beamsplitter, typically 50 % of the laser radiation is directed on one detector serving as a reference signal. The other 50 % is applied to the tissue sample. A second detector behind the sample and on the optical axis of the beam measures the transmitted intensity. By subtracting this intensity from the intensity measured by the reference detector, the attenuation coefficient[6] of the tissue can be obtained. The reader is reminded that total attenuation is due to both absorption and scattering. Thus, with this measurement no distinction can be made between these two processes.

In Fig. 2.15b, a setup for the evaluation of the absorbance is shown. Its basic component is called an *integrating sphere* and was discussed in theoretical studies by Jacquez and Kuppenheim (1955), Miller and Sant (1958), and Goebel (1967). The sphere has a highly reflecting coating. An integrated detector only measures light that has not been absorbed by a sample placed inside the sphere. Usually, the experiment consists of two measurements: one with and the other without the sample. The difference in both detected intensities is the amount absorbed by the sample. Thus, when taking the geometrical dimensions of the sample into account, its absorption coefficient can be obtained. Frequently, a baffle is positioned between sample and detector to prevent specular reflection from being detected.

A third experiment is illustrated in Fig. 2.15c. Here, the angular dependence of scattering can be measured by moving the detector on a 360° rotary stage around the sample. From the detected signals, the corresponding phase function of scattering is obtained. By fitting this result to a given function, e.g. the Henyey–Greenstein phase function, the coefficient of anisotropy can also be evaluated.

---

[6] Actually, this method does not lead to the same attenuation coefficient as defined by (2.30), since reflection on the tissue surface should also be taken into account. Specular reflection can be measured by placing a third detector opposite to the beamsplitter. Diffuse reflection can be evaluated only when using two integrating spheres as will be discussed below.

a

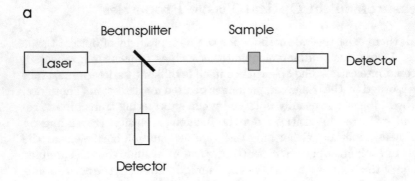

b

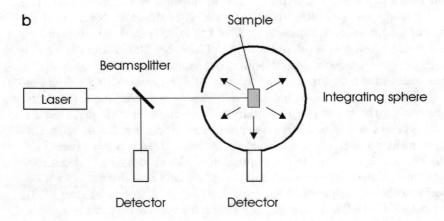

c

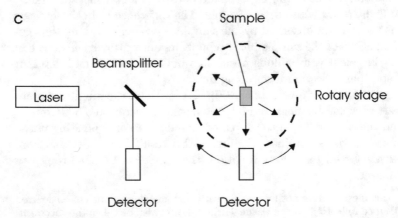

Fig. 2.15. (a) Experimental setup for measuring total attenuation. **(b)** Experimental setup for measuring absorption. **(c)** Experimental setup for measuring the angular dependence of scattering

The main disadvantage of the techniques shown in Fig. 2.15a–c is that they cannot be performed simultaneously. However, it is well known that the optical properties of biological tissue are altered during heating which is often associated with exposure to laser radiation. Hence, it is more accurate to measure these properties in the same experimental arrangement and at the same time. One commonly used setup satisfying this requirement is the double-integrating sphere geometry shown in Fig. 2.16. It was first applied to the measurement of optical tissue properties by Derbyshire et al. (1990) and Rol et al. (1990), and has meanwhile turned into an unsurpassed diagnostic tool.

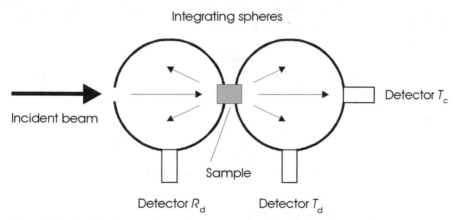

Fig. 2.16. Double-integrating sphere geometry for the simultaneous measurement of different optical tissue properties. The detectors measure the transmitted coherent intensity (T_c), the transmitted diffuse intensity (T_d), and the reflected diffuse intensity (R_d)

The double-integrating sphere geometry incorporates two more or less identical spheres which are located in front and behind the sample to be investigated. One sphere integrates all radiation which is either reflected or scattered backwards from the sample. Transmitted and forward scattered radiation is absorbed in the second sphere. With three detectors, all required measurements can be performed simultaneously. In Fig. 2.16, these detectors are labeled T_c for coherent transmittance, T_d for diffuse transmittance, and R_d for diffuse reflectance, respectively. Specular reflection can again be avoided by placing baffles between the sample and each detector. However, a small fraction of light in either sphere might penetrate again through the sample, thereby reaching the other sphere. The possibility of such an effect was first recognized by Pickering et al. (1992) who then developed an improved model by taking the multiple exchange of light between both spheres into account. First results regarding optical tissue properties with this more accurate theory were published by Pickering et al. (1993).

Since the Kubelka–Munk theory is one of the most frequently used methods to obtain data on optical tissue properties, we shall briefly discuss how the Kubelka–Munk coefficients introduced in Sect. 2.5 relate to measured values of diffuse reflectance and transmission. According to Kottler (1960), these expressions are given by

$$R_d = \frac{\sinh(S_{KM}yD)}{x\cosh(S_{KM}yD) + y\sinh(S_{KM}yD)} \; , \tag{2.46}$$

$$T_d = \frac{y}{x\cosh(S_{KM}yD) + y\sinh(S_{KM}yD)} \; , \tag{2.47}$$

where D is the optical thickness of the slab to be considered, and S_{KM} is the Kubelka–Munk coefficient for scattering. The parameters x and y can be expressed in terms of

$$x = \frac{1 + R_d^2 - T_d^2}{2\,R_d} \; , \tag{2.48}$$

$$y = \sqrt{x^2 - 1} \; . \tag{2.49}$$

According to Kottler (1960), the convenience of applying the Kubelka–Munk method arises from the fact that scattering and absorption coefficients may be directly calculated from measured reflection and transmission coefficients, i.e. by means of

$$S_{KM} = \frac{1}{yD} \, \ln\left[\frac{1 - R_d(x - y)}{T_d}\right] \; , \tag{2.50}$$

$$A_{KM} = (x - 1)\,S_{KM} \; . \tag{2.51}$$

The simplicity of the Kubelka–Munk model has made it a popular method for measuring optical properties of matter. In general, the model is based on the propagation of a uniform and diffuse radiance in a one-dimensional geometry. This is equivalent to the requirement of a forward and backward peaked phase function. Unfortunately, though, the assumptions of a purely diffuse radiance and a one-dimensional geometry are not satisfied by biological tissues. Thus, several attempts have been made to extend the Kubelka–Munk theory to partially collimated radiance, e.g. by Kottler (1960) and van Gemert et al. (1987), and to the case of anisotropic scattering, e.g. by Meador and Weaver (1979) and Jacques and Prahl (1987a). Despite these improvements, this method remains a rather simple approximation for the propagation of laser light in biological tissues.

Nearly all optical properties can be classified into either transport coefficients (α, α_s, g) or Kubelka–Munk coefficients (A_{KM}, S_{KM}), depending on the theory used to obtain them. It is not surprising that transport properties are derived from theories based on the transport equation (2.36), whereas Kubelka–Munk properties are calculated from (2.50) and (2.51).

Before providing a detailed list of optical tissue parameters, it should be noted that absorption and scattering coefficients may change during laser exposure. When discussing thermal effects in Sect. 3.2, we will find that carbonization, in particular, leads to increased absorption. However, the occurrence of carbonization is usually avoided during any kind of clinical surgery. Scattering, on the other hand, is already affected at lower temperatures, e.g. when tissue is coagulated[7]. In Fig. 2.17, the absorption and scattering coefficients are shown for white matter of human brain as calculated by Roggan et al. (1995a), using the Kubelka–Munk theory. Coagulation was achieved by keeping the tissue in a bath of hot water at approximately 75°C.

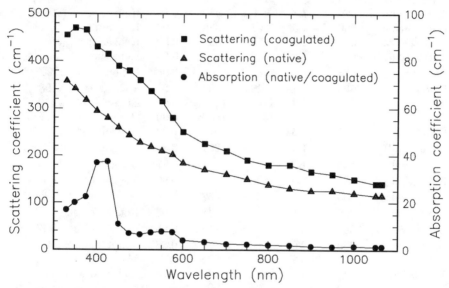

Fig. 2.17. Optical properties of white matter of human brain in its native and coagulated state. Data according to Roggan et al. (1995a)

We have now discussed different theoretical approaches and experimental setups for determining optical tissue properties. It is important to realize that both theory and experiment are necessary in order to yield valuable results. During the past decade, several research groups have been investigating all kinds of biological tissue. A short summary of the most significant results is given in Table 2.3. Type of tissue, laser wavelength, absorption coefficient, scattering coefficient, attenuation coefficient, coefficient of anisotropy, reference, and the applied theoretical method are listed. Additional data can be found in a review paper published by Cheong et al. (1990).

---

[7] Coagulation is a thermal effect which is very useful for many types of surgery, especially in achieving hemostasis and in laser-induced interstitial thermotherapy (LITT). Further details are given in Sect. 3.2.

Table 2.3. Optical properties of human tissues in vitro

Tissue	λ (nm)	α (cm^{-1})	α_s (cm^{-1})	α_t (cm^{-1})	g	Reference
Aorta advent.	476	18.1	267	285	0.74	Keijzer et al. (1989)[c]
Aorta advent.	580	11.3	217	228	0.77	Keijzer et al. (1989)[c]
Aorta advent.	600	6.1	211	217	0.78	Keijzer et al. (1989)[c]
Aorta advent.	633	5.8	195	201	0.81	Keijzer et al. (1989)[c]
Aorta intima	476	14.8	237	252	0.81	Keijzer et al. (1989)[c]
Aorta intima	580	8.9	183	192	0.81	Keijzer et al. (1989)[c]
Aorta intima	600	4.0	178	182	0.81	Keijzer et al. (1989)[c]
Aorta intima	633	3.6	171	175	0.85	Keijzer et al. (1989)[c]
Aorta media	476	7.3	410	417	0.89	Keijzer et al. (1989)[c]
Aorta media	580	4.8	331	336	0.90	Keijzer et al. (1989)[c]
Aorta media	600	2.5	323	326	0.89	Keijzer et al. (1989)[c]
Aorta media	633	2.3	310	312	0.90	Keijzer et al. (1989)[c]
Bladder	633	1.4	88.0	89.4	0.96	Cheong et al. (1987)[b]
Bladder	633	1.4	29.3	30.7	0.91	Splinter et al. (1989)[b]
Blood	665	1.3	1246	1247	0.99	Reynolds et al. (1976)[a]
Blood	685	2.65	1413	1416	0.99	Pedersen et al. (1976)[c]
Blood	960	2.84	505	508	0.99	Reynolds et al. (1976)[a]
Bone (skull)	488	1.4	200	201	0.87	Roggan et al. (1995a)[d]
Bone (skull)	514	1.3	190	191	0.87	Roggan et al. (1995a)[d]
Bone (skull)	1064	0.5	120	121	0.90	Roggan et al. (1995a)[d]
Brain (white)	633	1.58	51.0	52.6	0.96	Splinter et al. (1989)[b]
Brain (white)	850	0.8	140	141	0.95	Roggan et al. (1995a)[d]
Brain (white)	1064	0.4	110	110	0.95	Roggan et al. (1995a)[d]
Brain (grey)	633	2.63	60.2	62.8	0.88	Splinter et al. (1989)[b]
Breast	635	<0.2	395	395	–	Marchesini et al. (1989)[a]
Gallbladder	633	4.0	182	186	0.94	Maitland et al. (1993)[c]
Liver	515	18.9	285	304	–	Marchesini et al. (1989)[a]
Liver	635	2.3	313	315	0.68	Marchesini et al. (1989)[a]
Liver	850	0.3	200	200	0.95	Roggan et al. (1995a)[d]
Liver	1064	0.3	150	150	0.93	Roggan et al. (1995a)[d]
Lung	515	25.5	356	382	–	Marchesini et al. (1989)[a]
Lung	635	8.1	324	332	0.75	Marchesini et al. (1989)[a]
Muscle	515	11.2	530	541	–	Marchesini et al. (1989)[a]
Myocardium	1064	0.4	175	175	0.97	Splinter et al. (1993)[d]
Prostate	850	0.6	130	131	0.96	Roggan et al. (1995a)[d]
Prostate	1064	0.4	110	110	0.96	Roggan et al. (1995a)[d]
Skin (white)	633	2.7	187	190	0.81	Jacques et al. (1987b)[c]
Skin (white)	700	2.7	237	240	0.91	Graaff et al. (1993b)[d]
Skin (dark)	700	8.1	229	237	0.91	Graaff et al. (1993b)[d]
Uterus	635	0.35	394	394	0.69	Marchesini et al. (1989)[a]

[a] Lambert's law
[b] Kubelka–Munk or 3-flux theory
[c] Diffusion approximation
[d] Monte Carlo simulations

Finally, the reader is reminded that biological tissue is something very inhomogeneous and fragile. The inhomogeneity makes it difficult to transfer experimental data from one sample to another. Usually, it is taken into account by applying generous error bars. Throughout this book, however, error bars are often neglected to facilitate readability of the figures. Fragility, on the other hand, is something the experimenter is confronted with. It was pointed out by Graaff et al. (1993b) that optical properties determined *in vitro* may differ extremely from those valid *in vivo*. There are several reasons to be considered: First of all, living tissue does not have the same morphologic structure as excised tissue. One typical example is corneal tissue which turns into a turbid material within a few hours after dissection. Secondly, alterations are induced by unavoidable deformation and handling of the tissue such as drying, freezing, or just soaking in saline. According to Cilesiz and Welch (1993), it is dehydration in particular that leads to a gross effect on the optical properties of tissue[8]. During all these processes, scattering is extremely enhanced. Always keep in mind that published data are valid only for the documented type of tissue preparation. When needing data for clinical applications, a careful analysis of the specific setup should be performed instead of blindly relying on measurements reported elsewhere.

2.7 Questions to Chapter 2

Q2.1. A laser beam is refracted on its transition from air to corneal tissue. The angle of incidence is θ. Which angle θ'' is a potential angle of refraction?
A: $\theta'' < \theta$.   B: $\theta'' = \theta$.   C: $\theta'' > \theta$.

Q2.2. A biological tissue has the optical albedo $a = 0.9$. Its coefficients of absorption and scattering are α and α_s, respectively. Which is correct?
A: $\alpha = 0.9\alpha_s$.   B: $\alpha = 9\alpha_s$.   C: $\alpha_s = 9\alpha$.

Q2.3. The diffusion approximation may be applied to a biological tissue with an optical albedo
A: $a = 0.05$.   B: $a = 0.5$.   C: $a = 0.95$.

Q2.4. Lambert's law describes the loss in intensity due to
A: absorption.   B: scattering.   C: total attenuation.

Q2.5. If g is the coefficient of anisotropy, then isotropic scattering is characterized by
A: $g = 0$.   B: $g = 0.5$.   C: $g = 1$.

Q2.6. The index of refraction is approximately 1.5 for most glasses at visible wavelengths. How much light is lost due to reflection, when a red laser beam accidentally exits a laboratory through an observation window?

Q2.7. A 1 mm thick optical filter has an absorption coefficient of $10\,\mathrm{cm}^{-1}$ for a He-Ne laser at a wavelength of 633 nm. A collinear 5 mW beam is attenuated

---

[8] Theoretical and experimental results regarding the optical response of laser-irradiated tissue are summarized in the excellent book edited by Welch and van Gemert (1995).

by the filter with the beam propagation being perpendicular to the filter surface. What is the intensity of the attenuated laser beam when assuming that attenuation is due to absorption only?

Q2.8. Compare the intensities of scattered light in Rayleigh scattering for a 1 W Nd:YAG laser at a wavelength of 1064 nm and a frequency-doubled 1 W Nd:YAG laser at a wavelength of 532 nm.

Q2.9. The aortic wall has an absorption coefficient of $2.3\,\mathrm{cm}^{-1}$ and a scattering coefficient of $310\,\mathrm{cm}^{-1}$ for a He-Ne laser at a wavelength of 633 nm. What is the optical albedo of the tissue?

Q2.10. When cooking an egg "sunny side up" in a pan, the egg white turns from transparent to white at a temperature of approximately 60°C. Why?

3. Light Acting on Matter

In the previous chapter we learned that matter is acting on light in manifold ways. Now we are going to address the huge variety of how light is acting on matter. Specific tissue parameters as well as laser parameters contribute to this variety. The tissue parameters are classified into optical tissue parameters (such as the coefficients of reflection, absorption, and scattering) and thermal tissue parameters (such as heat conduction and heat capacity, defined in this chapter). On the other hand, we now have to take the whole set of laser parameters into account too: wavelength, exposure time, applied energy, focal spot size, energy density, and power density[1]. Among these, the exposure time is a very crucial parameter when selecting a certain type of interaction, as we will find later on.

During the first decades after the invention of the laser by Maiman (1960), many studies were conducted investigating potential interaction effects using various types of lasers and tissues. Although the number of possible combinations of experimental parameters is unlimited, mainly five categories of interaction types are classified today: *photochemical interactions*, *photothermal interactions*, *photoablation*, *plasma-induced ablation*, and *photodisruption*. Each of these interaction types will be thoroughly discussed in this chapter. In particular, the physical principles governing these interactions are reviewed. We will pay special attention to the microscopic mechanisms controlling various processes of laser energy conversion. Each type of interaction will be introduced by common macroscopic observations including typical experimental data and/or histology of tissue samples after laser exposure. At the end of each discussion, a comprehensive summary of the specific interaction type is given.

Before going into the details, I wish to draw your attention to one very interesting observation: all these obviously different interaction types share a single common datum—the associated energy densities are all within the rather narrow range from approximately $1 \, \mathrm{J/cm^2}$ to $1000 \, \mathrm{J/cm^2}$. This is very

---

[1] In the literature, the terms used for radiometric parameters such as fluence, irradiance, intensity, and energy dose are often somewhat confusing. Throughout this book, the following agreements are met: *power* is expressed in units of W, *energy* in units of J, the synonyms *power density*, *intensity*, and *irradiance* in units of $\mathrm{W/cm^2}$, the synonyms *energy density*, *fluence*, and *radiant exposure* in units of $\mathrm{J/cm^2}$, and *energy dose* in units of $\mathrm{J/cm^3}$ (see also the Appendix A.2).

© Springer Nature Switzerland AG 2019
M. H. Niemz, *Laser-Tissue Interactions*,
https://doi.org/10.1007/978-3-030-11917-1_3

surprising, since the power density itself varies over 15 orders of magnitude! It is just one parameter which controls all of these processes: the duration of laser exposure which is actually identical with the interaction time itself.

A double-logarithmic map with the five basic interaction types is shown in Fig. 3.1 as found in several experiments. The ordinate expresses the applied power density or irradiance in W/cm^2. The abscissa represents the exposure time in seconds. Two diagonals show constant energy fluences at $1\,J/cm^2$ and $1000\,J/cm^2$, respectively. According to this chart, the time scale can roughly be divided into five sections: continuous wave or exposure times $> 1\,s$ for *photochemical interactions*, 1 min down to approximately 1 µs for *photothermal interactions*, 100 ns down to 10 ns for *photoablation*, 100 ns down to 100 fs for *photodisruption*, and 500 ps down to 100 fs for *plasma-induced ablation*. The last two types of interaction are related to each other, but the higher fluence in photodisruption causes additional mechanical effects.

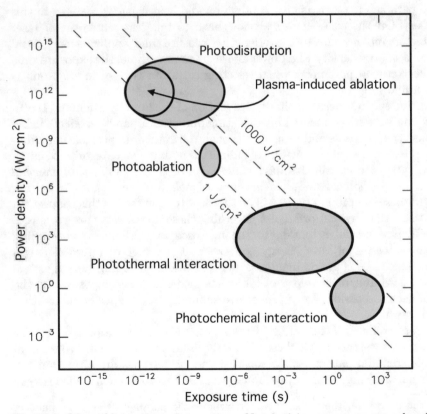

Fig. 3.1. Map of laser–tissue interactions. Non-ionizing processes are colored red, ionizing processes are colored blue. The ellipses give only rough estimates of the associated laser parameters. Photoablation covers a very small area because excimer lasers (which cause photoablation) only offer a very limited range of pulse durations. Modified from Boulnois (1986)

No matter whether you are a physician, a research scholar, an engineer, or a graduate student—please always think of these three questions first when searching for a laser that best suits your application:

---

1. Which is my target? $\Longrightarrow$ **WAVELENGTH (COLOR)**
The wavelength determines where the photon energy is absorbed, for example in water molecules or in hemoglobin. My advice: Think of each wavelength as a specific hormone that binds to some target!

2. Which interaction do I want? $\Longrightarrow$ **EXPOSURE TIME**
The exposure time determines which interaction type takes place, for example photochemical interaction or photothermal interaction.

3. Which strength do I want? $\Longrightarrow$ **FLUENCE**
The fluence determines how strong the interaction will be, for example weak shock waves or strong shock waves.

Also, please keep in mind that the five interaction types cannot always so strictly be distinguished as sketched in Fig. 3.1. For example, ultrashort laser pulses – each of them having no thermal effect – may add up to a measurable increase in temperature if applied at high repetition rates.

3.1 Photochemical Interaction

The group of photochemical interactions stems from empirical observations that light can trigger chemical reactions within macromolecules or tissues. One of the most popular examples was created by evolution itself: the release of energy during photosynthesis where sunlight is absorbed by chlorophyll. In medical applications, photochemical interactions play a significant role in wound healing, pain relief, disinfection, and the treatment of tumors. If light is the only input needed to trigger the chemical reaction (as in wound healing and pain relief), we nowadays speak of *photobiomodulation (PBM)*. If an additional chemical is needed to absorb the incident light, the treatments are referred to as *photoactivated disinfection (PAD)* and *photodynamic therapy (PDT)*. Historically, research on PAD and PDT was performed much earlier than on PBM. The reason is simple: It is much easier to study the combined action of light and an exogenous chemical rather than of light and an endogenous chemical, especially if the latter one had not been identified yet.

3.1.1 Photoactivated Disinfection (PAD)

During PAD, killing of bacteria is achieved by adding a chromophore (light absorbing molecule) to the tissue which turns toxic only upon absorbing light. A chromophore compound which is capable of causing light-induced reactions in other non-absorbing molecules is called a *photosensitizer*. After resonant excitation by laser irradiation, the photosensitizer performs several simultaneous or sequential decays which result in intramolecular transfer reactions. At the end of these diverse reaction channels, highly cytotoxic reactands are released causing an irreversible oxidation of essential cell structures. Thus, the main idea of photochemical treatment is to use a chromophore receptor acting as a catalyst. Its excited states are able to store energy transferred from resonant absorption. The release of this energy produces toxic compounds, setting the photosensitizer back to its original state.

Most photosensitizers belong to the group of organic dyes. Their electronic states are referred to as singlet states (total electron spin momentum $s = 0$) and triplet states ($s = 1$). Each state subdivides into a band of vibrational states. Intersystem crossing is permitted but comes with an increased lifetime. Potential reaction kinetics are listed in Table 3.1.

Table 3.1. Kinetics of photosensitization (S: photosensitizer, RH: substrate with H-bond, CAR: carotenoid). As a catalyst, the photosensitizer (colored red) is finally set back to its original state. Modified from Boulnois (1986)

Excitation	
• Singlet state absorption	$^1\mathrm{S} + h\nu \Longrightarrow\, ^1\mathrm{S}^*$
Decays	
• Radiative singlet decay	$^1\mathrm{S}^* \Longrightarrow\, ^1\mathrm{S} + h\nu'$ (*fluorescence*)
• Non-radiative singlet decay	$^1\mathrm{S}^* \Longrightarrow\, ^1\mathrm{S}$
• Intersystem crossing	$^1\mathrm{S}^* \Longrightarrow\, ^3\mathrm{S}^*$
• Radiative triplet decay	$^3\mathrm{S}^* \Longrightarrow\, ^1\mathrm{S} + h\nu''$ (*phosphorescence*)
• Non-radiative triplet decay	$^3\mathrm{S}^* \Longrightarrow\, ^1\mathrm{S}$
Type I reactions	
• Hydrogen transfer	$^3\mathrm{S}^* + \mathrm{RH} \Longrightarrow \mathrm{SH}^\bullet + \mathrm{R}^\bullet$
• Electron transfer	$^3\mathrm{S}^* + \mathrm{RH} \Longrightarrow \mathrm{S}^{\bullet-} + \mathrm{RH}^{\bullet+}$
• Formation of hydrogen dioxide	$\mathrm{SH}^\bullet + {}^3\mathrm{O}_2 \Longrightarrow\, ^1\mathrm{S} + \mathrm{HO}_2^\bullet$
• Formation of superoxide anion	$\mathrm{S}^{\bullet-} + {}^3\mathrm{O}_2 \Longrightarrow\, ^1\mathrm{S} + {}^3\mathrm{O}_2^{\bullet-}$
Type II reactions	
• Intramolecular exchange	$^3\mathrm{S}^* + {}^3\mathrm{O}_2 \Longrightarrow\, ^1\mathrm{S} + {}^1\mathrm{O}_2^*$
• Cellular oxidation	$^1\mathrm{O}_2^* + \mathrm{cell} \Longrightarrow \mathrm{cell}_{\mathrm{ox}}$
Carotenoid protection	
• Singlet oxygen extinction	$^1\mathrm{O}_2^* + {}^1\mathrm{CAR} \Longrightarrow\, ^3\mathrm{O}_2 + {}^3\mathrm{CAR}^*$
• Deactivation	$^3\mathrm{CAR}^* \Longrightarrow\, ^1\mathrm{CAR} + \mathrm{heat}$

After the absorption of laser photons, the photosensitizer is first transferred to an excited singlet state $^1\mathrm{S}^*$. Then, three potential decay channels are available: radiative and non-radiative singlet decay to the singlet ground state, and intersystem crossing to an excited triplet state. The latter may also promote to the singlet ground state by either radiative or non-radiative triplet decay. The radiative singlet and triplet decays are called *fluorescence* and *phosphorescence*, respectively. Typical lifetimes of fluorescence emission are of the order of nanoseconds, whereas phosphorescence may last from a few microseconds up to several hours. According to Foote (1968), two alternative reaction mechanisms exist for the decay of an excited triplet state (compare Table 3.1): *type I reactions* generating free radicals, and *type II reactions* transferring excitation energy to oxygen molecules.

During Type I reactions, the triplet state next interacts with a target molecule other than oxygen, resulting in the release of free neutral or ionized radicals. Further reaction with triplet oxygen may lead to the formation of hydrogen dioxide or superoxide anions. In Type II reactions, the triplet state of the photosensitizer directly interacts with molecular triplet oxygen $^3O_2$ which is then transferred to an excited singlet state $^1O_2^*$. During this reaction, the electronic spins are flipped in the following manner:

$$\uparrow\uparrow \qquad \uparrow\uparrow \qquad \uparrow\downarrow \qquad \uparrow\downarrow$$
$$^3S^* \quad + \quad ^3O_2 \quad \Longrightarrow \quad ^1S \quad + \quad ^1O_2^* \,.$$

Excited singlet oxygen $^1O_2^*$ is very reactive, thus leading to cellular oxidation and killing of bacteria. Malik et al. (1990) have observed bactericidal effects of laser-activated porphyrins. Wilson et al. (1993) have investigated the effect of different photosensitizers on *streptococcus sanguis*, a common bacterium in dental plaques. Some of their findings are summarized in Figs. 3.2a–b. Obviously, only the combined action of photosensitizer and laser exposure significantly reduces the fraction of surviving bacteria.

Today, tolonium chloride is one of the most often applied photosensitizers for PAD, especially in dentistry. Lee et al. (2004) and Williams et al. (2006) reported on a successful disinfection of root canals using tolonium chloride and diode lasers.

3.1.2 Photodynamic Therapy (PDT)

Since the very beginning of the 20th century, certain dyes have been known to induce photosensitizing effects as described by von Tappeiner (1900). The first intended therapeutic application of dyes in combination with light was proposed by von Tappeiner and Jesionek (1903). Later, it was observed by Auler and Banzer (1942) that for certain porphyrins the clearance period in tumor cells is longer than in healthy cells. If these dyes were somehow transferred to a toxic state, the tumor cells would preferentially be treated. Kelly and Snell (1976) have reported on the first endoscopic application of a photosensitizer in the case of human bladder carcinoma..

One of the first photosensitizers ever used in photodynamic therapy was a hematoporphyrin derivative called HpD. It is derived from calf blood and is a complex collection of different porphyrins, mainly dihematoporphyrin, hydroxyethylvinyl-deuteroporphyrin, and protoporphyrin. Among those sub-

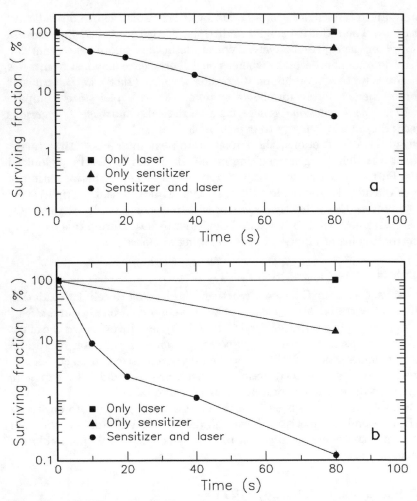

Fig. 3.2. (a) Effect of methylene blue and/or helium–neon laser (power: 7.3 mW) on the viability of streptococcus sanguis. **(b)** Effect of hematoporphyrin ester and/or helium–neon laser (power: 7.3 mW) on the viability of streptococcus sanguis. Data according to Wilson et al. (1993)

stances, dihematoporphyrin is the active constituent in providing the photo-sensitizing effect. Medical application of HpD was first performed and reported by Lipson and Baldes (1961). Proprietary names for HpD include Photofrin I and Photofrin II. Both of these agents are complex chemical mixtures with the latter one being strongly enriched in the tumor-localizing fraction during PDT. The chemical structure of dihematoporphyrin is shown in Fig. 3.3. It consists of two porphyrin rings connected by a C–O–C chain. According to Dolphin (1979), porphyrins are characterized by a high thermal stability and a dark color (from Greek: πορφυρα = purple).

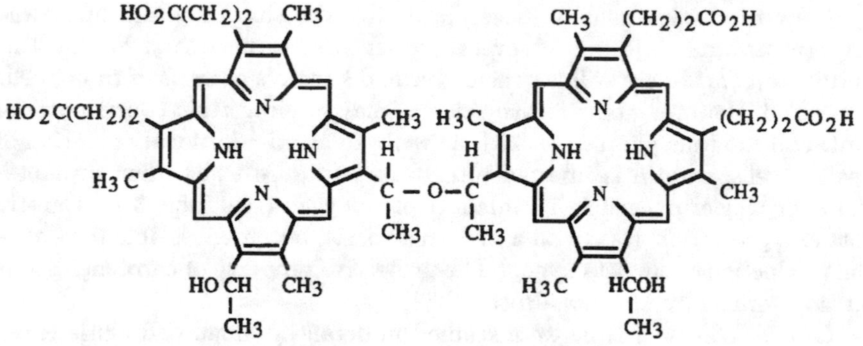

HO$_2$C(CH$_2$)$_2$ CH$_3$ CH$_3$ (CH$_2$)$_2$CO$_2$H

HO$_2$C(CH$_2$)$_2$ CH$_3$ H$_3$C   (CH$_2$)$_2$CO$_2$H

NH   HN   H   H   NH   HN

H$_3$C   C–O–C   CH$_3$

CH$_3$ CH$_3$

HO CH   CH$_3$ H$_3$C   HCOH

CH$_3$ CH$_3$

Fig. 3.3. Chemical structure of the active substance dihematoporphyrin which consists of two symmetric porphyrin rings

This is how photodynamic therapy is performed: first, a photosensitizer, e.g. hematoporphyrin derivative (HpD), is injected into a vein of the patient. In the case of HpD, 2.5–5 mg per kg body weight are applied. Within the next few hours, HpD is distributed among all soft tissues except the brain. The basic characteristic of a photosensitizer is that it remains inactive until irradiated. After 48–72 hours, most of it is cleared from healthy tissue, but its concentration in tumor cells has not decreased much even after a period of 7–10 days. Thus, HpD does not accumulate in tumor cells immediately after injection, but these cells show a longer storage ability (affinity) for HpD. The initial concentration is the same as in healthy cells, but the clearance is faster in the latter cells (see Fig. 3.4). After about three days, the concentration of HpD in tumor cells is about thirty times higher than in healthy cells.

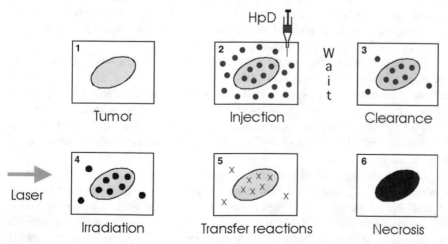

Fig. 3.4. Sequence of photodynamic therapy

Laser irradiation usually takes place after the third day and up to the seventh day after injection if several treatments are necessary. Within this period, tumor cells are still very sensitive and selective necrosis of tumor cells is enabled. However, some healthy tissues may retain certain constituents of HpD and are thus photosensitized, as well. To avoid additional oxidation of healthy cells, carotin is injected after laser exposure which then promotes the toxic singlet oxygen to harmless triplet oxygen (see Table 3.1). Carotin acts as a protection system on a molecular basis, because it is able to reverse the production of singlet oxygen. The protective property of carotenoids was already verified by Mathews-Roth (1982).

Cellular effects of HpD were studied in detail by Moan and Christensen (1981) and Berns et al. (1982). Usually, Type I and Type II reactions take place at the same time. Which mechanism is favored mainly depends on the concentration of available triplet oxygen and appropriate target molecules. Weishaupt et al. (1986) identified singlet oxygen as the primary toxic agent in the photodeactivation of tumor cells.

The absorption and fluorescence spectra of HpD are shown in Fig. 3.5, whereas Fig. 3.6 illustrates the corresponding energy level diagram. The very strong absorption at 350–400 nm originates from the broad excitation band $^1S_2$ of the sensitizer. On the other hand, the relative low absorption at wavelengths above approximately 600 nm coincides with the therapeutic window of biological tissues (see Sect. 2.2). More information on porphyrin photosensitizers and further studies on HpD are found in the book edited by Kessel and Dougherty (1983).

Primarily, the fluorescence spectrum of HpD is characterized by two peaks at 620 nm and 680 nm. They stem from the transitions $^1S_{1,0} \longrightarrow {}^1S_{0,0}$ and $^1S_{1,0} \longrightarrow {}^1S_{0,1}$, respectively. As in all macromolecules, the ground and the excited electronic states are further split into several vibrational states. After excitation, the macromolecule first relaxes to the lowest vibrational state belonging to the same excited electronic state. From there, it reaches the ground state by emitting fluorescence radiation.

The dependence of fluorescence on the concentration level of HpD is of considerable interest. Kinoshita (1988) observed that the emission peak at 620 nm decreases toward higher HpD concentrations. This effect is explained by self-absorption which dominates at concentrations higher than 10^{-3} mol/l. That's why comparing the relative fluorescence intensities is a good indicator for the concentration level of HpD and, thus, for distinguishing tumor cells from healthy cells.

An even more powerful technique is given by *time-resolved fluorescence*. It was observed by Yamashita (1984) that the duration of the fluorescence decay of HpD also depends on its respective concentration level. The lowest investigated concentration of 8.4×10^{-6} mol/l yields a fluorescence decay time in the nanosecond range, whereas higher concentrations such as 8.4×10^{-3} mol/l are characterized by decays as short as a few hundred picoseconds. Tumor

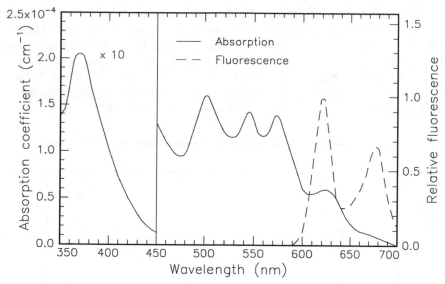

Fig. 3.5. Absorption and fluorescence spectra of hematoporphyrin derivative (HpD) dissolved in phosphate-buffered saline solution (PBS). Data according to Yamashita (1984)

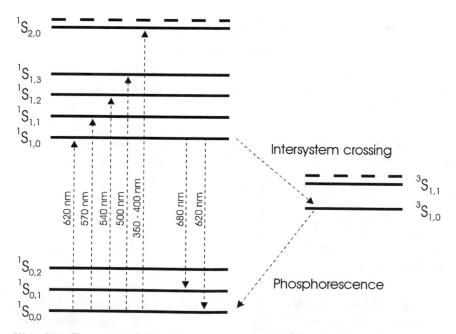

Fig. 3.6. Energy level diagram of HpD. Singlet (^{1}S) and triplet (^{3}S) states are shown. Dashed lines indicate higher excited states

diagnosis can thus be established by time-gated detection of the fluorescence. For instance, Fig. 3.7 compares the decay in fluorescence intensity of healthy cells and of tumor cells as observed by Kinoshita (1988). Since clearance of HpD is faster in healthy cells, the corresponding decay duration is significantly longer than in tumor cells. Today, this so-called *photodynamic diagnosis (PDD)* is a well established tool, especially in urology and in gastroenterology as reported by Jocham et al. (2008) and Koizumi et al. (2016).

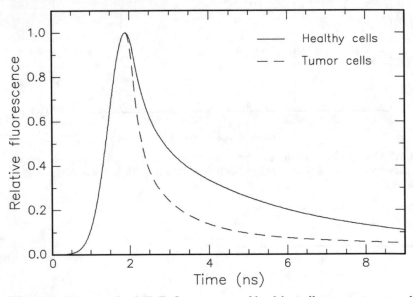

Fig. 3.7. Time-resolved HpD fluorescence of healthy cells versus tumor cells. The approximate decay durations are 2.5 ns and 1.0 ns, respectively. Data according to Kinoshita (1988)

The major disadvantage of HpD is the fact that treated patients need to remain in the dark during the first weeks of therapy, because HpD is distributed all over the body. So, during that time, healthy cells exposed to sun light or artificial light would be killed too. Other disadvantages are:

− since HpD absorbs very poorly in the red and near infrared spectrum, only tumors very close to the surface can be treated,
− the concentration gradient of HpD among tumor and healthy cells could be even steeper,
− the production of HpD from calf blood is very expensive.

In the early years of PDT, the initial isolation in the period between injection of HpD and laser irradiation was one of the biggest drawbacks. Carotenoid protection (as described above) could not help to overcome this problem, because the carotin had to be given *after* laser exposure.

Meanwhile, new photosensitizers with a much better performance have been developed such as phthalocyanins, naphthalocyanins, and phorbides (reduced porphyrins). But the best results today are achieved with meso-tetra-hydroxyphenyl-chlorin (mTHPC), a chemically well-defined substance unlike HpD, and with 5-aminolaevulinic acid (ALA) which itself is not even a photosensitizer but a precursor of porphyrins. It can be orally administered rather than being injected like most of the porphyrins.

Gossner et al. (1994) have compared the skin sensibility after application of mTHPC, DHE (dihematoporphyrin ester), and ALA. Their results are summarized in Fig. 3.8. ALA induces the least toxic damage: an increased sensibility is detectable only during the first two days after application. Although mTHPC is showing the best sensibility during the first week, a significant sensibility still remains after the second week. DHE performs worst with the sensibility still being strongly enhanced after four weeks.

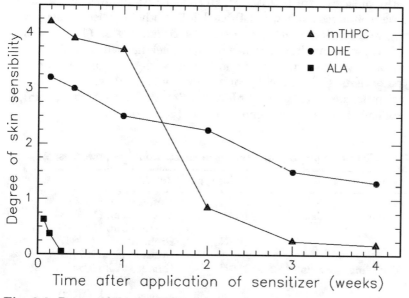

Fig. 3.8. Degree of skin sensibility after application of various photosensitizers as a function of time. Data according to Gossner et al. (1994)

According to Agostinis et al. (2011), an ideal photosensitizer would be a single pure compound with low manufacturing costs and good stability in storage. It should have a high absorption peak in the red spectrum as photons with longer wavelengths do not provide enough energy to excite oxygen to its singlet state. Moreover, it should have a rapid clearance from normal tissues, thereby minimizing phototoxic side-effects, but it should not have any toxicity on its own (in the dark).

3.1.3 Photobiomodulation (PBM)

Photobiomodulation is the newest of all scientific terms in this chapter. Of course, it was known for many decades that light induces biological effects in tissues, but the responsible mechanisms remained basically unclear until the beginning of the 21st century. Terms like "low-level light therapy" (LLLT) or "low-intensity laser therapy" (LILT) or "biostimulation" dominated for a long time. Yet none of these terms was well chosen, because LLLT would also include thermal effects, LILT would exclude any light from LEDs, and biostimulation would not include inhibitory effects caused by light.

Potential effects from very low power lasers (1–5 mW) on biological tissue have indeed been a subject of controversy, since they were first claimed by the Hungarian surgeon Mester at the end of the 1960s. At that time, hair growth, wound healing, and anti-inflammatory effects were reported to be caused by red light sources such as ruby lasers and helium–neon lasers. Typical energy fluences were 1–10 J/cm^2. In some cases, improvements for the patients were observed as shown in Table 3.2. But other studies came up with conflicting results. Here is just one example: Brunner et al. (1984) claimed to have observed wound healing when applying helium–neon lasers, whereas Hunter et al. (1984) reported no wound healing with the same laser and the same type of tissue (skin). Usually, only very few patients were treated, and no clinical protocols were established. According to Wilder-Smith (1988), it is quite difficult to rule out placebo effects in those cases.

Table 3.2. First attempts to prove biological effects from low power lasers

Observation	Target	Laser type	Reference
Hair growth	Skin	Ruby	Mester et al. (1968)
Wound healing	Skin	Ruby	Mester et al. (1969)
			Mester et al. (1971)
		He-Ne	Brunner et al. (1984)
			Lyons et al. (1987)
No wound healing	Skin	He-Ne	Hunter et al. (1984)
			Strube et al. (1988)
		Argon ion	Jongsma et al. (1983)
			McCaughan et al. (1985)
Stimulated collagen synthesis	Fibroblasts	Nd:YAG	Castro et al. (1983)
		He-Ne	Kubasova et al. (1984)
Suppressed collagen synthesis	Fibroblasts	Nd:YAG	Abergel et al. (1984)
Increased growth	Cells	Diode	Dyson and Young (1986)
Suppressed growth	Cells	He-Cd	Lin and Chan (1984)
		He-Ne	Quickenden et al. (1993)
Pain relief	Teeth	He-Ne	Carrillo et al. (1990)
		Diode	Taube et al. (1990)
No pain relief	Teeth	He-Ne	Lundeberg et al. (1987)
		Diode	Roynesdal et al. (1993)

Today, we know that photons from any light source can trigger chemical reactions in living cells which actually "modulate" a cell's biological behavior. That's why photobiomodulation is the appropriate term to describe these effects. Initially, it was sparsely used in just a few publications, for example by Eells et al. (2003) and Desmet et al. (2006). But then the term became more and more popular—and meanwhile, most professional societies are wholly devoted to "photobiomodulation". Introducing this term was a milestone for science as it is an umbrella term for a laser's ability to stimulate or to inhibit growth, to increase circulation, to reduce inflammation, to stimulate wound healing, and to relieve pain.

Photobiomodulation is a special type of photochemical interactions, where the photosensitizer is located in the cells themselves. One of the most investigated cell-inherent photosensitizers is cytochrome c oxidase. It absorbs light in the visible spectrum, preferably at 420 nm and at 550 nm according to data provided by Tanigawa et al. (2010). As illustrated in Fig. 3.9, cytochrome c oxidase is an enzyme located in the mitochondria[2] of cells. Upon absorbing photons, this enzyme can trigger any of these reactions according to Chung et al. (2012):

- causing an increase in adenosine triphosphate (ATP) which improves the cell's ability to fight infection,
- causing an increase in reactive oxygen species (ROS) which in turn activates transcription factors for cellular repair,
- causing the release of the vasodilator nitric oxide (NO) which in turn increases circulation.

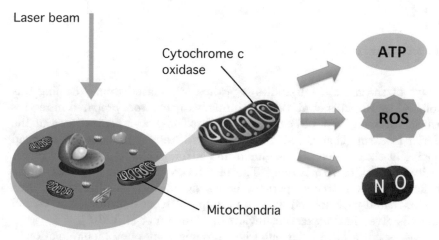

Fig. 3.9. Chemical reactions triggered by photoactivated cytochrome c oxidase

---

[2] Mitochondria are small organelles found in most eukaryotic organisms. One of their primary functions is to provide energy to the cell.

When talking about photobiomodulation, we must always keep in mind that any biological effect is influenced by both the degree of cell stimulation and the degree of tissue denaturation. For example, cytochrome c oxidase can activate transcription factors for cellular repair, but a too high laser fluence would wipe out this effect by causing harm to the cells. In general, this kind of behavior is known as *Arndt–Schulz Law*, illustrated in Fig. 3.10. It states that light stimulates biological effects (such as healing) only in a rather narrow range of fluence, another so-called *therapeutic window*[3]. If the fluence is too low, there is no effect at all. But if the fluence is too high, then any stimulating effect will turn into an inhibitory effect. So, lower fluences are often more beneficial than high fluences.

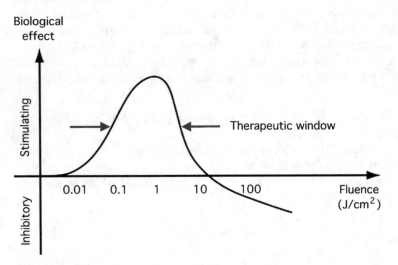

Fig. 3.10. Arndt–Schulz Law

One of the most cited studies regarding stimulated wound healing was published by Hopkins et al. (2004). Human subjects were prepared to receive artificial wounds (two $1.27\,cm^2$ circular abrasions) to a central area of the anterior forearm. Then the skin was irradiated using a 46-diode cluster head emitting at six wavelengths ranging from 660 nm to 820 nm. The total power density was set to $75\,mW/cm^2$. The results are shown in Fig. 3.11. From the fourth day on, treated patients had a significantly smaller wound area than untreated patients and the sham groups. Further information on wound healing is given in the excellent review by Hamblin et al. (2018).

Many studies have been published proving that photobiomodulation is also very successful in relieving pain, but pain is not as easy to measure as the size of a wound. A very reliable method was described by Bjordal et al. (2006).

---

[3] Not to be confused with the therapeutic window described in Sect. 2.2, which denotes the low absorption of biological tissues in the near infrared.

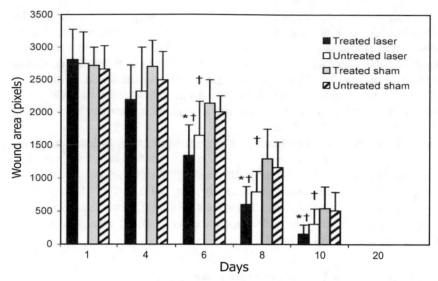

Fig. 3.11. Improved wound healing of human skin after exposure to a combination of red and NIR light (wavelengths: 660 nm to 820 nm, power density: $75\,\mathrm{mW/cm^2}$). Data according to Hopkins et al. (2004)

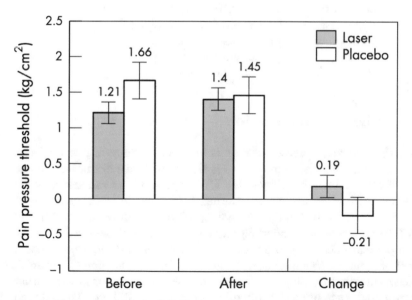

Fig. 3.12. Successful pain relief in human subjects after exposure to a GaAs diode laser (wavelength: 904 nm, power density: $20\,\mathrm{mW/cm^2}$). Data according to Bjordal et al. (2006)

They applied a 904 nm GaAs laser to reduce pain in patients with Achilles tendon symptoms. The total power density was set to only 20 mW/cm^2. Pain pressure was recorded by pressing a circular metallic tip of area 1 cm^2 with gradually increasing force to the most painful spot of the tendon. The results are summarized in Fig. 3.12. The mean pain pressure threshold significantly increased in the group of lasered patients from 1.21 kg/cm^2 to 1.4 kg/cm^2, whereas the placebo group even reported a decrease in pain pressure threshold from 1.66 kg/cm^2 to 1.45 kg/cm^2.

Photobiomodulation has become a very powerful tool in laser medicine, but one important question remains: Does it have to be a laser? Or would an incoherent light source be able to do the same job? In other words: Which of these properties – coherence, narrow bandwidth, polarization – is of primary importance for modulating biological effects? As long as we don't know the answer to this question, photobiomodulation will remain a field of research with a lot of speculation involved.

3.1.4 Summary of Photochemical Interaction

- *Main idea:*   using a photosensitizer acting as catalyst
- *Observable effects:*   stimulated or inhibited growth, increased circulation, reduced inflammation, improved wound healing, pain relief
- *Typical lasers:*   diode laser, He-Ne laser
- *Typical pulse durations:*   1 s ... CW
- *Typical power densities:*   0.01 W/cm^2 ... 10 W/cm^2
- *Special applications:*   photoactivated disinfection, photodynamic therapy, photobiomodulation

3.2 Photothermal Interaction

The term *photothermal interaction* stands for a large group of tissue effects, where the increase in local temperature is the significant parameter change. Thermal effects can be induced by either CW or pulsed laser radiation. While photochemical processes are often governed by a specific reaction pathway, thermal effects generally tend to be non-specific according to Parrish and Deutsch (1984). However, depending on the duration and peak value of the tissue temperature achieved, different effects like *coagulation*, *vaporization*, *carbonization*, and *melting* may be distinguished. In the following paragraphs, these effects shall first be visualized by selected photographs taken with either light microscopy or scanning electron microscopy (SEM). Thereafter, detailed models of heat generation, heat transport, and associated heat effects are presented. Finally, the principles of laser-induced interstitial thermotherapy are discussed.

Fig. 3.13a is one of the best self-explaining pictures in this book. It shows thermal effects in a chicken liver caused by various lasers. The Er:YAG laser vaporized the tissue, the CO_2 laser carbonized the tissue, and both the Nd:YAG and diode laser primarily coagulated the tissue. These apparently different effects are mainly due to the wavelength applied, because it's the wavelength which determines *where* the photon energy is absorbed. Radiation from the Er:YAG laser is strongly absorbed by water molecules causing vaporization. Radiation from the CO_2 laser is also absorbed by water molecules, but it penetrates deeper into the tissue—so, laser energy is efficiently heating up the tissue which eventually causes carbonization. Radiation from both the Nd:YAG and diode laser penetrates much deeper into the tissue and is primarily absorbed by hemoglobin causing coagulation and some carbonization. Yet there is a slight difference using these two lasers: Since light from the diode laser has a shorter wavelength, it is scattered more than the Nd:YAG laser radiation, resulting in a broader coagulation zone.

Coagulation. The histologic appearance of coagulated tissue is illustrated in Fig. 3.13b. A sample of uterine tissue was coagulated using a CW Nd:YAG laser. In a histologic section, the coagulated area can easily be detected when staining the tissue with hematoxylin and eosin. Coagulated tissue appears darker than other tissue. During the process of coagulation, temperatures reach at least 60°C, and the tissue becomes necrotic.

Vaporization. In Figs. 3.14a–b, a tooth was exposed to 20 pulses from an Er:YAG laser. During the ablation process, complete layers of tooth substance were removed, leaving stair-like structures behind. This effect is due to the existence of so-called *striae of Retzius* which are layers with a high content of water molecules. Water strongly absorbs the Er:YAG wavelength at 2.94 μm. The induced increase in pressure – water expands in volume as it vaporizes – leads to localized microexplosions. In the literature, vaporization is sometimes also referred to as *thermal decomposition*.

Carbonization. In Fig. 3.15a, a sample of skin is shown which was exposed to a CW CO_2 laser for the purpose of treating metastases. In this case, however, far too much energy was applied, and carbonization occurred. At temperatures above approximately 100°C, the tissue starts to carbonize, i.e. carbon is released, leading to a blackening in color. Coagulation, vaporization, and carbonization is seen in Fig. 3.15b, where cervical tissue was exposed to a CW CO_2 laser.

Melting. Finally, Figs. 3.16a–b show the surface of a tooth after exposure to 100 pulses from a Ho:YAG laser. Several cracks can be seen leaving the application spot radially. They originate from thermal stress induced by a local temperature gradient across the tooth surface. Melted and afterwards down-cooled tooth substance as well as gas bubbles are observed similar to solidified lava. The temperature must have reached a few hundred degrees Celsius to melt the tooth substance which mainly consists of hydroxyapatite, a chemical compound of calcium and phosphate.

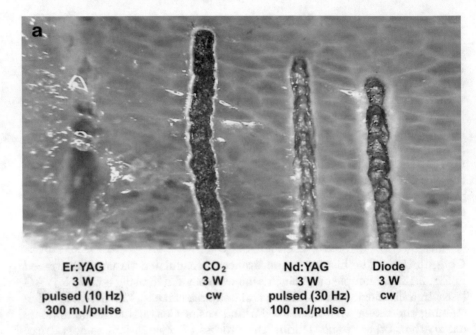

Er:YAG	CO$_2$	Nd:YAG	Diode
3 W	3 W	3 W	3 W
pulsed (10 Hz)	cw	pulsed (30 Hz)	cw
300 mJ/pulse		100 mJ/pulse	

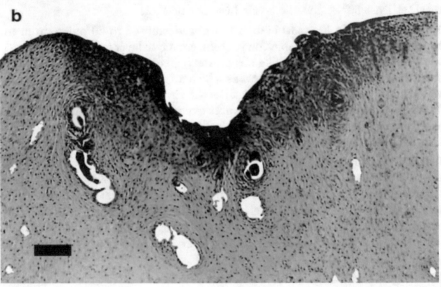

Fig. 3.13. (a) Chicken liver exposed to radiation from Er:YAG, CO$_2$, Nd:YAG, and diode laser (power: 3 W each). Vaporization appears transparent, carbonization appears dark, and coagulation appears pale. Photograph kindly provided by Dr. Kato (Tokyo), published by Aoki et al. (2015). **(b)** Uterine tissue of a wistar rat coagulated by a CW Nd:YAG laser (power: 10 W, bar: 80 μm). Photograph kindly provided by Dr. Kurek (Heidelberg)

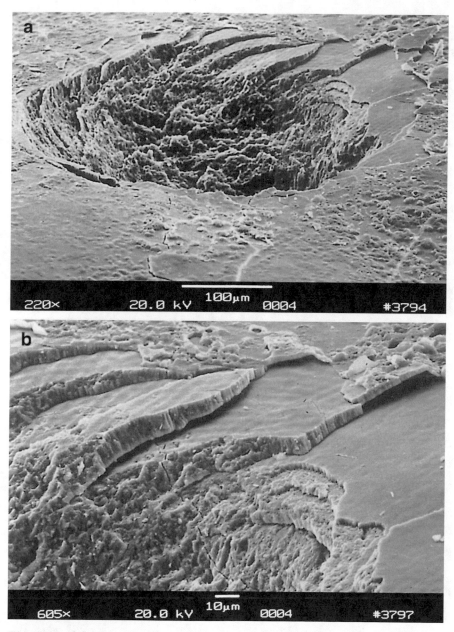

Fig. 3.14. (a) Human tooth vaporized by 20 pulses from an Er:YAG laser (pulse duration: 90 µs, pulse energy: 100 mJ, repetition rate: 1 Hz). (b) Enlargement showing the edge of ablation

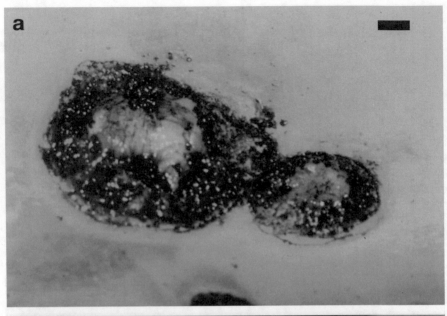

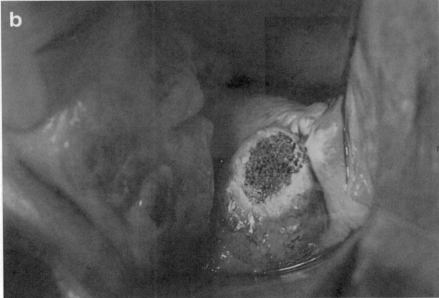

Fig. 3.15. (a) Tumor metastases on human skin carbonized by a CW CO_2 laser (power: 40 W, bar: 1 mm). **(b)** Cervical tissue coagulated, vaporized, and carbonized by a CW CO_2 laser (power: 10 W). Photographs kindly provided by Dr. Kurek (Heidelberg)

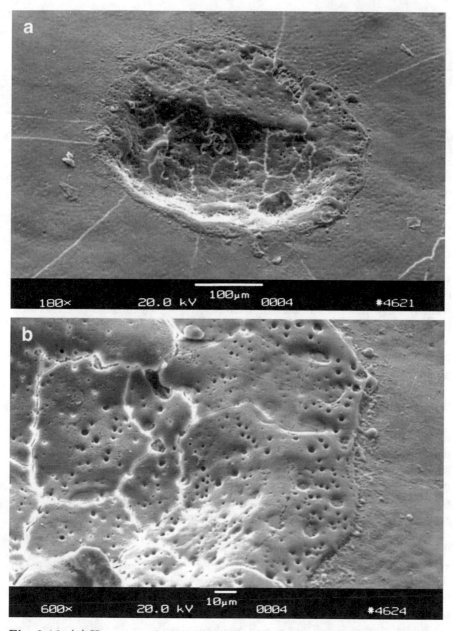

Fig. 3.16. (a) Human tooth melted by 100 pulses from a Ho:YAG laser (pulse duration: 3.8 µs, pulse energy: 18 mJ, repetition rate: 1 Hz). (b) Enlargement showing the edge of the melted zone

Temperature certainly is the governing parameter of all thermal laser–tissue interactions. And, for the purpose of predicting the thermal response, a model for the temperature distribution inside the tissue must be derived. Before we start working on this task, let us first look at the basics of what happens during thermal interaction.

At the microscopic level, thermal effects originate from bulk absorption occurring in molecular vibration–rotation bands followed by non-radiative decay. The reaction with a target molecule A can be considered as a two-step process. First, absorption of a photon with an energy $h\nu$ promotes the molecule to an excited state A^*; and second, inelastic collisions with some partner M of the surrounding medium lead to a deactivation of A^* and a simultaneous increase in the kinetic energy of M. So, temperature rises due to the transfer of photon energy to kinetic energy. This two-step process can be written as

- absorption: $A + h\nu \longrightarrow A^*$,
- deactivation: $A^* + M(E_{kin}) \longrightarrow A + M(E_{kin} + \Delta E_{kin})$.

How efficient is this two-step process? To answer this question, we have to consider both steps separately. First, absorption is facilitated due to the extremely large number of accessible vibrational states of most biomolecules. Second, the channels available for deactivation and thermal decay are also numerous, because typical energies of laser photons (Er:YAG laser: 0.35 eV, Nd:YAG laser: 1.2 eV, ArF laser: 6.4 eV) exceed by far the kinetic energy of a molecule at room temperature which is only about 0.025 eV. Thus, both of these steps are highly efficient provided that the duration of laser exposure is long enough.

The spatial extent and degree of tissue damage primarily depend on magnitude, exposure time, and placement of deposited heat inside the tissue. The deposition of laser energy, however, is not only a function of *laser parameters* such as wavelength, power density, exposure time, spot size, and repetition rate. It also strongly depends on *optical tissue properties* like absorption and scattering coefficients. For the description of storage and transfer of heat, *thermal tissue properties* are of primary importance such as heat capacity and thermal conductivity.

In biological tissue, absorption is mainly due to the presence of free water molecules, proteins, pigments, and other macromolecules as discussed in Sect. 2.3. It is governed by Lambert's law which we already encountered in (2.13). The absorption coefficient strongly depends on the wavelength of the incident laser radiation. In thermal interactions, absorption by water molecules plays a significant role. Therefore, the absorption spectrum of water – one important constituent of most tissues – is plotted in Fig. 3.17. In the visible range, the absorption coefficient of water is extremely small. In this section of the spectrum and in the UV, absorption in tissue is higher than shown in Fig. 3.17, depending on the relative content of macromolecules such as melanin and hemoglobin. Toward the IR range of the spectrum, however,

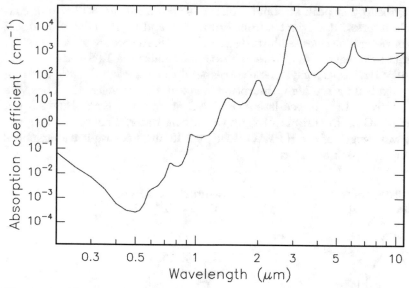

Fig. 3.17. Absorption of water. Data calculated from Hale and Querry (1973)

water molecules are the dominant absorbers, since their absorption coefficient then increases by several orders of magnitude. Typical absorption coefficients α and the corresponding absorption lengths L for the most important laser wavelengths are summarized in Table 3.3. It should be borne in mind that the total attenuation in the UV is strongly enhanced by Rayleigh scattering as discussed in Sect. 2.3.

Table 3.3. Absorption coefficients α and absorption lengths L of water at different wavelengths. Data calculated from Hale and Querry (1973)

Wavelength (nm)	Laser type	α (cm^{-1})	L (cm)
193	ArF	0.1	10
248	KrF	0.018	55
308	XeCl	0.0058	170
351	XeF	0.0023	430
514	Argon ion	0.00029	3400
633	He-Ne	0.0029	340
694	Ruby	0.0056	180
800	Diode	0.020	50
1053	Nd:YLF	0.57	1.7
1064	Nd:YAG	0.61	1.6
2120	Ho:YAG	36	0.028
2940	Er:YAG	12 000	0.00008
10600	CO$_2$	860	0.001

The absorption peak at about $3\,\mu m$ – as shown in Fig. 3.17 – is of considerable interest. It originates from symmetric and asymmetric vibrational modes of water molecules as illustrated in Fig. 3.18. According to Pohl (1976), the resonance frequencies of these vibrational modes are $1.08 \times 10^{14}\,$Hz and $1.13 \times 10^{14}\,$Hz, respectively. These correspond to a wavelength close to $3\,\mu m$, thereby explaining the high absorption peak at this wavelength. Thus, the whole family of Er^{3+} doped lasers (Er:YAG at $2.94\,\mu m$, Er:YLF at $2.8\,\mu m$, and Er:YSGG at $2.79\,\mu m$) is strongly vaporizing water. The same holds true for the wavelength of the Ho:YAG laser at $2.12\,\mu m$, because it also matches an absorption peak of water.

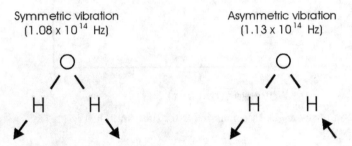

Fig. 3.18. Vibrational oscillations of water molecules

So far, we have encountered the basic origins of thermal effects, and we will now proceed with setting up a theoretical model based on thermodynamics. Our model consists of three parts: *heat generation*, *heat transport*, and *heat effects*. In order to derive such a model, several input parameters need to be taken into account: laser parameters, optical as well as thermal tissue parameters, and the type of tissue. The flow chart shown in Fig. 3.19 illustrates which parts of our model are affected by either of these parameters.

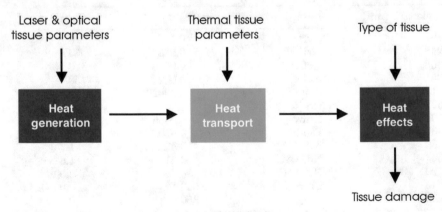

Fig. 3.19. Flow chart for modeling thermal interaction

Heat generation is determined by laser parameters and optical tissue parameters – primarily irradiance, exposure time, and the absorption coefficient – with the absorption coefficient itself being a function of the laser wavelength. Heat transport is solely characterized by thermal tissue parameters such as heat conductivity and heat capacity. Heat effects, finally, depend on the type of tissue and the temperature achieved inside the tissue.

We assume that a slab of tissue is exposed in air to a Gaussian-shaped laser beam as illustrated in Fig. 3.20. For the sake of simplicity, a cylindrical geometry is chosen with z denoting the optical axis, and r the distance from this axis. Then, the amplitude of the electric field and the corresponding intensity inside the tissue are given by

$$\boldsymbol{E}(r, z, t) = \boldsymbol{E}_0 \exp\left(-\frac{r^2}{w^2} - \frac{\alpha z}{2}\right) \exp\left(-\frac{4t^2}{\tau^2}\right) , \tag{3.1}$$

$$I(r, z, t) = I_0 \exp\left(-\frac{2r^2}{w^2} - \alpha z\right) \exp\left(-\frac{8t^2}{\tau^2}\right) , \tag{3.2}$$

where $\boldsymbol{E}_0$ and I_0 are the incident values of electric field and intensity, respectively, w is the beam waist, α is the absorption coefficient, and τ is the exposure time. From (3.2), we obtain that either at $r = w$ or at $t = \tau/2$ the intensity is cut down to $1/e^2$ of its incident value. The incident values $\boldsymbol{E}_0$ and I_0 are related to each other by the basic electrodynamic equation

$$I_0 = \frac{1}{2} \varepsilon_0 c \boldsymbol{E}_0^2 ,$$

where ε_0 is the dielectric constant, and c is the speed of light. Scattering inside the tissue is neglected in a first approximation. Experiments taking scattering effects into account will be found at the end of Sect. 3.2.4.

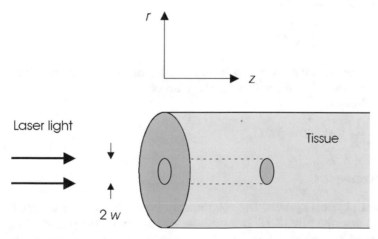

Fig. 3.20. Geometry of tissue irradiation

3.2.1 Heat Generation

By means of the two-step process described above, heat is generated inside the tissue during laser exposure. Deposition of heat in tissue is due only to light that is absorbed in the tissue. For a light flux in the z-direction in a non-scattering medium, the local heat deposition per unit area and time in a thickness Δz is given by

$$S(r,z,t) = \frac{I(r,z,t) - I(r,z+\Delta z,t)}{\Delta z} \quad \text{in units of} \quad \frac{\mathrm{W}}{\mathrm{cm}^3} \;.$$

And, as Δz approaches zero,

$$S(r,z,t) = - \frac{\partial I(r,z,t)}{\partial z} \;.$$

Therefore, under all circumstances, heat deposition is determined by

$$S(r,z,t) = \alpha I(r,z,t) \;. \tag{3.3}$$

Thus, the heat source $S(r,z,t)$ inside the exposed tissue is a function of the absorption coefficient α and the local intensity $I(r,z,t)$. Since α is strongly wavelength-dependent, the same applies for S. If phase transitions (vaporization, melting) or tissue alterations (coagulation, carbonization) do not occur, an alteration in heat content $\mathrm{d}Q$ induces a linear change in temperature $\mathrm{d}T$ according to a basic law of thermodynamics

$$\mathrm{d}Q = m\,c\,\mathrm{d}T \;, \tag{3.4}$$

where m is the tissue mass, and c is the specific heat capacity expressed in units of $\mathrm{kJ\ kg^{-1}\ K^{-1}}$. According to Takata et al. (1977), for most tissues, a good approximation is given by

$$c = \left(1.55 + 2.8\,\frac{\varrho_{\mathrm{w}}}{\varrho}\right) \frac{\mathrm{kJ}}{\mathrm{kg\ K}} \;,$$

where ϱ is the tissue density expressed in units of $\mathrm{kg/m^3}$, and ϱ_{w} is its water content expressed in units of $\mathrm{kg/m^3}$. In the case of water, i.e. $\varrho_{\mathrm{w}} = \varrho$, the last relation reduces to

$$c = 4.35\,\frac{\mathrm{kJ}}{\mathrm{kg\ K}} \quad \text{at} \quad T = 37°\mathrm{C} \;.$$

3.2.2 Heat Transport

Within a closed physical system, the relationship between temperature and heat content is described by (3.4). In real laser–tissue interactions, however, there are losses of heat to be taken into account, as well. They are based on either *heat conduction*, *heat convection*, or *heat radiation*. Usually, the latter two can be neglected for most types of laser applications. One typical example of heat convection in tissue is heat transfer due to blood flow. The perfusion rates of some human organs are summarized in Table 3.4. Due to the low perfusivity of most tissues, however, heat convection is negligible in a first approximation. Only during long exposures and in special cases such as laser-induced interstitial thermotherapy (LITT) does it play a significant role and should be considered by adding a negative heat loss S_{loss} to the source term S. Heat radiation is described by the *Stefan–Boltzmann law* which states that the radiated power is related to the fourth power of temperature. Due to the moderate temperatures achieved in most laser–tissue interactions, heat radiation can often be neglected.

Table 3.4. Blood perfusion rates of some selected human organs. Data according to Svaasand (1985)

Tissue	Perfusion rate (ml $min^{-1} g^{-1}$)
Fat	0.012–0.015
Muscle	0.02–0.07
Skin	0.15–0.5
Brain	0.46–1.0
Kidney	$\simeq 3.4$
Thyroid gland	$\simeq 4.0$

Heat conduction, though, is a considerable heat loss term and is the primary mechanism by which heat is transferred to unexposed tissue structures. The heat flow j_Q is proportional to the temperature gradient according to the general diffusion equation[4]

$$j_Q = -k \, \nabla T \, . \tag{3.5}$$

Herein, the constant k is called *heat conductivity* and is expressed in units of $W \, m^{-1} \, K^{-1}$. According to Takata et al. (1977), k can be approximated by

$$k = \left(0.06 + 0.57 \, \frac{\varrho_w}{\varrho} \right) \frac{W}{m \, K} \, .$$

In the case of water, i.e. $\varrho_w = \varrho$, the last relation reduces to

$$k = 0.63 \, \frac{W}{m \, K} \quad \text{at} \quad T = 37°C \, .$$

The dynamics of the temperature behavior of a certain tissue type can also be expressed by a combination of the two parameters k and c. It is called *temperature conductivity* and is defined by

---

[4] This equation is the analog to the electrodynamic equation $j = -\sigma \, \nabla \, \phi$, where j is the electric current density, σ is the electric conductivity, and ϕ is the electric potential.

$$\kappa = \frac{k}{\varrho c} \quad \text{in units of} \quad \frac{\text{m}^2}{\text{s}} \, . \tag{3.6}$$

The value of κ is approximately the same for liquid water and most tissues
– about $1.4 \times 10^{-7} \, \text{m}^2/\text{s}$ according to Boulnois (1986) –, since a decrease in
heat conductivity due to a lower water content is usually compensated by
a parallel decrease in heat capacity.

   With these mathematical prerequesites, we are able to derive the general
heat conduction equation. Our starting point is the *equation of continuity*
which states that the temporal change in heat content per unit volume[5], $\dot{q}$,
is determined by the divergence of the heat flow $\boldsymbol{j}_Q$:

$$\text{div} \, \boldsymbol{j}_Q = - \dot{q} \, . \tag{3.7}$$

Inserting (3.7) into (3.4) leads to

$$\dot{T} = \frac{1}{mc} \dot{Q} = \frac{1}{\varrho c} \frac{\dot{Q}}{V} = \frac{1}{\varrho c} \dot{q} = - \frac{1}{\varrho c} \text{div} \, \boldsymbol{j}_Q \, . \tag{3.8}$$

   The other important basic equation is the diffusion equation, i.e. (3.5).
Its combination with (3.8) yields

$$\dot{T} = \kappa \, \Delta T \, , \tag{3.9}$$

where Δ is the *Laplace operator*. This is the homogeneous *heat conduction
equation* with the temperature conductivity as defined by (3.6). With an
additional heat source S like the absorption of laser radiation, (3.8) and (3.9)
turn into the inhomogeneous equations

$$\dot{T} = - \frac{1}{\varrho c} \left(\text{div} \, \boldsymbol{j}_Q - S \right) , \tag{3.10}$$

$$\dot{T} = \kappa \, \Delta T + \frac{1}{\varrho c} S \, . \tag{3.11}$$

   Next, we want to solve the homogeneous part of the heat conduction equation, i.e. (3.9). It describes the decrease in temperature after laser exposure
due to heat diffusion. In cylindrical coordinates, (3.9) can be expressed by

$$\dot{T} = \kappa \left(\frac{\partial^2}{\partial r^2} + \frac{1}{r} \frac{\partial}{\partial r} + \frac{\partial^2}{\partial z^2} \right) T \, , \tag{3.12}$$

with the general solution

$$T(r, z, t) = T_0 + \frac{\chi_0}{(4\pi\kappa t)^{3/2}} \exp\left(- \frac{r^2 + z^2}{4\kappa t} \right) , \tag{3.13}$$

---

[5] Note that q in (3.7) is expressed in units of J/cm^3, while Q is in units of J. We
thus obtain according to *Gauss' theorem*: $\int dV \, \text{div} \, \boldsymbol{j}_Q = \oint df \, \boldsymbol{j}_Q = - \dot{Q} \, .$

where T_0 is the initial temperature, and χ_0 is an integration constant. The proof is straightforward. We simply assume that (3.13) represents a correct solution to (3.12) and find

$$\dot{T} = -\frac{3}{2}\frac{T - T_0}{t} + \frac{r^2 + z^2}{4\kappa t^2}\,(T - T_0)\,,$$

$$\frac{\partial^2}{\partial r^2}T = \frac{\partial}{\partial r}\left(-2r\frac{T - T_0}{4\kappa t}\right) = -\frac{T - T_0}{2\kappa t} + 4r^2\frac{T - T_0}{16\kappa^2 t^2}\,,$$

$$\frac{1}{r}\frac{\partial}{\partial r}T = -\frac{T - T_0}{2\kappa t}\,,$$

$$\frac{\partial^2}{\partial z^2}T = \frac{\partial}{\partial z}\left(-2z\frac{T - T_0}{4\kappa t}\right) = -\frac{T - T_0}{2\kappa t} + 4z^2\frac{T - T_0}{16\kappa^2 t^2}\,.$$

Hence,

$$\kappa\,\Delta T = -\frac{T - T_0}{2t} + r^2\frac{T - T_0}{4\kappa t^2} - \frac{T - T_0}{2t} - \frac{T - T_0}{2t} + z^2\frac{T - T_0}{4\kappa t^2}\,,$$

$$\kappa\,\Delta T = -\frac{3}{2}\frac{T - T_0}{t} + \frac{r^2 + z^2}{4\kappa t^2}\,(T - T_0) = \dot{T}\quad,\text{ q.e.d.}$$

The solution to the inhomogeneous heat conduction equation (3.11) strongly depends on the temporal and spatial dependences of $S(r, z, t)$. Usually, it is numerically evaluated assuming appropriate initial value and boundary conditions. Nevertheless, an analytical solution can be derived if the heat source function $S(r, z, t)$ is approximated by a delta-function

$$S(r, z, t) = S_0\,\delta(r - r_0)\,\delta(z - z_0)\,\delta(t - t_0)\,.$$

For the sake of simplicity, we assume that the heat conduction parameters are isotropic[6]. Thus,

$$S(z, t) = S_0\,\delta(z - z_0)\,\delta(t - t_0)\,.$$

In this case, the solution can be expressed by a one-dimensional *Green's function* which is given by

$$G(z - z_0, t - t_0) = \frac{1}{\sqrt{4\pi\kappa(t - t_0)}}\,\exp\left[-\frac{(z - z_0)^2}{4\kappa(t - t_0)}\right]\,. \tag{3.14}$$

By means of this function, the general solution for a spatially and temporally changing irradiation is determined by

$$T(z, t) = \frac{1}{\varrho c}\int_0^t\int_{-\infty}^{+\infty}S(z', t')\,G(z - z', t - t')\,\mathrm{d}z'\,\mathrm{d}t'\,. \tag{3.15}$$

---

[6] A valuable theoretical approach to the three-dimensional and time-dependent problem is found in the paper by Halldorsson and Langerholc (1978).

The spatial extent of heat transfer is described by the time-dependent *thermal penetration depth*

$$z_{\text{therm}}(t) = \sqrt{4\kappa t} \; . \tag{3.16}$$

The term "penetration depth" originates from the argument of the exponential function in (3.14), since (3.16) turns into

$$\frac{z_{\text{therm}}^2(t)}{4\kappa t} = 1 \; .$$

Thus, $z_{\text{therm}}(t)$ is the distance in which the temperature has decreased to $1/e$ of its peak value. In Table 3.5, the relationship expressed by (3.16) is evaluated for water ($\kappa = 1.4 \times 10^{-7} \, \text{m}^2/\text{s}$). We keep in mind that heat diffuses in water to approximately $0.7 \, \mu\text{m}$ within $1 \, \mu\text{s}$.

Table 3.5. Thermal penetration depths of water

Time t	Thermal penetration depth $z_{\text{therm}}(t)$
1 ns	24 nm
1 μs	0.7 μm
1 ms	24 μm
1 s	0.7 mm

For thermal decomposition of tissues, it is important to adjust the duration of the laser exposure in order to minimize thermal damage to adjacent structures. By this means, the least possible necrosis is obtained. The scaling parameter for this time-dependent problem is the so-called *thermal relaxation time* according to Hayes and Wolbarsht (1968) and Wolbarsht (1971). It is obtained by equating the optical penetration depth L as defined by (2.16) to the thermal penetration depth z_{therm}, hence

$$L = \sqrt{4\kappa\tau_{\text{therm}}} \; , \tag{3.17}$$

where τ_{therm} is the thermal relaxation time. One might argue the significance of τ_{therm}, because it is a theoretical parameter. The following considerations help us understand its meaning: for exposure durations $\tau < \tau_{\text{therm}}$, heat does not even diffuse to the optical penetration depth L. Hence, thermal damage to non-decomposed tissue is neglible. But for $\tau > \tau_{\text{therm}}$, heat can diffuse to a multiple of the optical penetration depth L, i.e. thermal damage to tissue adjacent to the decomposed volume is possible.

The thermal relaxation time also becomes important when coagulating tissues: we always get the most efficient heating, if the exposure duration τ is shorter than the thermal relaxation time τ_{therm}. And we always get the most efficient cooling through blood flow, if the "non-exposure time" (the time gap between laser pulses) is longer than the thermal relaxation time τ_{therm}.

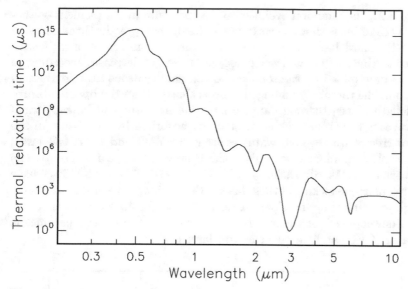

Fig. 3.21. Thermal relaxation times of water

Because of (3.17), the wavelength-dependence of L^2 is transferred to τ_{therm}. In Fig. 3.21, thermal relaxation times of water are shown as calculated from Fig. 3.17 and (3.17). We find that the shortest thermal relaxation time of approximately $1\,\mu s$ occurs at the absorption peak of water near $3\,\mu m$. We may thus conclude that laser pulse durations $\tau < 1\,\mu s$ are usually[7] not associated with thermal damage. This statement is also referred to as the "$1\,\mu s$ rule".

– *Case I:* $\tau < 1\,\mu s$. For nano- or picosecond pulses, heat diffusion during the laser pulse is negligible. If, in addition, we make the simplifying assumption that the intensity is constant during the laser pulse, we obtain from (3.3) that

$$S = \alpha I_0 \ .$$

For a quantitative relationship $T(t)$ at the tissue surface ($r = z = 0$), we may write

$$T = \left\{ \begin{array}{ll} T_0 + \frac{\alpha I_0}{\varrho c}\, t & \text{for } 0 \leq t \leq \tau \\ T_0 + T_{\max} \left(\frac{\tau}{t}\right)^{3/2} & \text{for } t > \tau \end{array} \right\} , \tag{3.18}$$

where $T_{\max}$ is the maximum increase in temperature given by

$$T_{\max} = \frac{\alpha I_0}{\varrho c}\, \tau \quad \text{at} \quad t = \tau \ .$$

---

[7] Laser pulses shorter than $1\,\mu s$ can also lead to thermal effects if they are applied at a high repetition rate as discussed later in this section.

In Fig. 3.22, the temporal evolution of temperature in the pigment epithelium of the retina is shown according to (3.18). By neglecting heat diffusion during the short laser pulse, the temperature first increases linearly with respect to time. After the laser pulse, i.e. $t > \tau$, it decreases according to $t^{-3/2}$ as determined by the solution to the homogeneous heat conduction equation. The thermally damaged zone is shorter than the optical absorption length. Thus, thermal damage to adjacent tissue can be kept small if a wavelength is selected that is strongly absorbed by the tissue. In the case of tissues with a high water content, Er:YAG and Er:YSGG lasers are potential candidates for this task. However, only a few groups like Andreeva et al. (1986), Eichler et al. (1992), and Pelz et al. (1994) have reported on mode locking of these lasers. But their operation was not stable enough for clinical applications. Alternatives might arise due to advances in the development of tunable IR lasers such as the optical parametric oscillator (OPO) and the free electron laser (FEL).

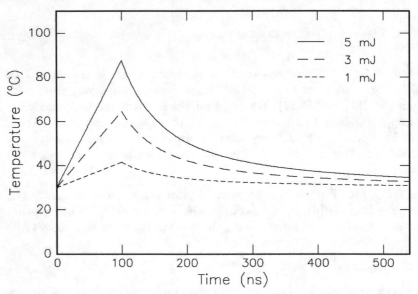

Fig. 3.22. Temporal evolution of temperature in the pigment epithelium of the retina during and after exposure to a short laser pulse ($\tau = 100$ ns, beam diameter: 2 mm, pulse energy: as labeled, $T_0 - 30°$C, $\alpha - 1587$ cm^{-1}, $\varrho - 1.35$ g cm^{-3}, and $c = 2.55$ J g^{-1} K^{-1}). Tissue parameters according to Hayes and Wolbarsht (1968) and Weinberg et al. (1984)

– *Case II:* $\tau > 1$ µs. For pulse durations during which heat diffusion is considerable, the thermally damaged zone is significantly broadened. In this case, the solution to the inhomogeneous heat conduction equation cannot be given analytically but must be derived numerically, for instance by using the methods of *finite differences* and *recursion algorithms*. This procedure becomes necessary because heat diffusion during the laser pulse can no longer be neglected. Thus, for this period of time, temperature does not linearly increase as assumed in (3.18) and Fig. 3.22. Detailed simulations were performed by Weinberg et al. (1984) and Roggan and Müller (1993). One example is found in Fig. 3.27 during the discussion of laser-induced interstitial thermotherapy.

A high repetition rate ν_{rep} of the laser pulses can evoke an additional increase in temperature if the rate of heat transport is less than the rate of heat generation. The dependence of temperature on repetition rate of the laser pulses was modeled by van Gemert and Welch (1989). The significance of the repetition rate becomes evident when looking at Figs. 3.23a–b. In this case, 1000 pulses from a picosecond Nd:YLF laser were focused on the same spot of a human tooth at a repetition rate of 1 kHz. Although, usually such short pulses do not evoke any thermal effect as discussed above, radial cracking and melting obviously occurred at the surface of the tooth. In particular, the enlargement shown in Fig. 3.23b demonstrates that the chemical compounds of the tooth had melted and recrystallized in a cubic structure. Thus, the temperature achieved must have reached a few hundred °C due to insufficient heat transport.

In order to get a basic feeling for the impact of typical laser parameters, the following simple calculations will be useful. We assume that a pulse energy of 2.5 μJ is absorbed within a tissue volume of 1000 μm^3 which contains 80 % water. The amount of water in the specified volume is equal to 8×10^{-10} cm^{-3} or 8×10^{-10} g, respectively. There are now three steps to be taken into account when heading for a rough approximation of the final temperature. First, energy is needed to heat the tissue up to 100°C. Second, energy is transferred to vaporization heat. And third, all the remaining energy leads to a further increase in temperature of the water vapor.

– Step 1: 37°C $\longrightarrow$ 100°C (assumed body temperature: 37°C)

$$Q_1 = mc\Delta T = 8 \times 10^{-10} \, \text{g} \; 4.35 \, \frac{\text{kJ}}{\text{kg} \, °\text{C}} \; 63°\text{C} = 2.2 \times 10^{-7} \, \text{J} \; .$$

– Step 2: Vaporization at 100°C

$$Q_2 = mQ_{vapor} = 8 \times 10^{-10} \, \text{g} \; 2253 \, \frac{\text{kJ}}{\text{kg}} = 1.8 \times 10^{-6} \, \text{J} \; .$$

– Step 3: 100°C $\longrightarrow T_{final}$

$$Q_3 = 2.5 \, \mu\text{J} - Q_1 - Q_2 = 0.48 \, \mu\text{J} \; ,$$

$$T_{final} = 100°\text{C} + \frac{Q_3}{mc_{vapor}} \; .$$

Thus, with c_{vapor} being equal to 1.87 kJ / (kg °C), the final temperature is approximately 420°C.

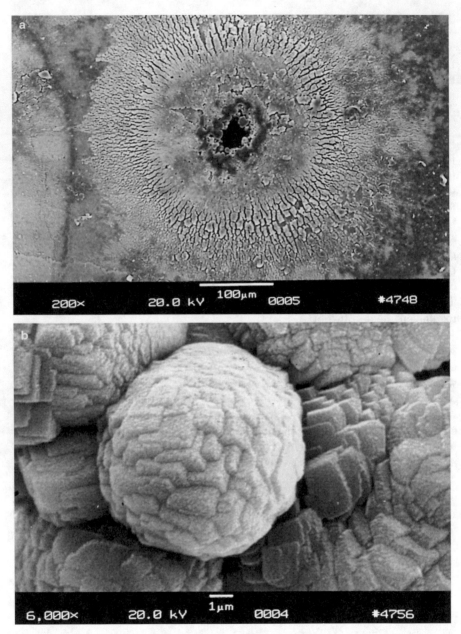

Fig. 3.23. (a) Hole in tooth created by focusing 1000 pulses from a Nd:YLF laser on the same spot (pulse duration: 30 ps, pulse energy: 1 mJ, repetition rate: 1 kHz). (b) Enlargement showing cubic recrystallization in form of plasma sublimations. Reproduced from Niemz (1994a). © 1994 Springer-Verlag

3.2.3 Heat Effects

The model developed above usually predicts the spatial and temporal distribution of temperature inside biological tissues very well if initial values and boundary conditions are properly chosen. This, however, isn't always an easy task. In general, though, approximate values of achievable temperatures can often be estimated. Therefore, the last topic in our model of thermal interaction deals with biological effects related to different temperatures inside the tissue. As already stated at the beginning of this section, these can be manifold, depending on the type of tissue. The most important and significant tissue alterations will be reviewed here.

Assuming a body temperature of 37°C, no measurable effects are observed for the next 5°C. The lowest temperatures by which biological tissue is thermally affected range from approximately 42–50°C. Conformational changes of molecules, accompanied by bond destruction and membrane alterations, are summarized in the single term *hyperthermia*. If such a hyperthermia lasts for several minutes, a significant percentage of the tissue will already undergo necrosis as described below by *Arrhenius' equation*. Beyond 50°C, a measurable reduction in enzyme activity is observed, resulting in reduced energy transfer within the cell and immobility of the cell. Furthermore, certain repair mechanisms of the cell are disabled. Thereby, the fraction of surviving cells is further reduced.

At 60°C, denaturation of proteins and collagen occurs which leads to coagulation of tissue and necrosis of cells. The corresponding macroscopic response is visible paling of the tissue. Several treatment techniques such as laser-induced interstitial thermotherapy (LITT) aim at temperatures just above 60°C. At even higher temperatures (> 80°C), the membrane permeability is drastically increased, thereby destroying the otherwise maintained equilibrium of chemical concentrations.

At 100°C, water molecules contained in most tissues start to vaporize. The large vaporization heat of water (2253 kJ/kg) is advantageous, since the vapor generated carries away excess heat and helps to prevent any further increase in the temperature of adjacent tissue. Due to the large increase in volume during this phase transition, gas bubbles are formed inducing mechanical ruptures and thermal decomposition of tissue fragments.

Only if all water molecules have been vaporized, and laser exposure is still continuing, does the increase in temperature proceed. At temperatures exceeding 100°C, carbonization takes place which is observable by the blackening of adjacent tissue and the escape of smoke. To avoid carbonization, the tissue is usually cooled with either water or gas. Finally, beyond 300°C, melting can occur, depending on the target material.

All these steps are summarized in Table 3.6, where the local temperature and the associated tissue effects are listed. For illustrating photographs, the reader is referred to Figs. 3.13–3.16 and Fig. 3.23.

Table 3.6. Thermal effects of laser radiation

Temperature	Biological effect
37°C	Normal
45°C	Hyperthermia
50°C	Reduction in enzyme activity, cell immobility
60°C	Denaturation of proteins and collagen, coagulation
80°C	Permeabilization of membranes
100°C	Vaporization, thermal decomposition (ablation)
> 100°C	Carbonization
> approx. 300°C	Melting

In general, the exact temperature for the onset of cell necrosis is rather difficult to determine. As a matter of fact, it was observed that not only the temperature achieved but also the temporal duration of this temperature plays a significant role for the occurrence of irreversible damage. It is illustrated in Fig. 3.24 how the critical temperature and the corresponding temporal duration relate to each other if irreversible damage is meant to occur. The curve is derived from several empirical observations. In the example selected in Fig. 3.24, a temperature of 60°C lasting for at least 6 s will lead to irreversible damage.

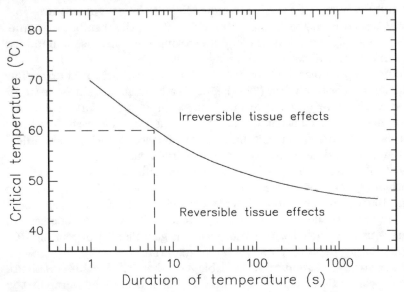

Fig. 3.24. Critical temperatures for the occurrence of cell necrosis. Data according to Henriques (1947) and Eichler and Seiler (1991)

Areas in which the temperature reaches values higher than 60°C are co-agulated, and irradiated tissue cells become necrotic. Areas with maximum temperatures less than 60°C are treated hyperthermically only, and the probability of cells staying alive depends on the duration and temporal evolution of the temperature obtained. For a quantitative approximation of the remaining active molecules and cells at a certain temperature level, *Arrhenius' equation* is very useful:

$$\ln \frac{C(t)}{C_0} = -A \int_0^t \exp\left(-\frac{\Delta E}{RT(t')}\right) dt' \equiv -\Omega \, , \tag{3.19}$$

where C_0 is the initial concentration of molecules or cells, $C(t)$ is the concentration at a time t, A is Arrhenius' constant, R is the universal gas constant, and ΔE and Ω are specific tissue properties. According to Welch (1984), the coefficient A can be approximated by

$$A \simeq \frac{kT}{h} \exp \frac{\Delta S}{R} \, ,$$

where ΔS is the activation entropy, k is Boltzmann's constant, and h is Planck's constant. But, according to Johnson et al. (1974), this relation has no simple significance.

The local degree of tissue damage is determined by the damage integral given in (3.19). The damage degree is defined as the fraction of deactivated molecules or cells given by

$$D_\mathrm{d}(t) = \frac{C_0 - C(t)}{C_0} = 1 - \exp\left(-\Omega\right) \, .$$

Thus, by inserting an appropriate value of the tissue constant Ω, we are able to calculate the probable damage degree $D_\mathrm{d}(t)$ as a function of time t. This was performed by Weinberg et al. (1984) in the case of retinal tissue.

Unfortunately, experimental data for the two parameters A and ΔE are very difficult to obtain due to the inhomogeneity of most tissues and the uncertainty in measuring the surviving fraction. However, in Table 3.7 some values of various tissues are listed as found in different studies.

Table 3.7. Arrhenius' constants of different tissues

Tissue	$A(s^{-1})$	ΔE (J/mol)	Reference
Retina	1.0×10^{44}	2.9×10^5	Vassiliadis et al. (1971) Weinberg et al. (1984)
Retina ($T < 50°$C)	4.3×10^{64}	4.2×10^5	Takata et al. (1977)
Retina ($T > 50°$C)	9.3×10^{104}	6.7×10^5	Takata et al. (1977)
Skin	3.1×10^{98}	6.3×10^5	Henriques (1947)
Liver	1.0×10^{70}	4.0×10^5	Roggan and Müller (1993)

Laser radiation acts thermally if power densities $\geq 10\,W/cm^2$ are applied from either CW radiation or pulse durations exceeding approximately 1 μs. Typical lasers for coagulation are Nd:YAG lasers or diode lasers. CO_2 lasers and Er:YAG lasers are very suitable for vaporization and the precise thermal cutting of tissue. Carbonization and melting can occur with almost any type of laser if sufficient power densities and exposures are selected.

Frequently, not only one but several thermal effects are induced in biological tissue, depending on the laser parameters. These effects might even range from carbonization at the tissue surface to hyperthermia a few millimeters inside the tissue. In most applications, though, only one specific effect is aimed at. Therefore, careful evaluation of the required laser parameters is essential. Reversible and irreversible tissue damage can be distinguished. Carbonization, vaporization, and coagulation certainly are irreversible processes, because they induce irreparable damage. Hyperthermia, though, can turn out to be either a reversible or an irreversible process, depending on the type of tissue and laser parameters. Since the critical temperature for cell necrosis is determined by the exposure time as shown in Fig. 3.24, no well-defined temperature can be declared which distinguishes reversible from irreversible effects. Thus, exposure energy, exposure volume, and exposure duration together determine the degree and extent of tissue damage. The coincidence of several thermal processes is illustrated in Fig. 3.25. The location and spatial extent of each thermal effect depend on the temperature locally achieved during and after laser exposure.

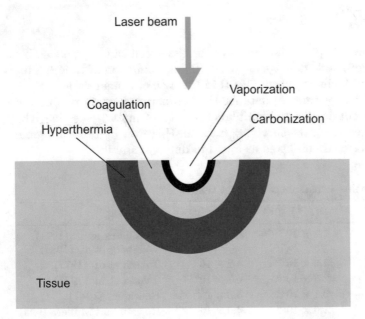

Fig. 3.25. Location of thermal effects inside biological tissue

3.2.4 Laser-Induced Interstitial Thermotherapy (LITT)

Our ability to precisely coagulate tissues is the cornerstone of a special tumor treatment called *laser-induced interstitial thermotherapy (LITT)*. Meanwhile, it has been applied to various types of tumors, especially in the prostate, the uterus, the liver, and the retina. LITT has become a well established tool in minimally invasive surgery (MIS). Detailed descriptions and clinical results can be found in the publications by Stafford et al. (2010), Roggan and Müller (1993), Wallwiener et al. (1994), Colin et al. (2011), Muschter et al. (1992), and Svaasand et al. (1989). The most significant applications of LITT reside in the disciplines of gynecology and urology, e.g. the treatment of malignant tumors in the uterus and the treatment of benign prostatic hyperplasia (BPH) as described in Sects. 4.3 and 4.4, respectively.

The principal idea of LITT is to position an appropriate laser applicator inside the tissue to be coagulated, e.g. a tumor, and to achieve necrosis by heating cells above 60°C. As stated in Table 3.6, denaturation of proteins and coagulation occur at these temperatures. Due to the associated coagulation of blood vessels, severe hemorrhages are less likely to occur than in conventional surgery. This is of particular importance in the case of tumors, because they usually are highly vascularized. Either Nd:YAG lasers at 1064 nm or different types of diode lasers at 800–900 nm are applied, since light deeply penetrates into tissue at these wavelengths. Thus, large volumes can be treated with temperature gradients not as steep as those associated with conventional thermotherapies based on heat conduction only. Typical parameters of the procedure are 1–5 W of CW laser power for a period of several minutes and coagulation volumes with diameters of up to 40 mm.

In order to achieve a safe LITT procedure, knowledge of the final damage zone is essential. In most medical applications, it is very important to prevent injury from adjacent healthy tissue and sensitive structures. The spatial extent of the damage zone primarily depends on laser power, laser exposure, geometry of the laser applicator, and on thermal and optical tissue properties. Since the optical penetration depth of laser light in the near infrared is very high, deeper zones are reached more easily by the mechanism of light scattering rather than by heat conduction. Therefore, laser applicators emitting their radiation through a scattering surface are favored compared to focusing optics. Actually, first applications of LITT had been performed using bare fibers which were directly guided into the tissue. The main disadvantage was the high power density at the end surface of the fiber, leading to immediate carbonization of adjacent tissue. Improved LITT applicators scatter laser light isotropically into all spatial directions by means of a frosted (very rough) surface as described by Roggan et al. (1995c).

The basic setup for LITT is illustrated in Fig. 3.26. The laser applicator usually consists of a flexible fiber and a transparent catheter through which the fiber is moved into the tissue. Frequently, fibers with typical diameters of 400 µm are chosen. In order to ensure isotropic scattering at the interface

of the applicator and tissue, the surface of the fiber is etched by a special
technique after removing the cladding. Continuous emitting surfaces with
active lengths up to 20 mm can be manufactured. Depending on size and
geometry of the treated volume, though, fibers with different active lengths
should be available during surgery. Laser light is emitted from the distal end
of the fiber on an active length which is approximately half the tumor size.
The specially designed catheter protects the sensitive fiber from mechanical
stress. Optional cooling of the catheter may help to prevent thermal damage
to the fiber tip at high power densities. Only then can large tissue volumes
be safely coagulated at a moderate gradient of temperature.

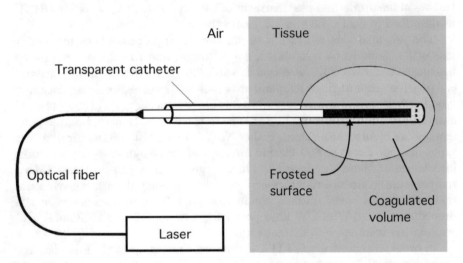

Fig. 3.26. Experimental setup for laser-induced interstitial thermotherapy. Laser
radiation is applied to the tissue through an optical fiber. The fiber is placed inside
the tissue by means of a specially designed, transparent catheter. Tissue necrosis
occurs in selected coagulated volumes only

At the start of any LITT treament, the catheter is placed inside the
tumor assisted by qualified monitoring. According to Roggan et al. (1994),
Eyrich et al. (2000), and Wacker et al. (2001), either ultrasound or magnetic
resonance imaging (MRI) can be used for this purpose[8]. The fiber is then
carefully guided through the catheter without actually being in contact with
the tissue. In an accidental break of the fiber, glass splinters will remain inside
the catheter without injuring the patient. After laser exposure, the fiber is
either completely pulled back or only by its active length if several treatments
are to be performed. Finally, that is if no further irradiation is required, the
fiber is withdrawn. And, after a short period of temperature equalization, the

---

[8] In MRI-controlled LITT treatments, a special marker is mounted at the distal
end of the fiber to facilitate its localization.

protecting catheter is removed. The same target may be treated several times by LITT to increase the spatial extent of tissue necrosis. For large lesions, the distance between adjacent puncturing canals should not exceed 1.5 mm to ensure an overlap of the coagulated tissue volumes.

The whole procedure can be performed either intraoperatively or percutaneously. In both cases, it should be preceded by suitable irradiation planning to obtain best surgical results. This can be achieved using appropriate computer simulations by considering a variety of input parameters. Most important among these are the position, extent, and topology of the diseased volume as well as optical and thermal tissue parameters, and the rate of blood perfusion. A choice of laser power and exposure duration should then be provided by the program. Detailed computer simulations for the laser treatment of liver tissue are illustrated in Figs. 3.27 and 3.28. They were obtained by applying the method of finite differences to the inhomogeneous heat conduction equation (3.11). This is performed by dividing the tissue into consecutive shells and using a mathematical algorithm which takes into account the heat conduction from adjacent layers only. After a few recursions of this algorithm, a steady-state solution is obtained expressing the desired temperature distribution. Further details on the simulation technique can be found in the paper by Roggan and Müller (1993). From Fig. 3.27, we conclude that adequate computer simulations fit very well to measured data, whereas Fig. 3.28 proves that cooled laser applicators enable the coagulation of larger tissue volumes at comparable peak temperatures.

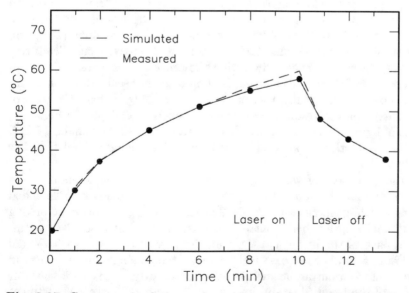

Fig. 3.27. Computer simulation and experimental results of LITT for liver tissue (laser power: 4 W, surface of applicator: 125 mm^2). The switch-off time of the laser is indicated. Data according to Roggan and Müller (1993)

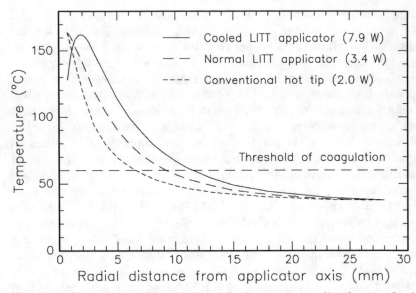

Fig. 3.28. Comparison of the calculated final temperature distributions for different types of thermotherapies (conventional hot tip, LITT without fiber cooling, LITT with fiber cooling). In the case of fiber cooling, the peak temperature does not occur at the applicator surface which is located at the *left* end of all curves. Data according to Roggan and Müller (1993)

The propagation of laser light in biological tissue and its transformation to thermal energy due to absorption is governed by the optical properties of the tissue. In our model of thermal interaction described above, only absorption was taken into account. During LITT, scattering also plays a significant role for reasons just stated. Especially during the process of coagulation, the optical properties of tissue are changed, leading to higher scattering but nearly the same absorption. Usually, this effect is observable by a change in color of the irradiated volume, resulting in a reduced penetration depth. In Fig. 3.29, a sample of liver tissue is shown which was irradiated using a standard LITT applicator as described above. The bright coagulated zone around the fiber is clearly visible.

For the treatment of very large tissue volumes with diameters of up to a few centimeters, the application of several fibers is strongly recommended according to Klingenberg et al. (2000). The laser output is distributed among several fibers by optical beamsplitters as illustrated in Fig. 3.30. Each beamsplitter separates the incident laser beam into two laser beams with equal power. Using a total of three optical beamsplitters, a single input beam can be divided into four output beams. Focusing lenses are used to couple the laser power out of the input fiber and into the output fibers. Further adjustment of the laser power within each output beam can be achieved by implementing optical filters.

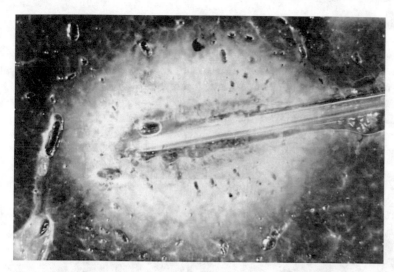

Fig. 3.29. Liver tissue after exposure to a Nd:YAG laser, using a standard LITT fiber applicator (laser power: 5.5 W, exposure duration: 10 min). The coagulated volume visibly pales. Photograph kindly provided by Dr. Roggan (Berlin)

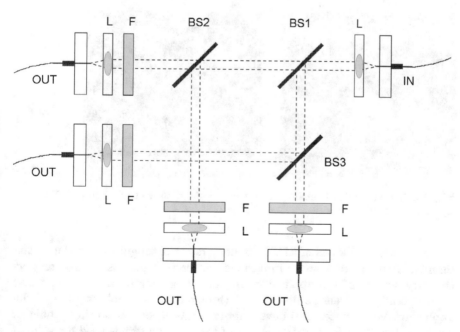

Fig. 3.30. LITT with four output fibers (BS: beamsplitter, F: filter, L: lens)

Depending on number and arrangement of the fibers, different geometries can be selected for coagulating biological tissue. As shown in Fig. 3.31, the surgeon may choose among either spherical, ellipsoidal, or trapezoidal lesions. He can thus adjust the coagulated volume to the geometry of a tumor. Some indications require the treatment of tumors closely located to vital structures. In these cases, it is extremely important to avoid the coagulation of organs at risk. This task can be achieved by applying trapezoidal lesions. Detailed discussion on this topic and on the advantages of a multi-fiber treatment is provided by Klingenberg et al. (2000).

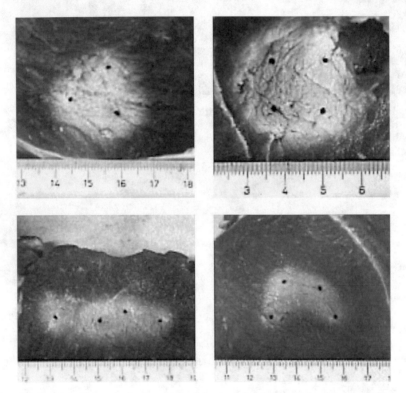

Fig. 3.31. Coagulations achieved in muscle tissue with up to four fibers (Nd:YAG laser, 4 W per fiber)

In general, two kinds of LITT applicators are distinguished: surface scatterers and volume scatterers. In surface scatterers, light is scattered only at the very surface of the applicator. In volume scatterers, light is scattered by tiny scattering centers distributed throughout the whole volume of the applicator. Volume scatteres have a more homogeneous scattering profile as shown in Fig. 3.32. Raw quartz is one of the basic materials used for volume scatterers. In raw quartz, the scattering centers consist of tiny gas bubbles.

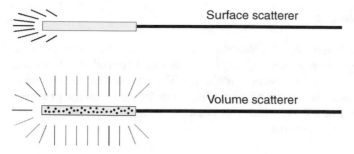

Fig. 3.32. Scattering profiles of surface scatterer and volume scatterer, respectively

As stated above, scattering is the dominant mechanism by which light is homogeneously distributed inside the tissue. In Sect. 2.5, we have already become acquainted with the *transport equation* of radiation (2.36) which takes scattering effects into account. It was mentioned that exact analytical solutions to this integro-differential equation do not exist. However, five approaches have been made to solve the equation:

- first-order scattering,
  Kubelka–Munk theory,
- diffusion approximation,
- Monte Carlo simulations,
- inverse adding–doubling.

In principle, the same five approaches can be chosen when trying to estimate the temperature field within laser-irradiated tissue. Detailed computer calculations based on Monte Carlo simulations are found in the paper by Roggan and Müller (1995b). For a further characterization of each method, the reader is referred to Sect. 2.5.

3.2.5 Summary of Photothermal Interaction

- *Main idea:* achieving a certain temperature which leads to the desired thermal effect
- *Observable effects:* coagulation, vaporization, carbonization, or melting
- *Typical lasers:* CO_2, Nd:YAG, Er:YAG, Ho:YAG, argon ion, diode laser
- *Typical pulse durations:* $1\,\mu s$ ... $1\,min$
- *Typical power densities:* $10\,W/cm^2$ ... $10^5\,W/cm^2$
- *Special applications:* hemostasis, thermal decomposition, laser-induced interstitial thermotherapy, hyperthermia

3.3 Photoablation

In Fig. 3.33, the cross-section of a human cornea is shown which was exposed to an ArF excimer laser at a wavelength of 193 nm. Material was removed very cleanly without any appearance of thermal damage such as coagulation or vaporization. Instead, evidence is given that the tissue was very precisely "etched". This kind of UV light-induced ablation is called *photoablation* and will be discussed in this section.

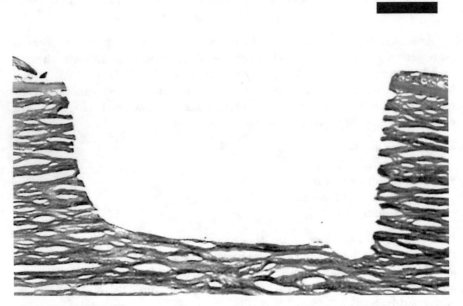

Fig. 3.33. Photoablation of corneal tissue achieved with an ArF excimer laser (pulse duration: 14 ns, energy density: 180 mJ/cm^2, bar: 100 µm). Photograph kindly provided by Dr. Bende (Tübingen)

Photoablation was first discovered by Srinivasan and Mayne-Banton (1982). They identified it as *ablative photodecomposition*, meaning that material is decomposed when exposed to high intense laser irradiation. Typical threshold values of this type of interaction are 10^7–10^9 W/cm^2 at laser pulse durations in the nanosecond range. The ablation depth, i.e. the depth of tissue removal per pulse, is determined by the pulse energy up to a certain saturation limit. The geometry of the ablation pattern itself is defined by the spatial parameters of the laser beam. The main advantages of photoablation are the very high precision of the etching process as demonstrated in Fig. 3.33, its excellent predictability, and the complete lack of thermal damage to adjacent tissue.

First studies regarding ablative photodecomposition were performed with polymethyl-metacrylate (PMMA), polyimide, Teflon, and other synthetic organic polymers[9]. Soon after, various biological tissues were also ablated. Then, photoablation became a very successful technique for refractive corneal surgery, where the refractive power of the cornea is altered in myopia, hyperopia, or astigmatism (see Sect. 4.1). In this section, for the sake of simplicity, we will follow the explanations given by Garrison and Srinivasan (1985) in the case of synthetic polymer targets. Due to the homogeneity of these materials, their behavior under certain experimental conditions is easier to understand. Thus, experimental data on the ablative process are very reliable and theoretical modeling is strongly facilitated. However, most of the theory applies for the more inhomogeneous biological tissues, as well.

Organic polymers are made up of large molecules consisting of more than 10^3 atoms, mainly carbon, hydrogen, oxygen, and nitrogen. A single molecular unit of up to 50 atoms is called a *monomer*. It is repeated several times along a well-defined axis to form the polymer chain. In Figs. 3.34a–b, the basic chemical structures of the two most popular polymers – PMMA and polyimide – are illustrated.

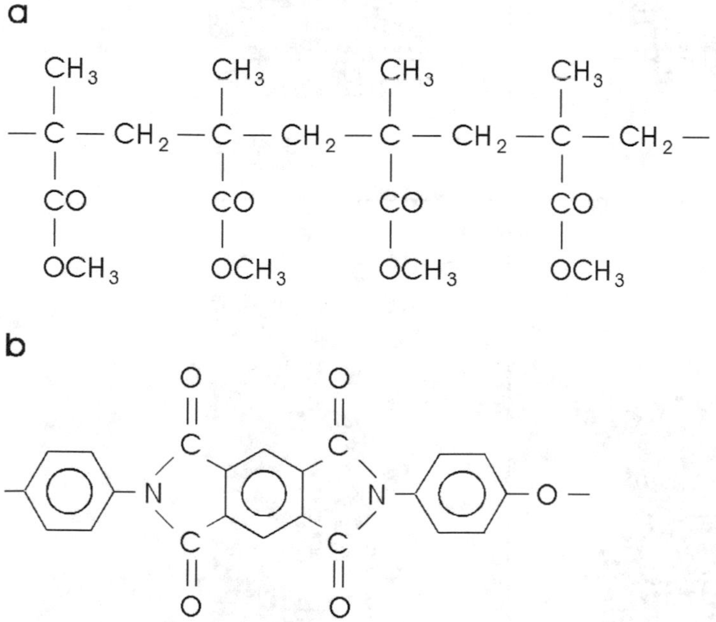

a

b

Fig. 3.34. (a) Chemical structure of four monomers of PMMA. (b) Chemical structure of one monomer of polyimide

---

[9] In the field of material processing, especially in the manufacture of miniaturized surface structures, photoablation by means of excimer lasers has meanwhile become a well-established technique.

For PMMA, the temporal progress of photoablation was modeled by Garrison and Srinivasan (1985) using *Newton's equation* of motion. The polymer is described by structureless monomer units held together by strong attractive forces. The interaction with laser radiation is simulated by allowing each monomer unit to undergo excitation directly from an attractive to a repulsive state. This promotion is associated with a change in volume occupied by each monomer, leading to a transfer of momentum and, thus, to the process of ablation. For the sake of simplicity, a face-centered cubic (fcc) crystalline array is assumed. With a density of $1.22\,\mathrm{g/cm^3}$ and a monomer mass of 100 amu, a lattice constant of 0.81 nm is calculated. The main attractive forces holding the monomer units together are the two carbon–carbon bonds along the chain. The strength of such a C–C bond is approximately 3.6 eV (see Table 3.8). The basic repulsive term is proportional to $1/r^{12}$, where r denotes a mean distance of two monomers. With these assumptions, the process of photoablation was simulated as shown in Fig. 3.35.

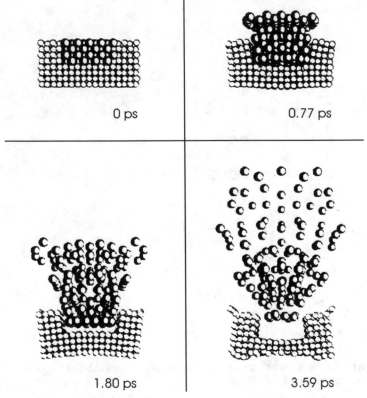

Fig. 3.35. Computer simulation of photoablation showing the movement of PMMA monomers as a function of time. Reproduced from Garrison and Srinivasan (1985) by permission. © 1985 American Institute of Physics

The character of the bonds within an organic polymer is primarily cova-
lent. In order to obtain a physical explanation of the photoablation process,
let us assume that two atoms A and B are bound by a common electron. The
corresponding energy level diagram of the ground and several excited states
is shown in Fig. 3.36. Due to the macromolecular structure, each electronic
level is split into further vibrational states. Absorption of a photon may promote the two atoms to an excited state $(AB)^*$. Usually, the excitation is
achieved by satisfying the *Franck–Condon principle*. It states that the radial
distance of the two nuclei of the atoms A and B will not be affected during
the process of excitation due to the small electron mass. Thus, transitions
characterized by a vertical line in the energy level diagram are favored as
indicated in Fig. 3.36. The probability of such a transition is increased if the
maxima of the corresponding *Schrödinger functions* of the initial and final
states coincide at the same radial distance.

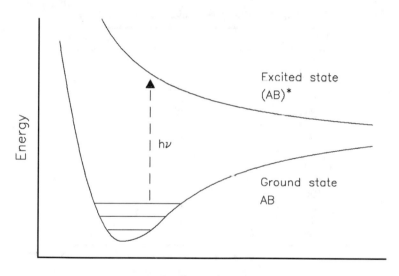

Fig. 3.36. Energy level diagram for photoablation

If a UV photon is absorbed, the energy gain is usually high enough to
access an electronic state which exceeds the bond energy. In this case, the two
atoms A and B may dissociate at the very next vibration. Thus, photoablation
can be summarized as a two-step process:

- excitation: $AB + h\nu \longrightarrow (AB)^*$,
- dissociation: $(AB)^* \longrightarrow A + B + E_{\mathrm{kin}}$.

The dissociation energies of some typical chemical bonds are listed in Table 3.8. Moreover, the wavelengths and the corresponding photon energies of
selected laser systems are given in Table 3.9. When comparing both tables,

Table 3.8. Dissociation energies of selected chemical bonds. Data according to Pauling (1962)

Type of bond	Dissociation energy (eV)
C=O	7.1
C=C	6.4
O–H	4.8
N–H	4.1
C–O	3.6
C–C	3.6
S–H	3.5
C–N	3.0
C–S	2.7

Table 3.9. Wavelengths and photon energies of selected laser systems

Laser type	Wavelength (nm)	Photon energy (eV)
ArF	193	6.4
KrF	248	5.0
Nd:YLF (4ω)	263	4.7
XeCl	308	4.0
XeF	351	3.5
Argon ion	514	2.4
Nd:YLF (2ω)	526.5	2.4
He-Ne	633	2.0
Diode	800	1.6
Nd:YLF	1053	1.2
Nd:YAG	1064	1.2
Ho:YAG	2120	0.6
Er:YAG	2940	0.4
CO_2	10600	0.1

we find that only photons from UV lasers – typically excimer lasers – provide an energy sufficient for dissociating such bonds. Therefore, the interaction mechanism of photoablation is limited to the application of UV light.

The ejected photoproducts of excimer laser ablation have been analyzed in several studies, e.g. by Srinivasan and Mayne-Banton (1982) and Brannon et al. (1985). Usually, a mixture of single atoms (C, N, H, O), molecules (C_2, CN, CH, CO), and stable fragments (MMA-monomer, HCN, benzene) were detected. It is interesting to add that the product composition was found to be wavelength-dependent. Srinivasan et al. (1986b) found that radiation at 248 nm generates fragments of a higher molecular weight than does 193 nm radiation. Thus, the corresponding etched surface at 248 nm is rougher and less predictable than at 193 nm.

Moreover, it was observed that a certain threshold intensity must be applied to achieve photoablation. Above this intensity, a well-defined depth is ablated, depending on the absorption coefficient and the incident intensity. At the same time, an audible report is heard and visible fluorescence is observed at the impact site. If the incident intensity is moderate and such that the ablation depth is smaller than the corresponding optical absorption length, subsequent pulses will enter partially irradiated tissue as well as unexposed tissue underlying it. Therefore, only the first few pulses are unique. After these, a linear relation between the number of applied pulses and the total etch depth is obtained. In practice, the etch depths are averaged over several pulses and noted as ablation depth per pulse. This value is reproducible within an uncertainty of about 10 % for most materials which is an excellent value when taking all the inhomogeneities of tissue into account. The physical principles of photoablation are summarized in Table 3.10.

Table 3.10. Principles of photoablation

Absorption of high-energy UV photons
⇓
Promotion to repulsive excited states
⇓
Dissociation
⇓
Ejection of fragments (no necrosis)
⇓
Ablation

We stated above that photoablation is a process which is restricted to the application of UV light. However, it is not limited to excimer lasers, since generating higher harmonics of other laser types can result in UV radiation, as well. The 4th harmonic of a solid-state laser, for instance, also induces photoablation as reported by Niemz et al. (1994b). In Figs. 3.37a–b and 3.38a–b, a sample of corneal tissue is shown which was partially exposed to either the second or the fourth harmonic of a Nd:YLF laser. In both cases, the pulse duration at the fundamental wavelength was set to 30 ps. The higher harmonics were induced by means of two BBO crystals. The pulse energies were approximately 150 μJ in the green and 20 μJ in the UV, respectively. The laser beam was scanned over the surface within a $1 \times 1\,\mathrm{mm}^2$ pattern. Whereas distinct impact sites of the focused laser beam are clearly visible in the section exposed to the second harmonic, a clean and homogeneous layer is ablated using the fourth harmonic. Only the latter belongs to the group of photoablation, whereas the other effect is attributed to plasma-induced ablation which will be discussed in Sect. 3.4.

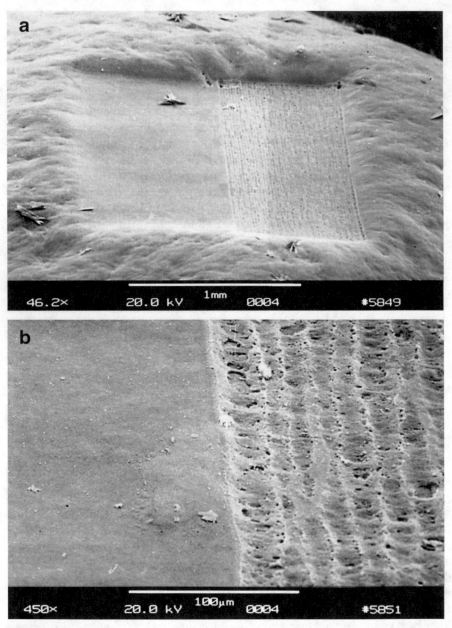

Fig. 3.37. (a) Human cornea exposed to the second (*right*) and fourth (*left*) harmonics of a Nd:YLF laser, respectively (pulse duration: 30 ps). (b) Enlargement of the boundary with the adjacent zones

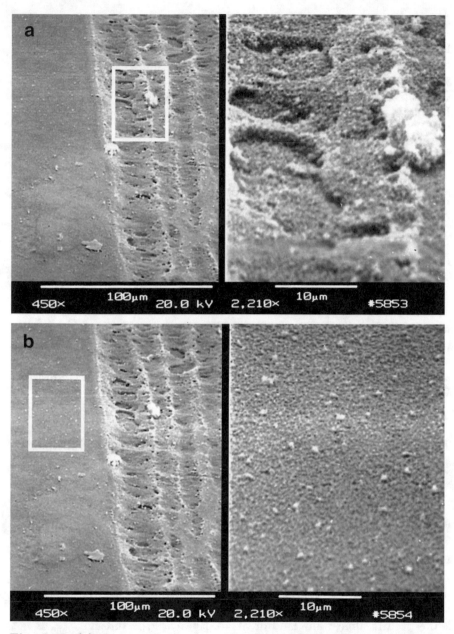

Fig. 3.38. (a) Human cornea exposed to the second and fourth harmonics of a Nd:YLF laser, respectively (*left*). Enlargement of area exposed to the second harmonic (*right*). **(b)** Same view (*left*). Enlargement of area exposed to the fourth harmonic (*right*)

3.3.1 Model of Photoablation

In order to come up with a model which describes the dependence of ablated depth on incident laser intensity, most research groups – such as Srinivasan and Mayne-Banton (1982), Andrew et al. (1983), Deutsch and Geis (1983), Garrison and Srinivasan (1985) – based their assumptions on the validity of Lambert's law of light absorption given by[10]

$$I(z) = I_0 \exp(-\alpha z) \,, \tag{3.20}$$

where z denotes the optical axis, I_0 is the incident laser intensity, and α is the absorption coefficient of the tissue. To evaluate the decrease in intensity, (3.20) has to be differentiated with respect to z which leads to

$$-\frac{\partial I}{\partial z} = \alpha \, I(z) \,. \tag{3.21}$$

Photoablation will take place only as long as the left side in (3.21) does not drop below a certain threshold value αI_{ph}, i.e.

$$\alpha I(z) \geq \alpha I_{ph} \,,$$

where I_{ph} is the threshold intensity of photoablation. This condition requires that a certain amount of energy must be absorbed per unit volume and time to achieve photoablation. The threshold intensity I_{ph} is determined by the minimal number of bonds that have to be dissociated to yield decomposition. We thus obtain

$$I_0 \exp(-\alpha z) \geq I_{ph} \,.$$

The ablation depth d, i.e. the depth at which $I(z) = I_{ph}$, should then be

$$d = \frac{1}{\alpha} \ln \frac{I_0}{I_{ph}} \simeq \frac{2.3}{\alpha} \log_{10} \frac{I_0}{I_{ph}} \,. \tag{3.22}$$

This simple model describes photoablation very well, except for the threshold I_{ph} at the onset of photoablation. The ablation curve of rabbit cornea is shown in Fig. 3.39 in a single-logarithmic plot. Usually, the ablation depth per pulse is given as a function of the incident energy density E_0, where $E_0 = I_0 \tau$ and τ is the pulse duration. The logarithmic dependence, i.e. the central and linear part in Fig. 3.39, is in good agreement with theoretical predictions based on (3.22). This section of the ablation curve is observed in almost any kind of tissue. However, the threshold I_{ph} is not as sharp as predicted by (3.22), i.e. the slope of the curve approaches zero when intercepting with the abscissa. This result most probably stems from the inhomogeneity in fragment sizes. The threshold varies around an average

---

[10] For a complete mathematical description of photoablation, the temporal shape of the applied laser pulses should also be taken into account. Further details are found in the paper by Srinivasan (1986a).

value $<I_{ph}>$ according to the size of the ablated fragment. Imagine that such a fragment was bound to several molecules prior to ablation. As soon as a certain ratio of dissociated molecules is reached, the fragment will be released. Because larger fragments have a smaller "surface" compared to their "volume", the threshold I_{ph} will decrease with increasing size of the ablated fragment. Consequently, averaging of various fragment sizes leads to a smooth intercept with the abscissa.

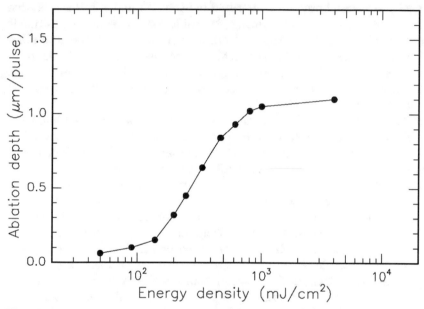

Fig. 3.39. Ablation curve of rabbit cornea obtained with an ArF excimer laser (pulse duration: 14 ns). Data according to Fantes and Waring (1989)

Above a second threshold I_{pl} – the threshold of plasma generation – the ablation depth per pulse obviously saturates as shown in Fig. 3.39. This effect stems from a well-known phenomenon called *plasma shielding*. Once a plasma is ignited at high power densities due to the generation of a high electric field, most of the succeeding laser radiation is absorbed by the plasma, thereby heating it up and leading to additional thermal effects. All abundant energy thus dissipates to heat and does not contribute to a further increase in ablation depth. Therefore, the ablation curve saturates at high energy densities. The absorption coefficient α_{pl} of the induced plasma is significantly higher than the original absorption coefficient α of the tissue as will be evaluated in Sect. 3.4. Hence, equations (3.20)–(3.22) are no longer valid. For a detailed discussion of plasma parameters and plasma shielding, the reader is referred to Sect. 3.4.

In the 1980s, the question was raised whether photoablation is based on a photochemical or a photothermal process. This discussion has led to much controversy. Andrew et al. (1983) and Brannon et al. (1985) claimed UV ablation to be exclusively of thermal character, whereas Garrison and Srinivasan (1985) tended to attribute it to photochemical effects. Today, it is well accepted that photoablation – or the synonym ablative photodecomposition in the sense of UV ablation – shall be considered as an interaction mechanism of its own that can certainly be distinguished from pure photochemical and thermal processes described in Sects. 3.1 and 3.2. However, due to the dissociation of molecules during photoablation, a chemical transition is involved which could also justify misleading terms like "photochemical ablation". But, with respect to the historical progress made in research, the term "photochemical" should be reserved for interactions with photosensitizers at long exposures. Only an ablation caused by UV photons should be regarded as photoablation[11].

In order to distinguish *photoablation* or *ablative photodecomposition* from *thermal interaction*, we take a closer look once more at the energy level diagram shown in Fig. 3.36. In the case of photoablation, we concluded that the energy of a single UV photon is sufficient to dissociate the former bound molecule AB. In thermal interactions, the situation is completely different. The photon energy is not high enough for the molecule to reach a repulsive state. The molecule is promoted only to a vibrational state within the ground level or to a rather low electronic state including any of its vibrational states. By means of non-radiative relaxation, the absorbed energy then dissipates to heat, and the molecule returns to its ground state. Hence, the crucial parameter for differentiating these two mechanisms – photoablation and thermal interaction – is the photon energy or laser wavelength. Only if $h\nu \geq 3.6\,\mathrm{eV}$, or in other measures $\lambda \leq 345\,\mathrm{nm}$, is the single photon dissociation of C–C bonds enabled. Of course, several photons with $h\nu < 3.6\,\mathrm{eV}$ may be absorbed. Then, all these photons can add up in energy and thus lead to a dissociated state, as well. However, during the time needed for such a multi-photon absorption process, other tissue areas become vibrationally excited, hence leading to a global increase in temperature and an observable thermal effect (usually either vaporization or melting). If this effect is associated with ablation, the whole process is called *thermal decomposition* and has to be distinguished from pure *ablative photodecomposition*. These statements hold true, unless ultrashort pulses with a pulse duration shorter than 500 ps are used at pulse energies high enough to induce a localized microplasma. Then, even VIS- and IR-lasers can interact non-thermally. But the discussion of this mechanism shall be deferred to Sect. 3.4.

---

[11] Actually, the term "photoablation" itself is not well defined. It only states that an ablation occurs which is caused by photons, but it does not imply any further details. Photoablation in the scientific sense, though, is a very precise ablation caused by UV photons.

In reality, pure photoablation is only observed for the 193 nm wavelength of the ArF excimer laser. Higher wavelengths are usually associated with a more or less apparent *thermal component*. Sutcliffe and Srinivasan (1986) even postulated that the thermal component of the radiation from a XeCl excimer laser at 308 nm is 100 % thermal if its radiation is incident on PMMA. In Fig. 3.40, the thermal components of three excimer lasers are compared with each other. Obviously, each laser induces thermal effects at low energy densities. In the cases of the ArF laser and the KrF laser, the slope of the curve strongly decreases above a certain threshold, i.e. the threshold of photoablation. Radiation from the XeCl laser, on the other hand, continues to act thermally even at higher energy densities.

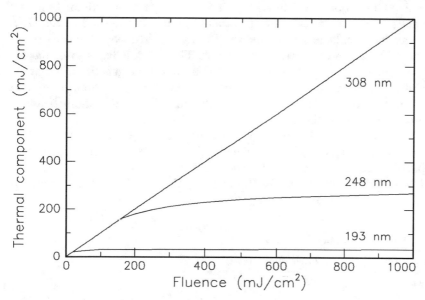

Fig. 3.40. Thermal component of UV radiation from three different excimer lasers (ArF laser at 193 nm, KrF laser at 248 nm, and XeCl laser at 308 nm) in PMMA. Data according to Sutcliffe and Srinivasan (1986)

The data presented in Fig. 3.40 are only valid for the target material PMMA. They cannot be generalized to any biological tissue, since the absorption characteristics of PMMA and biological tissue are quite different. However, the bottom line of the observation above – i.e. that the significance of the thermal component increases with higher wavelengths – will be true in biological tissue, as well. Therefore, we can conclude that ArF excimer lasers are our best choice when aiming at photoablation. XeCl lasers could be used for thermal decomposition, yet many solid-state lasers do a similar job. KrF lasers offer both photoablative and thermal decomposition but are less useful in medical applications as we will discuss next.

3.3.2 Cytotoxicity of UV Radiation

It has been argued that the application of UV radiation for photoablative purposes might induce mutagenic – and thus cytotoxic – effects within cells. This statement has to be taken seriously, since laser surgery, of course, should not evoke new maladies when eliminating others. It is a fact that DNA strongly absorbs UV radiation, especially at approximately 240–260 nm. And it is also well known that this radiation can cause mutagenic alterations of cells, e.g. exchanges of sister chromatids.

The effect of UV radiation on cells and biological tissue is initiated by photochemical reactions of chromophores contained by them. Absorption of UV photons by DNA causes alterations in the chemical structure of the DNA molecule. The major chemical change is the formation of a dimer from two adjacent pyrimidine bases. Other products are also synthesized in the DNA that may have biological consequences. Cells are frequently able to repair dimers before any adverse responses occur. This is an indispensable mechanism of protection, since the DNA contains important genetic information. Thus, if these photoproducts are not repaired, erroneous information may be passed on to progeny cells when the cell divides. This event, finally, leads to the process of mutagenesis.

Several studies have been done in order to evaluate potential hazards from UV laser radiation. Usually, the effects of one or two excimer lasers were compared to each other and to low-pressure Hg lamps. With these, most of the UV spectrum is covered. In Table 3.11, the wavelengths of commonly used UV sources are listed.

Table 3.11. Sources of UV radiation

Light source	Wavelength (nm)
ArF laser	193
KrF laser	248
Hg lamp	254
Nd:YLF laser (4ω)	263
Nd:YAG laser (4ω)	266
XeCl laser	308
XeF laser	351

Cytotoxic effects of UV radiation have been investigated by several groups, among them especially by Green et al. (1987), Kochevar (1989), and Rasmussen et al. (1989). In general, it can be concluded that the relative ability of excimer laser radiation to cause DNA defects decreases in the order of 248 nm > 193 nm > 308 nm. Only in some cases was radiation at 193 nm found to be less cytotoxic than at 308 nm. In Fig. 3.41, the surviving fraction of Chinese hamster ovary (CHO) cells after exposure to three different light sources is shown as a function of incident energy density. Whereas the Hg

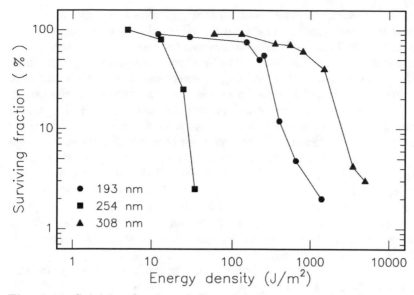

Fig. 3.41. Surviving fraction of Chinese hamster ovary (CHO) cells after UV irradiation. Light sources: ArF laser (193 nm), Hg lamp (254 nm), and XeCl laser (308 nm). Data according to Rasmussen et al. (1989)

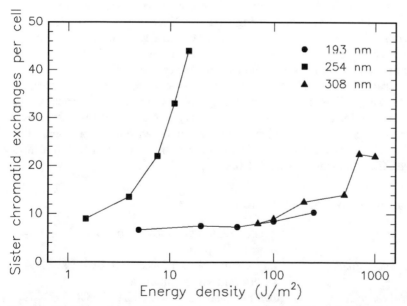

Fig. 3.42. Sister chromatid exchanges per cell after UV irradiation. Light sources: ArF laser (193 nm), Hg lamp (254 nm), and XeCl laser (308 nm). Data according to Rasmussen et al. (1989)

lamp induces measurable damage at $15\,\mathrm{J/m^2}$, it takes a few hundred $\mathrm{J/m^2}$ of excimer laser radiation to cause a similar effect. When comparing the ArF laser at 193 nm and the XeCl laser at 308 nm, the latter is less cytotoxic.

In the same study, the exchange of sister chromatids was measured. These can either be repaired by the cell or, in severe cases, lead to irreversible defects. In the latter case, either cell necrosis or uncontrolled cell proliferation, i.e. certain types of cancer, can occur. The extent of sister chromatid exchange was investigated for the same UV sources and is shown in Fig. 3.42. All three radiations caused an increase in the incidence of sister chromatid exchange but with different effectiveness. The required energy density at 254 nm, for instance, is much less than at 193 nm or 308 nm. Moreover, the slope of the curve at 254 nm is much steeper than the others. Thus, another proof is given that radiation from a Hg lamp can be considered as being more mutagenic than ArF or XeCl lasers. This is the main reason why there are no significant medical applications for KrF lasers, since its wavelength at 248 nm almost coincides with that of Hg lamps.

We conclude that radiation from excimer lasers is less mutagenic than UV light from Hg lamps. This observation can probably be explained by the existence of proteins in the cell matrix which strongly absorb radiation at 193 nm, before it reaches the cell nucleus containing the DNA. According to Green et al. (1987), about 60 % of incident radiation is already absorbed by only 1 µm of cytoplasm. Hence, the sensitive DNA inside the nucleus is shielded by the surrounding cytoplasm. However, potential risks should never be ignored when using ArF or XeCl lasers. Actually, a few altered cells might be enough to induce cancer in tissue. And, according to Rasmussen et al. (1989), there is some preliminary evidence that fluorescence emission around 250 nm occurs following absorption at 193 nm.

In general, the difficulty in judging the severity of mutagenic effects is due to the long follow-up periods during which maladies can develop. For instance, a haze inside the cornea is frequently observed within a few years after refractive corneal surgery has been performed with excimer lasers. The origin of this haze is yet unknown. It might well be attributed to cell alterations, even if corneal tumors do not occur.

3.3.3 Summary of Photoablation

- *Main idea:*   direct breaking of molecular bonds by high-energy UV photons
- *Observable effects:*   very clean ablation, associated with audible report and visible fluorescence
- *Typical lasers:*   excimer laser, e.g. ArF, KrF, XeCl, XeF
- *Typical pulse durations:*   10 ns ... 100 ns
- *Typical power densities:* $10^7\,\mathrm{W/cm^2}$ ... $10^9\,\mathrm{W/cm^2}$
- *Special applications:*   refractive corneal surgery

3.4 Plasma-Induced Ablation

When obtaining power densities exceeding 10^{11} W/cm^2 in solids and fluids – or 10^{14} W/cm^2 in air – a special phenomenon called *optical breakdown* occurs. This effect is demonstrated in Fig. 3.43. Here a series of 30 ps laser pulses from a mode locked and amplified Nd:YLF laser was focused on an extracted human tooth. A bright spark stemming[12] from a physical plasma is clearly visible which is pointing toward the laser source. This plasma is ionizing the target and thereby ablating material from the tooth. That's why we call it a *plasma-induced ablation* If several laser pulses are applied, a typical sparking noise at the repetition rate of the pulses is heard.

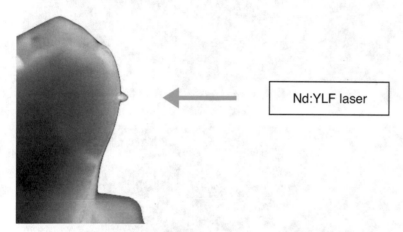

Fig. 3.43. Laser-induced plasma sparking on tooth surface caused by a series of ultrashort pulses from a Nd:YLF laser (pulse duration: 30 ps, pulse energy: 1 mJ, focal spot size: 30 μm). With these parameters, a power density of approximately 5×10^{12} W/cm^2 is obtained

By means of plasma-induced ablation, a clean and well-defined removal of tissue without evidence of thermal or mechanical damage can be achieved when choosing appropriate laser parameters. The effects of plasma-induced ablation on both soft and hard tissues are demonstrated in Figs. 3.44a–b. In one case, a sample of corneal tissue was incised using a picosecond Nd:YLF laser by applying the pulses in a straight line pattern. The cross-sectional view shows a very precise incision without mechanical ruptures. In the other case, pulses from the same Nd:YLF laser were scanned over the surface of a tooth in a 1×1 mm^2 square pattern. In particular, the fairly smooth bottom of the cavity and the steep side walls are very significant. Further details on potential applications are given in Sects. 4.1 and 4.2.

---

[12] The spark is caused by the recombination of positive calcium ions and free electrons, thereby releasing energy as visible light.

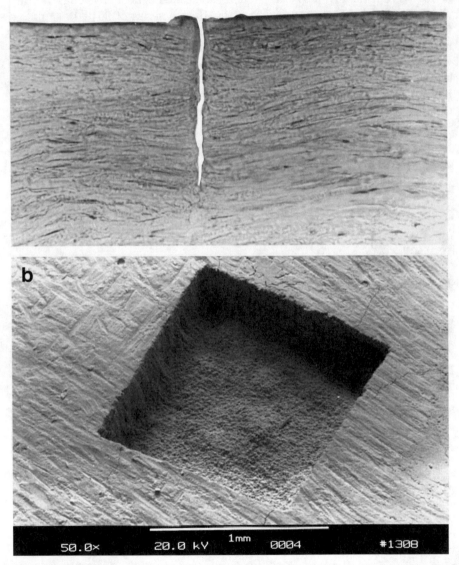

Fig. 3.44. (a) Incision in a human cornea achieved with a picosecond Nd:YLF laser (pulse duration: 30 ps, pulse energy: 200 μJ, bar: 50 μm). (b) Human tooth exposed to 16 000 pulses from a picosecond Nd:YLF laser (pulse duration: 30 ps, pulse energy: 1 mJ, pattern size: $1 \times 1\,\mathrm{mm}^2$). Superficial cracking occurred as an artefact during the preparation for electron microscopy

Plasma-induced ablation was thoroughly investigated and discussed by Teng et al. (1987), Stern et al. (1989), and Niemz et al. (1991). In the literature, plasma-induced ablation is sometimes referred to as *plasma-mediated ablation*[13]. Both synonyms express a generally well accepted interpretation that this kind of ablation is primarily caused by the plasma ionization itself. This is in contrast to a more mechanical process called *photodisruption* which will be described in Sect. 3.5.

The most important parameter of plasma-induced ablation is the local electric field strength E which defines when optical breakdown is achieved. If E exceeds a certain threshold value, i.e. if the applied electric field forces the ionization of molecules and atoms, breakdown occurs. The electric field strength itself is related to the local power density I by the basic electrodynamic equation

$$I(r, z, t) = \frac{1}{2} \varepsilon_0 c E^2 \,, \tag{3.23}$$

where ε_0 is the dielectric constant, and c is the speed of light. For picosecond pulses, typical threshold intensities of optical breakdown are $10^{11}\,\mathrm{W/cm^2}$. Hence, the corresponding electric field amounts to approximately $10^7\,\mathrm{V/cm}$. This value is comparable to the average atomic or intramolecular *Coulomb* electric fields, thus providing the necessary condition for plasma ionization. Within a few hundred picoseconds, an extremely large free electron density with typical values of $10^{18}/\mathrm{cm^3}$ and more is created within the focal volume of the laser beam. In general, plasma generation due to an intense electric field is referred to as *optical breakdown*. The term *optical breakdown* is derived from plasma physics and emphasizes that there is no optics anymore, once such a physical plasma is formed. This is because all of the electromagnetic radiation is strongly absorbed by the plasma itself. The physical principles of breakdown have been investigated in several theoretical studies, for example by Seitz (1949), Molchanov (1970), Yablonovitch and Bloembergen (1972), Bloembergen (1974), Epifanov (1981), and Sacchi (1991).

Fradin et al. (1973b) have measured the intensity of ruby laser pulses transmitted through NaCl. They used a $\mathrm{TEM_{00}}$ mode ruby laser with pulse energies of 0.3 mJ and pulse durations of 15 ns which was focused with a lens of 14 mm focal length. In Figs. 3.45a–c, the transmitted intensities are shown as a function of time. Obviously, the shape of the transmitted pulse is not altered in Fig. 3.45a, whereas the rear part of the pulse is strongly absorbed in Figs. 3.45b–c. Fradin et al. (1973b) concluded that the intensity as in Fig. 3.45a was just below the threshold of optical breakdown, while it was above in Fig. 3.45b, thereby damaging the NaCl sample. In Fig. 3.45c, breakdown was achieved at the very peak of the laser pulse.

---

[13] Throughout this book we will use the term *plasma-induced ablation* rather than *plasma-mediated ablation*, since only the first term emphasizes the primary cause of the ablation.

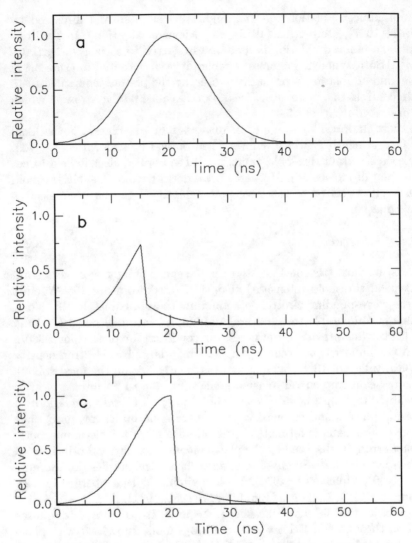

Fig. 3.45. (a–c) Ruby laser pulses transmitted through NaCl. Data according to Fradin et al. (1973b)

The initiation of plasma generation can be two-fold and was described in detail by Puliafito and Steinert (1984). It was observed that either Q-switched pulses in the nanosecond range or mode locked laser pulses in the picosecond or femtosecond range can induce a localized microplasma. In Q-switched pulses, the initial process for the generation of free electrons is supposed to be *thermionic emission*, i.e. the release of electrons due to thermal ionization as shown in Fig. 3.46. In mode locked pulses, *multi-photon ionization* may occur due to the high electric field induced by the intense laser pulse. In general, the term multi-photon ionization denotes processes in which coherent absorption of several photons provides the energy needed for ionization. Due to the requirement of coherence, multi-photon ionization is achievable only during high peak intensities as in picosecond or femto-second laser pulses. Plasma energies and plasma temperatures, though, are usually higher in Q-switched laser pulses because of the associated increase in threshold energy of plasma formation. Thus, optical breakdown achieved with nanosecond pulses is often accompanied by non-ionizing side effects and will be deferred to Sect. 3.5.

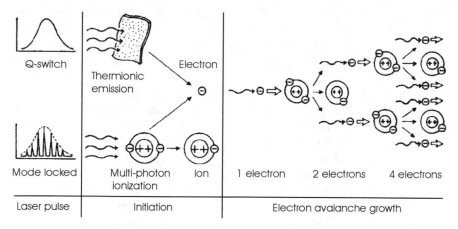

Fig. 3.46. Initiation of ionization with subsequent electron avalanche. Reproduced from Puliafito and Steinert (1984) by permission. © 1984 IEEE

In either case, however, a few "lucky" electrons initiate an avalanche effect, leading to the accumulation of free electrons and ions. A free electron absorbs a photon and accelerates. The accelerated electron collides with another atom and ionizes it, resulting in two free electrons, each of them with less individual kinetic energy than the initial electron. Again, these free electrons may absorb incoming photons, accelerate, strike other atoms, release two more electrons, and so forth. The basic process of photon absorption and electron acceleration taking place in the presence of an atom is called *inverse Bremsstrahlung*[14]. It is based on a free-free absorption, i.e. a transition, where a free electron is present in the initial and final states. In order to fulfill the conservation laws of energy and momentum, this event must necessarily take place in the electric field of an ion A^+ or a neutral atom. The process is schematically written as

$$h\nu + e + A^+ \longrightarrow e + A^+ + E_{kin} \ .$$

---

[14] In ordinary *Bremsstrahlung*, the opposite effect occurs, i.e. electrons are accelerated in the electromagnetic field of an atom, thereby emitting radiation.

In particular, the important feature of optical breakdown is that it makes possible the deposition of energy not only in pigmented tissues but also in nominally weakly absorbing media. This is due to the increased absorption coefficient of the induced plasma. Thereby, the field of potential medical laser applications is considerably widened. Especially in ophthalmology, transparent tissues – like cornea and lens – become potential targets of laser surgery. It is also worthwile noting that during the process there is no restriction of the photon energy involved. The electron may increase its kinetic energy by means of absorbing any amount of energy, thus leading to a very short risetime of the free electron density of the order of picoseconds. In order to achieve optical breakdown, though, the irradiance must be intense enough to cause rapid ionization so that losses do not quench the electron avalanche. According to Smith and Haught (1966), inelastic collisions and diffusion of free electrons from the focal volume are the main loss mechanisms during avalanche ionization.

As already emphasized, the electric field strength $\boldsymbol{E}$ is the crucial parameter for the onset of plasma-induced ablation. In the following paragraphs, a model of plasma-induced ablation is developed which is based on fundamental relations of electrodynamics and plasma physics. It derives a general expression for the dielectric factor ε and culminates in a relation for the absorption coefficient α_{pl} of the plasma. For visible and near infrared laser wavelengths, the original absorption coefficient α of most tissues is quite low compared to using UV light. However, calculations derived from the following model will prove that $\alpha_{\mathrm{pl}} \gg \alpha$, thus leading to enhanced absorption and efficient ablation.

3.4.1 Model of Plasma-Induced Ablation

Our model is based on *Maxwell's equations* of electrodynamics relating the electric and magnetic field strengths $\boldsymbol{E}$ and $\boldsymbol{H}$ to the electromagnetic inductions $\boldsymbol{D}$ and $\boldsymbol{B}$, respectively. In order to account for free electrons and currents associated with plasma formation, the inhomogeneous version of Maxwell's equations is considered. In their differential form, these equations are given by

$$\mathrm{rot}\, \boldsymbol{E} = -\frac{\partial \boldsymbol{B}}{\partial t} \ , \tag{3.24}$$

$$\mathrm{rot}\, \boldsymbol{H} = \frac{\partial \boldsymbol{D}}{\partial t} + \boldsymbol{j}_{\mathrm{f}} \ , \tag{3.25}$$

$$\mathrm{div}\, \boldsymbol{D} = \varrho_{\mathrm{f}} \ , \tag{3.26}$$

$$\mathrm{div}\, \boldsymbol{B} = 0 \ , \tag{3.27}$$

where ϱ_{f} denotes the density of all free electric charges, and $\boldsymbol{j}_{\mathrm{f}}$ the density of all free electric currents of the plasma.

The laser beam itself is approximated by a plane electromagnetic wave

$$\boldsymbol{E}(r, t) = \boldsymbol{E}_0 \, \exp[\,i(\omega t - \boldsymbol{kr}\,)] \;, \tag{3.28}$$

$$\boldsymbol{H}(r, t) = \boldsymbol{H}_0 \, \exp[\,i(\omega t - \boldsymbol{kr}\,)] \;, \tag{3.29}$$

where $\boldsymbol{k}$ is the propagation vector of the electromagnetic wave. Assuming the relative magnetic permeability of the tissue to be $\mu = 1$, the electromagnetic inductions can be expressed by

$$\boldsymbol{B} = \mu_0 \boldsymbol{H} \;, \tag{3.30}$$

$$\boldsymbol{D} = \varepsilon\varepsilon_0 \boldsymbol{E} \;, \tag{3.31}$$

where μ_0 and ε_0 are the electromagnetic constants, and ε is the dielectric factor of the plasma. Inserting (3.28)–(3.31) into the first two of Maxwell's equations leads to

$$\boldsymbol{k} \times \boldsymbol{E} = \omega\mu_0 \boldsymbol{H} \;,$$

$$\boldsymbol{k} \times \boldsymbol{H} = -\,\omega\varepsilon\varepsilon_0 \boldsymbol{E} + i\,\boldsymbol{j}_{\mathrm{f}} = -\omega\varepsilon\varepsilon_0 \boldsymbol{E} + i\,\sigma \boldsymbol{E} \;,$$

where σ is the electric conductivity of the plasma. Hence,

$$\boldsymbol{k} \times (\boldsymbol{k} \times \boldsymbol{E}) = -\,\omega^2 \varepsilon\mu_0\varepsilon_0 \boldsymbol{E} + i\,\omega\mu_0\sigma \boldsymbol{E} \;,$$

$$\boldsymbol{k} \times (\boldsymbol{k} \times \boldsymbol{E}) = \left(-\frac{\omega^2}{c^2}\,\varepsilon + i\,\frac{\omega\sigma}{c^2\varepsilon_0} \right) \boldsymbol{E} \;, \tag{3.32}$$

where $c - \sqrt{1/\varepsilon_0\mu_0}$ is the speed of light. In electrodynamics, (3.32) is referred to as the *telegraph equation*. Since electromagnetic waves are transversal, i.e. $\boldsymbol{kE} = 0$, the amount $k = |\boldsymbol{k}|$ can be expressed by

$$k^2 = \frac{\omega^2}{c^2}\,\varepsilon' \;, \tag{3.33}$$

with the complex dielectric factor

$$\varepsilon' = \varepsilon - i\,\frac{\sigma}{\omega\varepsilon_0} \;. \tag{3.34}$$

In the case of negligible absorption, (3.33) reduces to the simple relation of dispersion

$$k = \frac{\omega}{c}\,n \;, \tag{3.35}$$

where n is the index of refraction. However, since the plasma absorbs incident radiation, (3.35) is replaced by

$$k = \frac{\omega}{c}\,(n + i\tilde{\alpha}) \;, \tag{3.36}$$

where $\tilde{\alpha}$ is the index of absorption. Combining (3.33) and (3.36) leads to the general expression

$$\varepsilon' = (n + i\tilde{\alpha})^2 \, . \tag{3.37}$$

Moreover, the following two relations are derived from (3.36)

$$n = \frac{c}{\omega} \, \mathrm{Re}(k) \, ,$$

$$\tilde{\alpha} = \frac{c}{\omega} \, \mathrm{Im}(k) \, .$$

The absorption coefficient α – which is not to be mistaken for $\tilde{\alpha}$ – was defined in the previous section by (3.20). Together with (3.23) and (3.28), we thus obtain

$$\alpha_{\mathrm{pl}} = -\, 2 \, \mathrm{Im}(k) = -\, \frac{2\omega}{c} \, \tilde{\alpha}_{\mathrm{pl}} \, ,$$

which expresses the plasma absorption coefficient α_{pl} and the index of plasma absorption $\tilde{\alpha}_{\mathrm{pl}}$ in terms of each other. We may separate (3.37) into

$$\mathrm{Re}(\varepsilon') = n^2 - \tilde{\alpha}_{\mathrm{pl}}^2 \, ,$$

$$\mathrm{Im}(\varepsilon') = 2n\tilde{\alpha}_{\mathrm{pl}} \quad \Longrightarrow \quad \tilde{\alpha}_{\mathrm{pl}} = \frac{1}{2n} \, \mathrm{Im}(\varepsilon') \, .$$

Hence,

$$\alpha_{\mathrm{pl}} = -\, \frac{\omega}{nc} \, \mathrm{Im}(\varepsilon') \, , \tag{3.38}$$

stating that the existence of an imaginary part of ε' evokes absorption.

Next, we will consider the equation of motion of a plasma electron. It is given by

$$m_{\mathrm{e}} \frac{\partial \boldsymbol{v}_{\mathrm{e}}}{\partial t} = -e\boldsymbol{E} - \nu_{\mathrm{ei}} m_{\mathrm{e}} \boldsymbol{v}_{\mathrm{e}} \, , \tag{3.39}$$

where m_{e} is the electron mass, $\boldsymbol{v}_{\mathrm{e}}$ is the electron velocity, e is the electron charge, and ν_{ei} is the mean collision rate of free electrons and ions. Assuming that the plasma electrons are performing oscillations induced by the incident electromagnetic field at its frequency ω, we obtain

$$\mathrm{i}\omega m_{\mathrm{e}} \boldsymbol{v}_{\mathrm{e}} = -e\boldsymbol{E} - \nu_{\mathrm{ei}} m_{\mathrm{e}} \boldsymbol{v}_{\mathrm{e}} \, ,$$

$$\boldsymbol{v}_{\mathrm{e}} = -\, \frac{e}{m_{\mathrm{e}}(\nu_{\mathrm{ei}} + \mathrm{i}\omega)} \, \boldsymbol{E} \, . \tag{3.40}$$

Together with the basic relation for the free electric current

$$\boldsymbol{j}_{\mathrm{f}} = -N e \boldsymbol{v}_{\mathrm{e}} \, ,$$

where N is the density of free electrons, (3.40) turns into

$$\dot{j}_f = \frac{Ne^2}{m_e(\nu_{ei} + i\omega)} \, E \, ,$$

$$\sigma = \frac{Ne^2}{m_e(\nu_{ei} + i\omega)} \, . \tag{3.41}$$

Inserting (3.41) into (3.34) leads to

$$\varepsilon' = \varepsilon - i\frac{1}{\omega\varepsilon_0} \frac{Ne^2}{m_e(\nu_{ei} + i\omega)} = \varepsilon - i\frac{\omega_{pl}^2}{\omega(\nu_{ei} + i\omega)} \, ,$$

$$\varepsilon' = \varepsilon - \frac{\omega^2\omega_{pl}^2}{\omega^2\nu_{ei}^2 + \omega^4} - i\frac{\omega\nu_{ei}\omega_{pl}^2}{\omega^2\nu_{ei}^2 + \omega^4} \, , \tag{3.42}$$

where ω_{pl} is the *plasma frequency* defined by

$$\omega_{pl}^2 = \frac{Ne^2}{\varepsilon_0 m_e} \, . \tag{3.43}$$

From (3.38) and (3.42), the plasma absorption coefficient is derived as

$$\alpha_{pl} = \frac{\omega}{nc} \frac{\omega\nu_{ei}\omega_{pl}^2}{\omega^2\nu_{ei}^2 + \omega^4} = \frac{\nu_{ei}}{nc} \frac{\omega_{pl}^2}{\omega^2 + \nu_{ei}^2} \, . \tag{3.44}$$

In cold laser plasmas, i.e. $\nu_{ei} \ll \omega$, (3.44) can be simplified to

$$\alpha_{pl} = \frac{\nu_{ei}}{nc} \frac{\omega_{pl}^2}{\omega^2} \, . \tag{3.45}$$

Therefore, plasma absorption is enhanced in the IR region of the spectrum. And, since $\omega_{pl}^2 \sim N$ and $\nu_{ei} \sim N$, we obtain the important relation

$$\alpha_{pl} \sim N^2 \, ,$$

stating that the absorption is a non-linear function of the free electron density and, thus, of the absorbed energy itself.

In Fig. 3.47, typical absorption coefficients α_{pl} for various incident power densities have been collected. The plasma was induced in a cuvette filled with distilled water by picosecond pulses from a Nd:YLF laser at a wavelength of 1.053 μm. The absorption coefficient has already gained more than a factor of 30 once the incident energy density is twice the threshold value of plasma generation. In the example shown in Fig. 3.47, the absorption coefficient increases from 0.1 cm^{-1} at low fluences to approximately 90 cm^{-1} at 23 J/cm^2. Therefore, the induced plasma serves as a shield for succeeding laser photons. This phenomenon is referred to as the *plasma shielding effect*. At higher pulse energies, its significance increases even more. Primarily, the enhanced absorption of the plasma is caused by the high density of free electrons capable of absorbing laser photons. Thus, by means of plasma generation, a very efficient type of ablation is created.

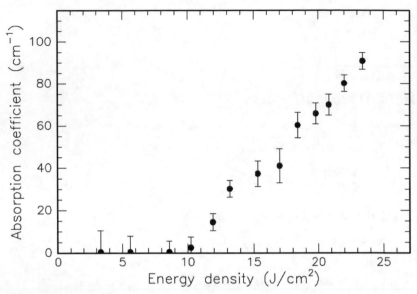

Fig. 3.47. Absorption of laser plasmas induced in distilled water by a Nd:YLF laser (wavelength: 1.053 μm, pulse duration: 30 ps). Unpublished data

The condition for plasma growth and sustainment is that losses – such as inelastic collisions and free electron diffusion – do not quench the avalanche process. As the free electron density N increases, photon scattering is enhanced. Consequently, laser photons are not only absorbed but also scattered by the plasma. The critical density N_{crit} at which a net amount of energy is not converted any further to plasma energy is obtained when the plasma frequency becomes equal to the frequency of the incident electromagnetic wave[15]. Mathematically, this condition is written as $\omega_{pl} = \omega$ and, when inserting (3.43),

$$N_{crit} = \frac{\varepsilon_0 m_e}{e^2} \, \omega^2 \; . \tag{3.46}$$

For visible laser radiation, i.e. $\omega \simeq 10^{15}\,\mathrm{Hz}$, the electron density may thus reach values of up to a few $10^{20}/\mathrm{cm}^3$.

In the remaining part of this section, we will evaluate the threshold dependence of optical breakdown on pulse duration of the laser. In several studies – as discussed below – a square root dependence of the threshold power density on pulse duration was observed when using picosecond or nanosecond laser pulses. This relationship, however, was only known as an *empirical scaling law* as stated by Koechner (1992). On the following pages, a model is developed which illuminates its theoretical background.

---

[15] Plasma oscillations are characterized by density-dependent eigenfrequencies. The highest amplitude occurs at resonance, i.e. at $\omega_{pl} = \omega$. At $\omega_{pl} > \omega$, the plasma oscillation is decelerated due to the phase shift of the external force.

The temporal behavior of the free electron density $N(t)$ is described by

$$\frac{\partial N}{\partial t} = [\beta - \gamma N(t)] \, N(t) - \delta N(t) \ , \tag{3.47}$$

with the rate parameters β for *avalanche ionization*, γ for *inelastic collision*, and δ for *electron diffusion*. Avalanche ionization is a two-step process. First, electrons gain energy in the electromagnetic field of the laser pulse, then further electrons are dissociated by subsequent collisions with either atoms or ions. According to Smith and Haught (1966), inelastic collisions and diffusion of free electrons from the focal volume are the main loss mechanisms during avalanche ionization. They are accounted for in (3.47) by negative signs. The avalanche parameter β primarily depends on the incident intensity $I(t)$. A *reduced avalanche rate* can be defined by

$$\beta' = \beta - \gamma \, N(t) \ ,$$

which takes into consideration that some of the collisions are inelastic, i.e. no additional electrons are released but kinetic energy is transferred to the collision partners. This energy is lost for further avalanche processes and thus reduces the effective rate parameter β'. The rate parameter γ is multiplied by $N(t)$, since the cross-section for the occurrence of inelastic collisions scales linearly with $N(t)$. By substituting $u(t) = 1/N(t)$, (3.47) turns into

$$\frac{\partial u}{\partial t} + (\beta - \delta) \, u(t) - \gamma = 0 \ .$$

This differential equation has the following general solution

$$u(t) = c(t) \, \exp\left[-\int_0^t (\beta - \delta) \, dt' \right] \ , \tag{3.48}$$

with

$$c(t) = c_0 + \gamma \int_0^t \exp\left[\int_0^{t'} (\beta - \delta) \, dt'' \right] dt' \ .$$

The initial condition is expressed by $u(0) = 1/N_0$. Hence, it follows that $c_0 = 1/N_0$, where N_0 is the initial electron density. After resubstituting $N(t)$ in (3.48), the following general solution is derived

$$N(t) = \frac{\exp \int_0^t (\beta - \delta) \, dt'}{\frac{1}{N_0} + \gamma \int_0^t \exp\left[\int_0^{t'} (\beta - \delta) \, dt'' \right] dt'} \ . \tag{3.49}$$

We will now assume a laser pulse with duration τ and constant intensity I_0. According to Molchanov (1970), the parameter β can be approximated by

$$\beta = \left\{ \begin{array}{ll} \eta I_0 & \text{for} \quad 0 \leq t \leq \tau \\ 0 & \text{else} \end{array} \right\} \ , \tag{3.50}$$

where η is called the *ionization probability* and is expressed in units of $(J/cm^2)^{-1}$. In our model, multi-photon processes are not taken into account. It is rather assumed that the initial electrons for the avalanche are provided by thermal ionization as stated by Bloembergen (1974) who estimated an initial electron density of $N_0 \simeq 10^8$–10^{10} cm^{-3} and a breakdown density of $N_{th} \simeq 10^{18}$ cm^{-3}.

– *Case I:* $0 \leq t \leq \tau$. By making the simplifying assumptions that the parameters η and δ are not time-dependent, (3.49) reduces to

$$N(t) = N_0 \, \frac{e^{(\beta-\delta)t}}{1 + \frac{\gamma N_0}{\beta-\delta} \left[e^{(\beta-\delta)t} - 1\right]} \; .$$

According to Bloembergen (1974), a certain threshold ratio s can be defined for the occurrence of optical breakdown

$$s = \ln \frac{N_{th}(t)}{N_0} = \ln \frac{e^{(\beta-\delta)t}}{1 + \frac{\gamma N_0}{\beta-\delta} \left[e^{(\beta-\delta)t} - 1\right]} \; .$$

When assuming $\beta \gg \delta$, we obtain

$$s = (\beta - \delta)t - \ln \left[1 + \frac{\gamma N_0}{\beta - \delta} e^{(\beta-\delta)t}\right] \; . \tag{3.51}$$

It is now supposed that optical breakdown occurs near the end of the laser pulse, i.e. at $t = \tau$, and that the free electron density saturates at this threshold condition. Hence,

$$\frac{\partial N}{\partial t} \bigg|_{t=\tau} = \beta N_\tau - \gamma N_\tau^2 - \delta N_\tau \simeq 0 \quad \text{with} \quad N_\tau = N(\tau) \; ,$$

$$\frac{\gamma N_\tau}{\beta - \delta} \simeq 1 \; ,$$

$$1 + \frac{\gamma N_0}{\beta - \delta} e^{(\beta-\delta)\tau} \simeq 1 + \frac{\gamma N_0}{\beta - \delta} e^{\beta\tau} \simeq 1 + \frac{\gamma N_\tau}{\beta - \delta} \simeq 2 \; . \tag{3.52}$$

Therefore, (3.51) reduces to

$$s = (\beta - \delta)\tau - \ln \left(1 + \frac{\gamma N_\tau}{\beta - \delta}\right) \; . \tag{3.53}$$

Because of (3.52), the logarithm in (3.53) is evaluated in a series at the argument "2" up to the second term

$$s = (\beta - \delta)\tau - \ln 2 - \frac{1}{2} \left[\left(1 + \frac{\gamma N_\tau}{\beta - \delta}\right) - 2\right] \; ,$$

with the general solution

$$\beta - \delta = \frac{s + \ln 2 - 0.5}{2\tau} + \sqrt{\left(\frac{s + \ln 2 - 0.5}{2\tau}\right)^2 + \frac{\gamma N_\tau}{2\tau}} \; .$$

The rate parameters for inelastic collision and diffusion are defined by $\gamma = (N_\tau \tau_c)^{-1}$ and $\delta = \tau_d^{-1}$, where τ_c and τ_d are the respective mean time constants. Then, together with (3.50), the following two relations are obtained for the threshold parameters of plasma generation

$$\eta E_{th} = \frac{s}{2} + \sqrt{\left(\frac{s}{2}\right)^2 + \frac{\tau}{2\tau_c} + \frac{\tau}{\tau_d}} \,, \qquad (3.54)$$

$$\eta I_{th} = \frac{s}{2\tau} + \sqrt{\left(\frac{s}{2\tau}\right)^2 + \frac{1}{2\tau\tau_c} + \frac{1}{\tau_d}} \,, \qquad (3.55)$$

where E_{th} and I_{th} are the threshold values of incident energy density and intensity, respectively. The term $(\ln 2 - 0.5)$ has been neglected, because Bloembergen (1974) estimated a typical ratio of $s = \ln(N_{th}/N_0) \simeq 18$, where N_{th} is the electron density at the threshold of optical breakdown.

– *Case II: $t > \tau$.* The integration in (3.49) is performed by means of

$$N(t) = \frac{\exp\left[\int_0^\tau (\beta - \delta)\,dt' + \int_\tau^t (\beta - \delta)\,dt'\right]}{\frac{1}{N_0} + \gamma \int_0^\tau \exp\left[\int_0^{t'} (\beta - \delta)\,dt''\right]dt' + \gamma \int_\tau^t \exp\left[\int_0^{t'}(\beta - \delta)\,dt''\right]dt'} \,. \qquad (3.56)$$

Keeping in mind that $\beta = 0$ for any $t' > \tau$, (3.56) can be written as

$$N(t) = N_0 \frac{e^{(\beta-\delta)\tau}\,e^{-\delta(t-\tau)}}{1 + \gamma N_0 \int_0^\tau e^{(\beta-\delta)t'}\,dt' + \gamma N_0 \int_\tau^t e^{(\beta-\delta)\tau - \delta(t'-\tau)}\,dt'} \,, \qquad (3.57)$$

with the final solution

$$N(t) = N_0 \frac{e^{\beta\tau - \delta t}}{1 + \frac{\gamma N_0}{\beta - \delta}\left[e^{(\beta-\delta)\tau} - 1\right] - \frac{\gamma N_0}{\delta}e^{\beta\tau}(e^{-\delta t} - e^{-\delta\tau})} \,. \qquad (3.58)$$

The last equation generally describes the decrease in electron density after the laser pulse due to collision and diffusion processes.

In Figs. 3.48 and 3.49, the threshold dependences of incident energy density and intensity on pulse duration are illustrated, respectively, assuming $\tau_c = 1\,fs$ according to Bloembergen (1974) and $\tau_d = 500\,ps$ according to Zysset et al. (1989). It is supposed that the electron diffusion time is of the order of plasma lifetimes achieved with picosecond pulses. With these values, numerical evaluation of (3.54) and (3.55) yields that $E_{th} \sim \tau^x$ and $I_{th} \sim \tau^{x-1}$ with

$$\begin{array}{lll} x < 0.1 & \text{for} & \tau < 100\,fs \,, \\ x = 0.4 \ldots 0.6 & \text{for} & 4\,ps < \tau < 8\,\mu s \,, \\ x > 0.8 & \text{for} & \tau > 300\,\mu s \,. \end{array}$$

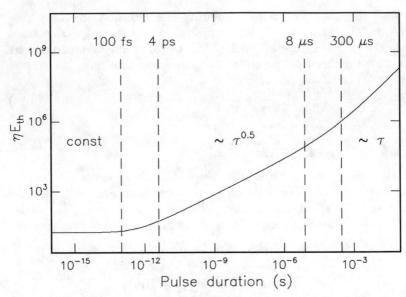

Fig. 3.48. Threshold dependence of incident energy density on laser pulse duration. A square root dependence is postulated for pulse durations between 4 ps and 8 μs. The parameter η is tissue-dependent and is expressed in units of $(\text{J/cm}^2)^{-1}$

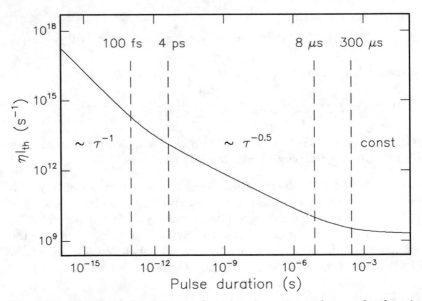

Fig. 3.49. Threshold dependence of incident intensity on laser pulse duration. An inverse square root dependence is postulated for pulse durations between 4 ps and 8 μs. The parameter η is tissue-dependent and is expressed in units of $(\text{J/cm}^2)^{-1}$

Primarily, the two limits within the three time domains are determined by the time constants of inelastic collision and diffusion modified by the empirical threshold parameter s. There remains, of course, an uncertainty in the absolute value of these time constants. This uncertainty is thus carried forward to the domain limits, i.e. insertion of longer time constants would shift the two limits to longer pulse durations, as well.

Indeed, the results of a series of experiments indicate a relationship of the form $E_{th} \sim \sqrt{\tau}$ for pulse durations in the picosecond and nanosecond range. These studies have been made with a variety of anorganic target materials such as NaCl and SiO_2 as performed by Fradin et al. (1973a), Bloembergen (1974), Vaidyanthan et al. (1980), van Stryland et al. (1981), Taylor et al. (1987), and Du et al. (1994). A similar threshold dependence was obtained for various biological tissues as investigated by Stern et al. (1989), Niemz et al. (1993a), and Vogel et al. (1994b).

In Table 3.12, the exponent in the relationship $E_{th} \sim \tau^x$ is listed for various targets and pulse durations. They are calculated from threshold data which were published by different authors. Obviously, most of the values fit very well to the postulated relationship $E_{th} \sim \sqrt{\tau}$, i.e. $x \simeq 0.5$, except the data obtained by Zysset et al. (1989). The latter discrepancy might be caused by performing the experiments at extremely small focal spot sizes and the associated errors in threshold determination.

Table 3.12. Exponents in the relationship $E_{th} \sim \tau^x$ as calculated from published threshold data. Pulse durations shorter than 1 ps were not considered to remain in the time domain of the square root dependence

Target	Reference	Considered pulse durations	x
H_2O (distilled)	Docchio et al. (1986)	30 ps -7 ns	0.49
	Zysset et al. (1989)	40 ps -10 ns	0.71
	Vogel et al. (1994a)	30 ps -6 ns	0.54
Cornea	Stern et al. (1989)	1 ps -8 ns	0.44
	Niemz et al. (1993a)	1 ps -200 ps	0.52
	Vogel et al. (1994b)	30 ps -6 ns	0.47
Lens	Vogel et al. (1994b)	30 ps -6 ns	0.46
Vitreous	Vogel et al. (1994b)	30 ps -6 ns	0.48
Tooth	unpublished	1.5 ps -30 ps	0.46

In Figs. 3.50a–b, the threshold dependence of the incident energy density is illustrated in corneal tissue and teeth. The data were obtained with two different laser systems, a Nd:YLF laser and a Rhodamine 590 dye laser, respectively. The dye laser was necessary to provide pulse durations in the femtosecond range. From the slope of each curve, the corresponding exponents $x = 0.52$ and $x = 0.46$ are calculated, respectively. At the shortest pulse durations, both curves tend to decrease in slope.

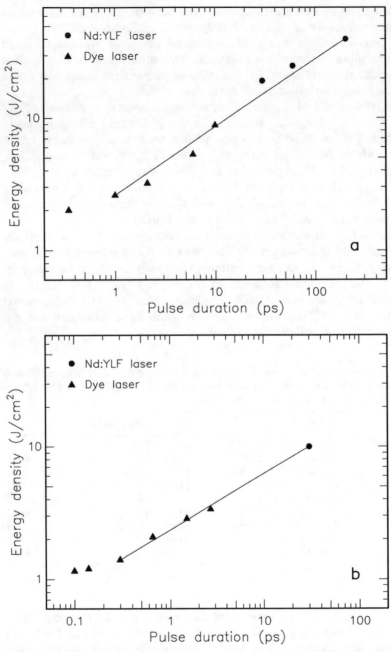

Fig. 3.50. (a) Threshold dependence of incident energy density on laser pulse duration for corneal tissue. Measured using a Nd:YLF laser and a Rhodamine 590 dye laser, respectively. The slope of the linear section is 0.52. Data according to Niemz et al. (1993). **(b)** Same dependence for teeth. The slope of the linear section is 0.46. Unpublished data

The ionization probability η was defined by (3.50). It is a tissue-dependent parameter and must thus be evaluated from experimental data. Its significance becomes evident when comparing different media with each other. In Table 3.13, some values of η are listed as calculated from published threshold data and (3.54). Obviously, the ionization probabilities of cornea and tooth are significantly higher than the corresponding values of water, lens, or vitreous. Values measured by different authors and under various experimental conditions match each other fairly well.

Table 3.13. Ionization probabilities of various targets as calculated from published threshold data

Target	Reference	Pulse duration	$\eta\ (\mathrm{J/cm^2})^{-1}$
H_2O (distilled)	Docchio et al. (1986)	30 ps	6.0
		7 ns	6.4
	Zysset et al. (1989)	40 ps	2.2
		10 ns	0.71
	Vogel et al. (1994a)	30 ps	1.3
		6 ns	0.95
Cornea	Stern et al. (1989)	1 ps	16.2
		30 ps	12.9
	Niemz et al. (1993a)	10 ps	8.0
		30 ps	6.6
		60 ps	7.3
		200 ps	8.1
	Vogel et al. (1994b)	30 ps	8.8
		6 ns	9.7
Lens	Vogel et al. (1994b)	30 ps	2.7
		6 ns	3.1
Vitreous	Vogel et al. (1994b)	30 ps	3.8
		6 ns	3.9
Tooth	unpublished	1.5 ps	14.0
		30 ps	13.2

According to (3.58), the temporal evolution of the free electron density after the laser pulse can be approximated by

$$N(t) \simeq N_{\mathrm{max}}\ \exp(-\,\delta t) = N_{\mathrm{max}}\ \exp\left(-\frac{t}{\tau_{\mathrm{d}}}\right)\ .$$

For the same $\tau_{\mathrm{d}} = 500\,\mathrm{ps}$ as chosen above and a typical laser pulse duration of 1 ns, the temporal decrease in electron density is illustrated in Fig. 3.51. In theory, the lifetime of the plasma is thus a few nanoseconds. Time-resolved measurements of the plasma were performed by Docchio et al. (1988a-b) and Zysset et al. (1989). The photon emission of a plasma induced by a 30 ps pulse from a Nd:YLF laser is captured in Fig. 3.52. The upper and lower traces represent laser pulse and plasma emission, respectively. Both signals

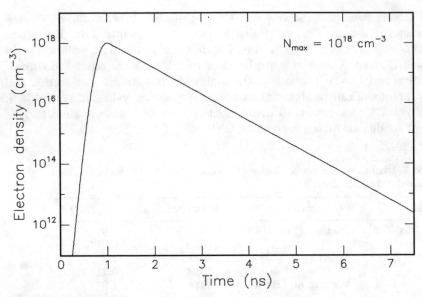

Fig. 3.51. Temporal evolution of free electron density in a laser-induced plasma. The simulation is based on a laser pulse duration of 1 ns, i.e. the maximum value N_{max} of the electron density is reached after 1 ns

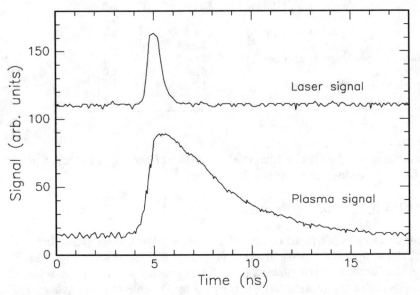

Fig. 3.52. Temporal traces of a Nd:YLF laser pulse (pulse duration: 30 ps, pulse energy: 500 μJ) and plasma emission. The plasma signal was detected by a photodiode with a risetime of approximately 1 ns

were detected by a photodiode with a risetime of 1 ns. Even though the laser signal is broadened in time, the lifetime of the plasma can be estimated to be approximately 3 ns.

The model presented above finally explains the observed square root dependence for the time domain $4\,\mathrm{ps} < \tau < 8\,\mu\mathrm{s}$. It also indicates that the threshold parameters of optical breakdown do not decrease any further when going below 100 femtoseconds. So, shorter pulses do not provide any significant advantage to several medical laser applications.

3.4.2 Analysis of Plasma Parameters

The interaction type of plasma-induced ablation can be used for diagnostic purposes, as well. By means of a spectroscopic analysis of the induced plasma spark, the free electron density and the temperature of the plasma can be evaluated. Moreover, detailed information on the chemical consistency of the target can be obtained, thereby allowing certain conclusions to be drawn regarding the state of health of the investigated tissue volume.

One example for the analysis of laser-induced plasmas is the simultaneous caries diagnosis and therapy developed by Niemz (1994a) which is based on the detection of tooth demineralization and is enabled by the evaluation of plasma spectra. A typical setup for such an experiment is shown in Fig. 3.53. Optical breakdown is induced on the tooth surface by a picosecond Nd:YLF laser. Then, the plasma is imaged on the entrance pupil of a spectrometer and analyzed. Finally, the data are fed into a computer and processed for further evaluation.

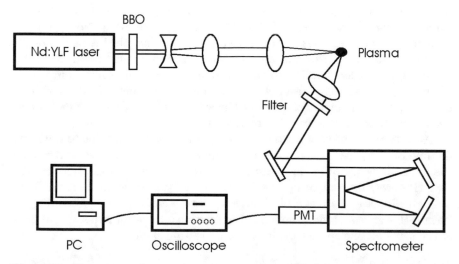

Fig. 3.53. Experimental setup for a computer-controlled spectroscopic analysis of laser-induced plasmas

Two typical plasma spectra are shown in Figs. 3.54a–b, one of which was obtained from healthy tooth substance, the other from carious substance. Calcium in neutral and singly ionized states and the major doublet of neutral sodium is found in the spectra according to emission wavelengths listed by Weast (1981). They are the minerals occurring most in dental hydroxyapatite with the chemical formula $Ca_{10}(PO_4)_6(OH)_2$. Phosphorus is also expected, but neither of its strongest emission lines at 602.4 nm and 603.4 nm is observed. This result most probably stems from the fact that the plasma temperature is not high enough to transfer phosphate to its atomic constituents, i.e. phosphorus and oxygen. The two lines between 390 nm and 400 nm probably result from neutral calcium (394.8 nm, 397.3 nm) and the major emission of singly ionized calcium (393.3 nm, 396.8 nm). One strong peak can be seen at about 526.5 nm that partly arises from a multiplet of calcium lines and from external second-harmonic generation (SHG). A few percent of the incident laser beam are converted to the second harmonic by means of a BBO crystal as seen in Fig. 3.53. Therefore, in the spectra, this peak is labeled "Ca & 2ω". SHG at the tooth surface is also conceivable, but its amplitude is expected to be surpassed by the calcium multiplet. The detected signal at 526.5 nm is normalized to unity. This normalization is useful when comparing spectra of healthy and carious teeth. The latter spectra show a strong decrease for all mineral lines due to the demineralization process induced by caries. Slight deviations in the ratio of calcium and sodium intensities might be observed because of different demineralization stages.

The normalization realized in this study is correct only if the diffuse reflectivities of healthy and carious teeth are the same for the second harmonic at 526.5 nm. This was tested by filtering out the fundamental wavelength, thereby preventing plasma generation, and measuring the reflected signal while moving different areas of the tooth through the focus. No significant difference in the intensities was observed. However, since the intensities of the calcium multiplet at 526.5 nm are also decreased in carious teeth, the ratio of SHG photons contributing to the normalization signal at 526.5 nm varies. Therefore, a direct comparison of absolute intensities cannot be derived from these spectra. For further details on the subject of simultaneous caries diagnosis and therapy, the reader is referred to Sect. 4.2.

An exact evaluation of the plasma temperature is difficult due to the short lifetime of the plasma. Temporal measurements are required to understand the dynamic behavior of the plasma temperature and the free electron density. However, by comparing the intensities of two different calcium lines, a rough estimate of a mean plasma temperature can be given. According to Lochte-Holtgreven (1968), the following relation applies for two spectral lines of the same atomic species in the same ionization stage

$$\frac{I_1}{I_2} = \frac{A_1 g_1 \lambda_2}{A_2 g_2 \lambda_1} \exp\left(-\frac{E_1 - E_2}{kT}\right) , \qquad (3.59)$$

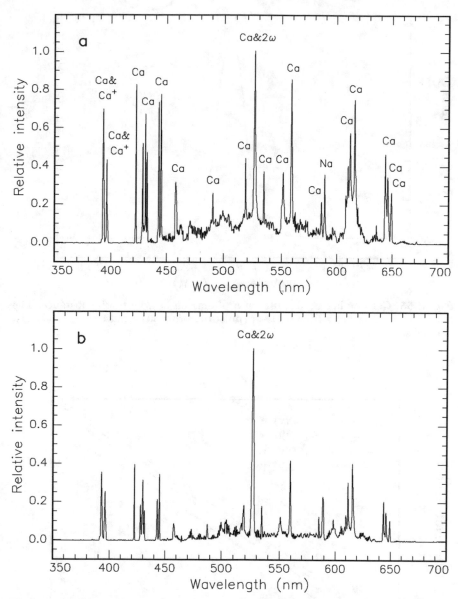

Fig. 3.54. (a) Spectrum of plasma spark on healthy tooth substance induced by a Nd:YLF laser (pulse duration: 30 ps, pulse energy: 500 μJ). Lines of neutral calcium (Ca), singly ionized calcium (Ca⁺), and neutral sodium (Na) are visible. The signal at 526.5 nm partly originates from calcium and from second harmonic generation (SHG) of the laser beam. **(b)** Spectrum of plasma spark on carious tooth substance. Due to the process of demineralization, the intensity of all mineral lines is reduced. Data according to Niemz (1994a)

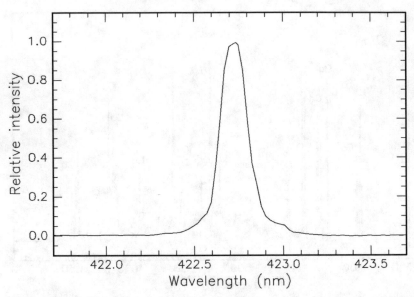

Fig. 3.55. Calcium line at 422.7 nm of a plasma on healthy tooth substance. The plasma was induced by a Nd:YLF laser (pulse duration: 30 ps, pulse energy: 500 μJ). Unpublished data

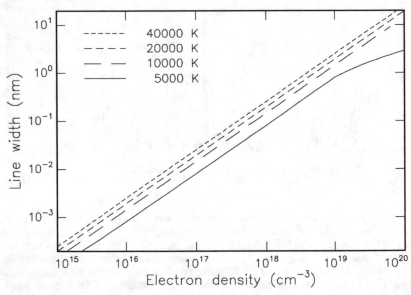

Fig. 3.56. Dependence of spectral width on plasma density and temperature according to Saha's equation (plasma temperature: as labeled). Given values apply for the calcium line at 422.7 nm. Unpublished data

where I is the detected intensity, A is the transition probability, g is the statistical weight of the upper energy level, λ is the detected wavelength, E is the upper energy level, k is Boltzmann's constant, and T is the plasma temperature. By differentiating (3.59), we obtain

$$\frac{\Delta T}{T} = \frac{kT}{E_1 - E_2} \frac{\Delta(I_1/I_2)}{I_1/I_2} . \tag{3.60}$$

Hence, the accuracy of temperature determination can be improved by choosing two upper energy levels far apart from each other. The accuracy of the measured intensities is better for singlets. Because of these two reasons, it is convenient to compare the intensities at the two calcium wavelengths of $\lambda_1 = 422.7\,\mathrm{nm}$ and $\lambda_2 = 585.7\,\mathrm{nm}$. According to Weast (1981), the corresponding parameters are

$$A_1 = 2.18 \times 10^8\,\mathrm{s}^{-1}, \quad g_1 = 3, \quad E_1 = 2.94\,\mathrm{eV},$$
$$A_2 = 0.66 \times 10^8\,\mathrm{s}^{-1}, \quad g_2 = 5, \quad E_2 = 5.05\,\mathrm{eV}.$$

On average, the intensity ratio of the spectra shown in Figs. 3.54a–b is about $I_1/I_2 \simeq 4$. Thus, the mean plasma temperature corresponding to these spectra is about $5\,\mathrm{eV}$ or $60\,000\,\mathrm{K}$ which is in good agreement with theoretical predictions by Mulser et al. (1973). This plasma temperature is not to be mistaken for the local tissue temperature, since it represents the kinetic energy of plasma electrons only.

In order to evaluate the free electron density of the plasma, the spectral widths of singlet lines need to be measured. These experiments were performed by Niemz (1994a) and yielded typical spectral widths of $0.15\,\mathrm{nm}$ for the calcium line at $422.7\,\mathrm{nm}$ as shown in Fig. 3.55. From these, the free electron density can be determined by applying *Saha's equation* of plasma physics. The derivation and significance of Saha's equation is discussed in detail by Griem (1964). Its physical interpretation is summarized in Fig. 3.56, where the spectral width at a given wavelength – in this case the calcium line at $422.7\,\mathrm{nm}$ – is plotted as a function of electron density and plasma temperature. In the example shown, a measured spectral width of $0.15\,\mathrm{nm}$ corresponds to a free electron density of approximately $10^{18}/\mathrm{cm}^3$. Moreover, Fig. 3.56 illustrates that the free electron density shows a rather weak dependence on plasma temperature.

3.4.3 Summary of Plasma-Induced Ablation

- *Main idea:*   ablation by creating an ionizing plasma
- *Observable effects:*   very clean ablation, associated with audible report and plasma sparking
- *Typical lasers:*   Nd:YAG, Nd:YLF, Ti:Sapphire
- *Typical pulse durations:*   100 fs ... 500 ps
- *Typical power densities:* $10^{11}\,\mathrm{W/cm}^2$ ... $10^{13}\,\mathrm{W/cm}^2$
- *Special applications:*   refractive corneal surgery, caries therapy

3.5 Photodisruption

The physical effects associated with optical breakdown are *plasma formation* and *shock wave generation*. If breakdown occurs inside soft tissues or fluids, *cavitation* and *jet formation* may additionally take place. The meaning of these terms will be discussed in this section. The ablative process due to plasma ionization has already been described in Sect. 3.4. However, when discussing plasma-induced ablation, we neglected any secondary effects of the plasma. At higher pulse energies – and thus higher plasma energies – shock waves and other mechanical side effects become more significant and might even determine the global effect upon the tissue. Primarily, this is due to the fact that mechanical effects scale linearly with the absorbed energy. Then, because of the mechanical impact, the term *disruption* (from Latin: *ruptus* = ruptured) is more appropriate.

The effect of rupturing becomes evident when looking at Figs. 3.57a–b. These two photographs show a 90 µm thick glass plate on which picosecond pulses from a Nd:YLF laser were focused. Since glass is a simple detector for shocks – it splinters above a certain threshold value – the mechanical impact on the anterior and posterior surfaces is easily seen.

Cavitation is an effect that occurs when focusing the laser beam not on the surface of a tissue but into the tissue. For instance, a cavitation within a human cornea is seen in Fig. 3.58. It was generated by focusing a single picosecond pulse from a Nd:YLF laser underneath the epithelium. Immediately after laser exposure, the tissue was fixated to prevent the cavitation bubble from collapsing. The cross-section of the cavitation is elongated on an axis determined by the orientation of the collagen fibrils, because shear forces can readily split the tissue in this direction. Cavitation bubbles consist of gaseous vapors – mainly water vapor and carbon oxides – which eventually diffuse again into the surrounding tissue.

Photodisruption has become a well-established tool of minimally invasive surgery (MIS), since it was introduced by Krasnov (1973) and then further investigated by Aron-Rosa et al. (1980) and Fankhauser et al. (1981). Two of the most important applications of photodisruptive interaction are posterior capsulotomy of the lens – frequently being necessary after cataract surgery – and laser-induced lithotripsy of urinary calculi.

During photodisruption, the tissue is split by mechanical forces. Whereas plasma-induced ablation is spatially confined to the breakdown region, shock wave and cavitation effects propagate into adjacent tissue, thus limiting the localizability of the interaction zone. For pulse durations in the nanosecond range, the spatial extent of the mechanical effects is already of the order of millimeters even at the very threshold of breakdown. Actually, purely plasma-induced ablation is not observed for nanosecond pulses, because the threshold energy density of optical breakdown is higher compared to picosecond pulses as shown in Fig. 3.48, and the pressure gradient scales with plasma energy. Hence, for nanosecond pulses, optical breakdown is always associated with

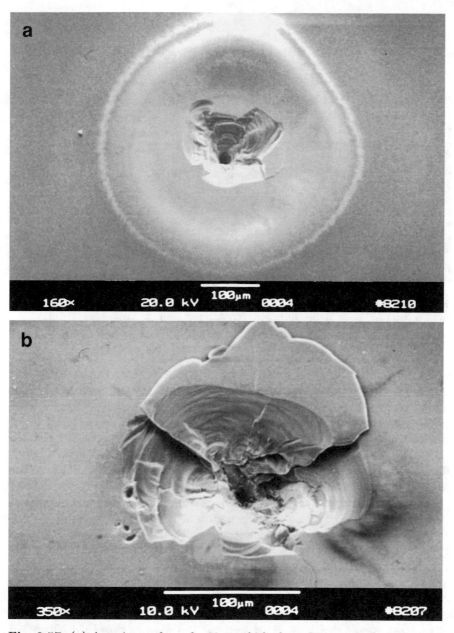

Fig. 3.57. (a) Anterior surface of a 90 μm thick glass plate exposed to ten pulses from a Nd:YLF laser (pulse duration: 30 ps, pulse energy: 1 mJ, focal spot size: 30 μm). (b) Posterior surface of a glass plate after a similar experiment. Reproduced from Niemz (1994a). © 1994 Springer-Verlag

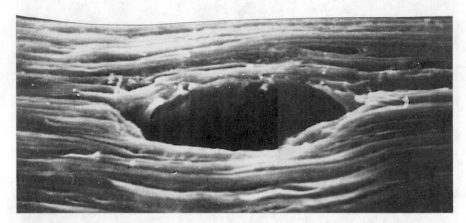

Fig. 3.58. Cavitation bubble within a human cornea induced by a single pulse from a Nd:YLF laser (pulse duration: 30 ps, pulse energy: 1 mJ, bar: 30 µm)

shock wave formation even at the very threshold. Since adjacent tissue can be damaged by disruptive forces, the presence of these effects is often an undesired but associated symptom. In contrast, picosecond or femtosecond pulses permit the generation of high peak intensities with considerably lower pulse energies. With these extremely short pulse durations, optical breakdown may still be achieved while significantly reducing plasma energy and, thus, disruptive effects. Moreover, spatial confinement and predictability of the laser–tissue interaction is strongly enhanced.

Since both interaction mechanisms – plasma-induced ablation as well as photodisruption – rely on plasma generation, it is not always easy to distinguish between these two processes. Actually, in the 1970s and 1980s, all tissue effects evoked by ultrashort laser pulses were attributed to photodisruption. It was only because of intense research that a differentiation between ablations solely due to ionization and ablations owing to mechanical forces seems justified. For instance, it was found by Niemz (1994a) that, in the case of picosecond pulses, ablation without mechanical side effects takes place at incident power densities of a few times the plasma threshold. Based on these findings and the theory describing the dependence of threshold parameters (see Figs. 3.48 and 3.49 in Sect. 3.4), approximate thresholds of both interaction types are listed in Table 3.14. The data represent estimated values for corneal tissue, assuming a mean ionization probability of about $\eta = 10\,(\mathrm{J/cm^2})^{-1}$ as obtained from Table 3.13.

Table 3.14. Estimated parameters for the onset of plasma-induced ablation and photodisruption in corneal tissue

Pulse duration	Onset of plasma-induced ablation Energy density (J/cm^2)	Onset of photodisruption Energy density (J/cm^2)
100 fs	2.0	50
1 ps	3.3	50
10 ps	8.0	50
100 ps	23.0	50
1 ns	–	72
10 ns	–	230
100 ns	–	730
	Power density (W/cm^2)	Power density (W/cm^2)
100 fs	2.0×10^{13}	5.0×10^{14}
1 ps	3.3×10^{12}	5.0×10^{13}
10 ps	8.0×10^{11}	5.0×10^{12}
100 ps	2.3×10^{11}	5.0×10^{11}
1 ns	–	7.2×10^{10}
10 ns	–	2.3×10^{10}
100 ns	–	7.3×10^{9}

The values listed in Table 3.14 are graphically presented in Fig. 3.59. Obviously, plasma-induced ablation is limited to a rather narrow range of pulse durations up to approximately 500 ps. At longer pulse durations, the energy density necessary for achieving breakdown always induces significant mechanical side effects.

In general, photodisruption may be regarded as a bundle of mechanical effects initiated by optical breakdown. The primary mechanisms are shock wave generation and cavitation, completed by jet formation if cavitations collapse in fluids and near a solid boundary. In Fig. 3.60, a schematic sequence of these processes is illustrated indicating their relations to each other. Moreover, the distinction between photodisruption and plasma-induced ablation is emphasized.

The four effects – plasma formation, shock wave generation, cavitation, and jet formation – all take place at a different time scale. This is schematically illustrated in Fig. 3.61. Plasma formation begins during the laser pulse and lasts for a few nanoseconds afterwards as already mentioned when discussing Fig. 3.52. This is basically the time needed by the free electrons to diffuse into the surrounding medium. Shock wave generation is associated with the expansion of the plasma and, thus, already starts during plasma formation. However, the shock wave then propagates into adjacent tissue and leaves the focal volume. Approximately 30–50 ns later, it has slowed down to an ordinary acoustic wave. Cavitation, finally, is a macroscopic effect start-

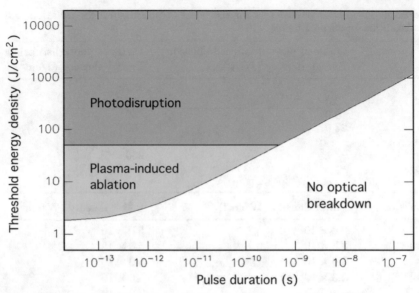

Fig. 3.59. Distinction of plasma-induced ablation and photodisruption according to applied energy density

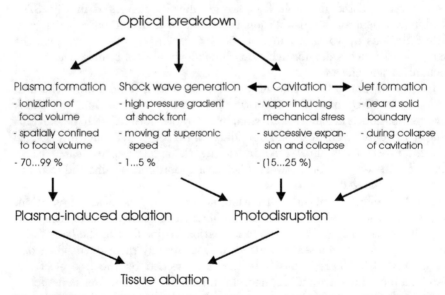

Fig. 3.60. Physical processes associated with optical breakdown. Percentages given are rough estimates of the approximate energy transferred to each effect (incident pulse energy: 100 %). Cavitation occurs in soft tissues and fluids only. In fluids, part of the cavitation energy might be converted to jet formation

ing roughly 50–150 ns after the laser pulse. The time delay is caused by the water molecules being vaporized. Usually, the cavitation bubble performs several oscillations of expansion and collapses within a period of a few hundred microseconds as will be shown below. Since the pressure inside the bubble increases again during each collapse, every rebound of the cavitation bubble is accompanied by another shock wave. Furthermore, every collapse can induce a jet formation if the bubble is generated in the vicinity of a solid boundary. Each of these effects contributing to photodisruption will be discussed in detail in the following paragraphs, because they all have their own physical significance.

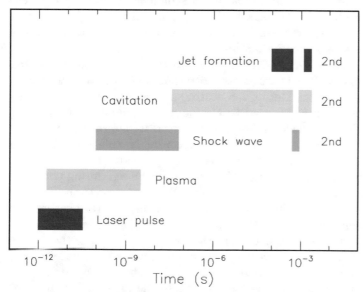

Fig. 3.61. Approximate time scale for all processes contributing to photodisruption. Assumed is a 30 ps laser pulse. The first and second occurrences of shock wave, cavitation, and jet formation are indicated

3.5.1 Plasma Formation

The principles of laser-induced plasma formation have already been considered in Sect. 3.4. It should be emphasized, however, that the amount of energy absorbed during photodisruption is typically two or more orders of magnitude higher than during plasma-induced ablation. This is an immediate consequence of the different energy densities associated with either process as already emphasized in Fig. 3.1. Thus, the free electron density and the plasma temperature are also higher than for purely plasma-induced ablation. Therefore, in photodisruptive laser–tissue interactions, the following three effects are enabled or become more significant:

- plasma shielding,
- Brillouin scattering,
- multiple plasma generation.

Once formed, the plasma absorbs and scatters further incident light. This property "shields" underlying structures which are also in the beam path. The importance of the *plasma shielding* effect for medical laser applications was first recognized by Steinert et al. (1983) and Puliafito and Steinert (1984). In ophthalmology, the retina is considerably protected by this plasma shield during laser surgery of the lens or the vitrous. In Fig. 3.47, we have already encountered an increased absorption coefficient of the plasma. However, under the conditions of plasma-induced ablation, a significant amount of laser energy is still transmitted by the plasma. During photodisruptive interactions, thus at denser plasmas, the absorption coefficient is even enhanced, and the plasma serves as a very effective shield.

In *Brillouin scattering*, incident light is scattered by thermally excited acoustic waves and shifted in frequency corresponding to potential phonon frequencies of the material. During the heating process of the plasma, acoustic waves are generated which lead to Brillouin scattering. When applying even higher irradiances, the laser light itself may create alterations in optical density by which, in turn, it is scattered. Accordingly, this effect is called *stimulated Brillouin scattering*. It was described by Ready (1971).

Finally, at the very high electric field strengths achieved during photodisruption, *multiple plasma generation* is enabled. Whereas close to the ablation threshold only one spark is induced at the very focus, several plasmas can be ignited at higher pulse energies. In the latter case, only the first section of the laser pulse will induce a plasma at the focal spot. Then, as the fluence increases during the pulse, succeeding radiation may also generate optical breakdown before reaching the smallest beam waist. Thus, a cascade of plasmas is initiated pointing from the focal spot into the direction of the laser source. This effect is schematically illustrated in Fig. 3.62.

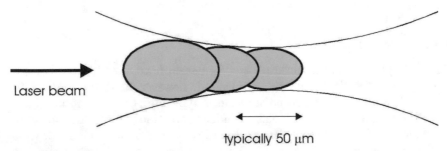

Fig. 3.62. Cascade of multiple laser-induced plasmas. A Gaussian-shaped laser beam is incident from the left

Plasma formation in distilled water has been extensively studied by Docchio et al. (1986), Zysset et al. (1989), and Vogel et al. (1994a). Some of their results were presented in Sect. 3.4 when discussing the threshold behavior of laser-induced optical breakdown. Moreover, Vogel et al. (1994a) have determined plasma sizes with time-resolved photography. They have observed that the length of plasmas is strongly related to the pulse duration. Their measurements are summarized in Fig. 3.63.

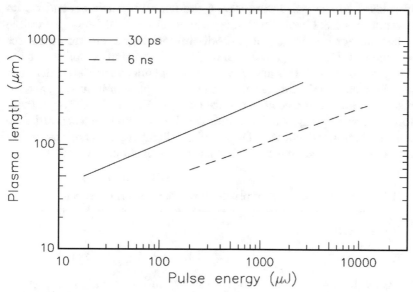

Fig. 3.63. Plasma length as a function of incident pulse energy. Measured using a Nd:YAG laser (pulse duration: as labeled, focal spot diameter: 4 μm) in distilled water. Data according to Vogel et al. (1994a)

Obviously, plasmas induced by 30 ps pulses are approximately 2.5 times as long as plasmas induced by 6 ns pulses of the same energy. Only at the respective thresholds of breakdown are the latter slightly longer. These different plasma lengths – and thus volumes – result in a considerably lower energy density of plasmas induced by picosecond pulses. Indeed, Vogel et al. (1994a) have observed a significant difference in the corresponding intensities of visible plasma fluorescence. Moreover, the plasma volume determines the fraction of incident energy to be converted to shock waves or cavitations. If the plasma volume is larger – as in plasmas induced by picosecond pulses – more energy is required for ionization and vaporization of matter. Hence, this amount of energy can no longer contribute to the generation of potential shock waves or cavitations. Therefore, we can conclude that plasmas induced by picosecond pulses are less likely to cause mechanical tissue damage than plasmas from nanosecond pulses.

The overall sequence of plasma formation is summarized in Table 3.15. In order to distinguish the physical parameters of plasma-induced ablation and photodisruption, two typical pulse durations of 10 ps and 100 ns are considered. These correspond to typical values of mode locked or Q-switched pulses, respectively. Energy densities and power densities at the threshold of plasma formation apply for corneal tissue and are taken from Table 3.14. Associated electric field strengths are calculated using (3.23). The critical electron density at the plasma threshold is derived from (3.46). It is not directly related to the pulse duration of the laser but does depend on its wavelength. In Sect. 3.4, we estimated typical densities of $10^{18}/cm^3$ for visible laser radiation. For the process of photodisruption, higher electron densities were obtained by Boulnois (1986). Finally, the absorption coefficient of the plasma is given by (3.45). It is also wavelength-dependent and determines the extent of the plasma shielding effect. In the case of plasma-induced ablation, some measured absorption coefficients are listed in Fig. 3.47. Although these data apply for water only, similar values can be assumed for corneal tissue because of its high water content. During photodisruption, higher absorption coefficients are accessible due to the increased electron density.

Table 3.15. Physical parameters of plasma formation in corneal tissue

	10 ps	100 ns
Pulse duration	10 ps	100 ns
⇓		
Energy density	$8.0\,J/cm^2$	$730\,J/cm^2$
⇓		
Power density	$8.0 \times 10^{11}\,W/cm^2$	$7.3 \times 10^9\,W/cm^2$
⇓		
Electric field strength	$2.5 \times 10^7\,V/cm$	$2.3 \times 10^6\,V/cm$
⇓		
Electron density of plasma	10^{18}–$10^{19}/cm^3$	10^{18}–$10^{20}/cm^3$
⇓		
Non-linear absorption of plasma (Plasma shielding)	1–100/cm	1–10 000/cm

The values listed in Table 3.15 provide a good estimate of the physical parameters associated with optical breakdown. In a first approximation, they apply for other targets, as well. In order to achieve a similar plasma electron density, roughly 100 times the energy density is needed when applying 100 ns pulses rather than 10 ps pulses. Therefore, provided the same focus size is chosen, plasmas induced by nanosecond pulses contain significantly more energy. This additional amount of energy must somehow dissipate into the surrounding medium. It is partly converted to the generation of shock waves, cavitation, and jet formation as will be discussed next.

3.5.2 Shock Wave Generation

As calculated in Sect. 3.4, laser-induced optical breakdown is accompanied by a sudden adiabatic rise in plasma temperature to values of up to a few 10 000 K. Primarily, this temperature can be attributed to the kinetic energy of free electrons. Due to their high kinetic energy, the plasma electrons are not confined to the focal volume of the laser beam but rather diffuse into the surrounding medium. When the inert ions follow at a certain time delay, mass is moved which is the basic origin of shock wave generation. This shock wave soon separates from the boundary of the plasma. It initially moves at hypersonic speed and eventually slows down to the speed of sound. In Fig. 3.64, the geometry of shock wave generation is illustrated.

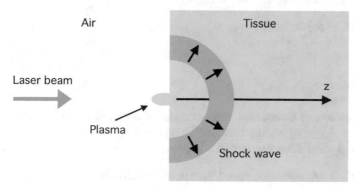

Fig. 3.64. Geometry of shock wave generation

Laser-induced shock waves in water were first investigated by Carome et al. (1966), Bell and Landt (1967), and Felix and Ellis (1971). Shock waves differ from sonic acoustic waves by their speed. Whereas the speed of sound in water, for instance, is 1483 m/s at 37°C, laser-induced shock waves typically reach speeds of up to 5000 m/s at the very focus. Both the hypersonic shock waves and sonic acoustic waves are referred to as *acoustic transients*. In order to derive a relation describing the pressure gradient at the shock front, let us consider a slab of tissue with a cross-section A_0 which is passed through by a shock front at a speed u_s, as seen in Fig. 3.65. During a time interval of $\mathrm{d}t$, the shock front moves a distance of $\mathrm{d}x_s$, thus

$$u_s = \frac{\mathrm{d}x_s}{\mathrm{d}t} \ .$$

The pressure inside the medium and its density are p_0 and ϱ_0, respectively. The shock front induces a sudden increase in local pressure from p_0 to p_1 and of the density from ϱ_0 to ϱ_1. The conservation of mass demands compensation by other particles intruding from the left side in Fig. 3.65. These particles move at a particle speed u_p which is usually lower than u_s. During a time

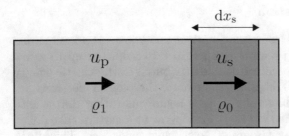

Fig. 3.65. Geometry of shock front moving through a slab of tissue

interval dt, a mass of $(\varrho_1 - \varrho_0)A_0 dx_s$ must be provided. This is achieved at the particle speed u_p from a zone with a higher[16] density ϱ_1:

$$u_p \varrho_1 A_0 dt = (\varrho_1 - \varrho_0) A_0 dx_s \ .$$

Hence,

$$u_p = \frac{\varrho_1 - \varrho_0}{\varrho_1} u_s \ . \tag{3.61}$$

Beside the conservation of mass, the conservation of momentum must also be fulfilled. A mass $A_0 \varrho_1 dx_s$ begins to move at a speed u_p and thus receives a momentum $A_0 \varrho_1 u_p dx_s$. This momentum is provided by two means:

- The mass $A_0 \varrho_1 u_p dt$ intrudes at a speed u_p, thereby supplying a momentum of $A_0 \varrho_1 u_p^2 dt$.
- The shock front induces an increase in pressure from p_0 to p_1. This pressure gradient induces a mechanical force $A_0(p_1 - p_0)$ which generates a momentum $A_0(p_1 - p_0)dt$ during the time interval dt.

The conservation law of momentum thus asks for

$$A_0 \varrho_1 u_p dx_s = A_0 \varrho_1 u_p^2 dt + A_0(p_1 - p_0)dt \ , \tag{3.62}$$

or

$$p_1 - p_0 = \varrho_1 u_p u_s - \varrho_1 u_p^2 \ . \tag{3.63}$$

Inserting (3.61) into (3.63) leads to a pressure increase

$$p_1 - p_0 = \varrho_0 u_p u_s \ . \tag{3.64}$$

An empirical relationship between the shock speed u_s and particle speed u_p was first determined by Rice and Walsh (1957). For water at high pressures exceeding 20 kbar, the following expression applies:

---

[16] In our model, the shock front has just passed the left area in Fig. 3.65, thus leaving a higher density ϱ_1 behind.

$$u_s = 1.483 + 25.306 \ \log_{10} \left(1 + \frac{u_p}{5.19} \right) , \tag{3.65}$$

where u_p and u_s must be inserted in units of km/s. For pressures lower than 20 kbar, an approximation was given by Doukas et al. (1991), i.e.

$$u_s = a + bu_p , \tag{3.66}$$

where a is the speed of sound and b is a dimensionless constant. In the case of water, $b = 2.07$ was estimated by Zweig and Deutsch (1992). Assuming a spherical shock wave with radius r, the conservation of momentum (3.62) leads to

$$4\pi r^2 \varrho_1 u_p u_s \Delta t = c_0 ,$$

where Δt is the risetime of the shock front, and c_0 is a constant denoting the overall momentum. Replacing u_p within the last equation by an expression obtained from (3.66) yields

$$u_s(u_s - a) = \frac{c_1}{r^2} ,$$

with

$$c_1 = \frac{b}{4\pi \varrho_1 \Delta t} \ c_0 ,$$

and the final solution

$$u_s(r) = \frac{a}{2} + \sqrt{\frac{a^2}{4} + \frac{c_1}{r^2}} . \tag{3.67}$$

The parameter c_1 can be empirically obtained. The particle speed is derived from (3.66)

$$u_p(r) = -\frac{a}{2b} + \frac{1}{b} \sqrt{\frac{a^2}{4} + \frac{c_1}{r^2}} . \tag{3.68}$$

Finally, the resulting pressure is obtained by inserting the expressions for particle speed u_p and shock speed u_s into (3.64), i.e.

$$p_1(r) = p_0(r) + \frac{\varrho_0 c_1}{b} \frac{1}{r^2} . \tag{3.69}$$

From (3.64) and (3.65), we can also derive two relationships of the form

$$u_s = u_s(p_1) ,$$

$$u_p = u_p(p_1) ,$$

respectively. In the case of water, they are graphically presented in Fig. 3.66. At $p_1 = 0$ kbar, the shock speed u_s approaches 1483 km/s, whereas the particle speed u_p remains at 0 km/s. However, these relationships can also be

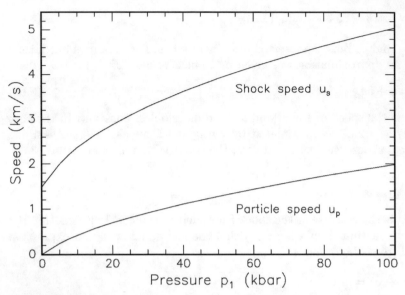

Fig. 3.66. Shock velocities and particle velocities in water as a function of shock wave pressure. Data according to Rice and Walsh (1957)

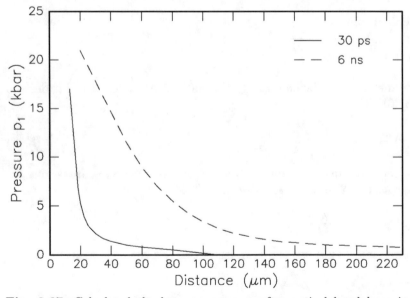

Fig. 3.67. Calculated shock wave pressures after optical breakdown in water. Breakdown was caused by pulses from a Nd:YAG laser (pulse duration: 30 ps, pulse energy: 50 µJ) and a Nd:YAG laser (pulse duration: 6 ns, pulse energy: 1 mJ), respectively. On the abscissa, the distance from the center of emission is given. Data according to Vogel et al. (1994a)

used to calculate the shock wave pressure p_1 as a function of shock speed u_s. By this means, shock wave pressures – which are usually very difficult to determine – can be derived from measured shock speeds.

Such calculations were performed by Vogel et al. (1994a) for shock waves induced by picosecond and nanosecond pulses. Their results are summarized in Fig. 3.67 where the shock wave pressure is shown in various distances from the center of emission. The initial pressure at the boundary of the laser plasma was 17 kbar for 50 µJ pulses with a duration of 30 ps, whereas it was 21 kbar for 1 mJ pulses with a duration of 6 ns. Although these values are quite similar, the pressure decay is significantly steeper for those shock waves which were induced by the picosecond pulses. In a distance of approximately 50 µm from the center of the shock wave emission, their pressure has already dropped to 1 kbar, whereas this takes a distance of roughly 200 µm when applying nanosecond pulses.

Moreover, it was observed by Vogel et al. (1994a) that the width of shock waves is smaller in the case of the picosecond pulses. They evaluated approximate widths of 3 µm and 10 µm for shock waves induced by either 30 ps or 6 ns pulses, respectively. Thus, the energies contained in these shock waves are not the same, because this energy is roughly given by

$$E_s \simeq (p_1 - p_0) A_s \, \Delta r \, , \tag{3.70}$$

with shock wave pressure p_1, shock wave surface area A_s, and shock wave width Δr. Due to different plasma lengths as shown in Fig. 3.63, we obtain initial values of $A_s \simeq 100 \, \mu m^2$ for 30 ps pulses and $A_s \simeq 2500 \, \mu m^2$ for 6 ns pulses when assuming a focal spot diameter of 4 µm. From (3.70), we then find that $E_s \simeq 0.5 \, \mu J$ for 30 ps pulses and $E_s \simeq 50 \, \mu J$ for 6 ns pulses. Thus, only 1–5 % of the incident pulse energy is converted to shock wave energy, and shock waves from picosecond pulses are significantly weaker than those induced by nanosecond pulses with comparable peak pressures. From the corresponding particle speeds, Vogel et al. (1994a) have calculated a tissue displacement of approximately 1.2 µm for 30 ps pulses and a displacement of roughly 4 µm for 6 ns pulses. These rather small displacements can cause mechanical damage on a subcellular level only, but they might induce functional changes within cells.

Primarily, there exist two types of experiments which are performed to investigate the dynamics of shock wave phenomena: optical and mechanical measurements. During optical measurements, the shock wave is detected by a weaker probe beam after being generated by the main laser beam. In some setups, an external helium–neon laser or dye laser is used as a probe beam. In other cases, the probe beam is extracted from the main beam by means of a beamsplitter and directed through an optical delay. A decrease in probe beam intensity is detected by a fast photodiode as long as the shock wave passes through the focus of the probe beam. By moving the focus of the probe beam with respect to the site of plasma generation, the propagation of the shock wave can be monitored on a fast digital oscilloscope. Fast photodiodes even enable a temporal analysis of the risetime of the shock front. Mechanical measurements rely on piezoelectric transducers transforming the shock wave pressure to a voltage signal. One commonly used detecting material is a thin foil made of polyvinyldifluoride (PVDF) which is gold coated on both sides for measuring the induced voltage by attaching two thin wires. The corresponding pressure is monitored on a fast digital oscilloscope. Both types of experiments are illustrated in Figs. 3.68 and 3.69.

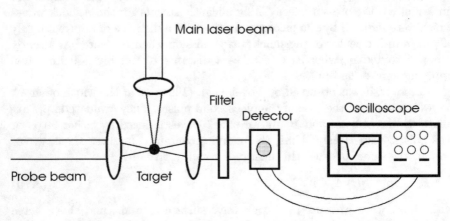

Fig. 3.68. Probe beam experiment for the detection of shock waves. A laser-induced shock wave deflects a second laser beam at the target. A fast photodiode measures the decrease in intensity of the probe beam

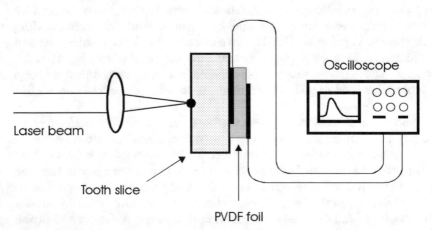

Fig. 3.69. PVDF experiment for the detection of shock waves. A laser-induced shock wave hits a piezoelectric transducer which converts pressure to a voltage signal. The voltage is measured by a digital oscilloscope

In Figs. 3.70 and 3.71, typical results are shown concerning the detection of acoustic transients. In Fig. 3.70, a shock wave was generated when inducing a plasma inside water by a 30 ps pulse from a Nd:YLF laser. A helium–neon laser served as a probe beam being focused at the depth of interest and detected by a fast photodiode. The probe beam is deflected as the shock wave passes through the focus of the probe beam, resulting in a decrease in detected intensity. A steep shock front is seen lasting for approximately 10 ns. After another 30 ns, the shock wave has completely passed through. During its overall duration of roughly 40 ns, the shock wave can hardly cause any gross tissue displacement. Thus, further evidence is given that shock wave damage is limited to a subcellular level.

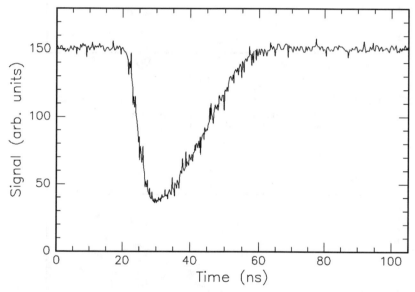

Fig. 3.70. Signal from a helium–neon laser serving as a detector for a shock wave generated in water by a Nd:YLF laser (pulse duration: 30 ps, pulse energy: 1 mJ)

In Fig. 3.71, on the other hand, the voltage signal from a PVDF transducer is shown. In this case, a plasma was induced at the surface of a tooth slice with a thickness of 0.5 mm. The detected PVDF signal arrives approximately 130 ns after the incident laser pulse, thus corresponding to a velocity of about 3800 m/s which is the speed of sound in teeth. The acoustic transient is reflected at the opposite surface of the tooth slice and is again detected after about 2×130 ns $= 260$ ns. Taking the round trip through the tooth slice into account, this time delay is related to the same speed of sound.

Since the shock wave loses energy when propagating through a medium, it eventually slows down until it finally moves at the speed of sound. With probe beam experiments as discussed above, traces of the shock wave at different

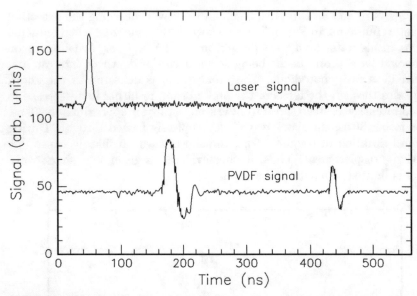

Fig. 3.71. Temporal trace of a Nd:YLF laser pulse (pulse duration: 30 ps, pulse energy: 500 μJ) and acoustic transient induced in a 0.5 mm thick tooth slice. The shock wave is detected by a PVDF transducer. Unpublished data

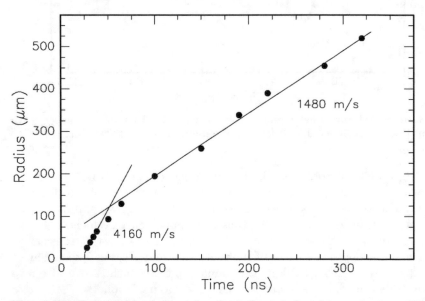

Fig. 3.72. Temporal evolution of a shock front induced in water by a Nd:YLF laser (pulse duration: 30 ps, pulse energy: 1 mJ). The shock speed decreases from 4160 m/s to 1480 m/s. Unpublished data

distances from its origin can be measured by moving the focus of the probe beam away from the plasma site. Typical results for such an experiment are shown in Fig. 3.72. From these data, a shock speed of approximately 4160 m/s is calculated for the first 30 ns. According to Fig. 3.66, this value corresponds to a shock wave pressure of approximately 60 kbar. The shock wave then slows down to about 1480 m/s – the speed of sound in water – after another 30 ns. Fig. 3.72 also illustrates that the spatial extent of this shock wave is limited to a radius of approximately 150 μm. Similar results were reported by Puliafito and Steinert (1984) and Teng et al. (1987).

3.5.3 Cavitation

Historically, interest in the dynamics of cavitation bubbles started to rise after realizing their destructive effect on solid surfaces such as ship propellers and other hydraulic equipment. Laser-induced cavitations occur if plasmas are generated inside soft tissues or fluids. By absorbing energy from the laser beam, the focal volume is vaporized. Thereby, work is done against the outer pressure of the surrounding medium, and kinetic energy is converted to potential energy being stored in the expanded cavitation bubble. Within less than a millisecond, the bubble implodes again as a result of the outer static pressure, whereby the bubble content – typically water vapor and carbon oxides – is strongly compressed. Thus, pressure and temperature rise again to values similar to those achieved during optical breakdown, leading to a rebound of the bubble. Consequently, a second transient is emitted, and the whole sequence may repeat a few times, until all energy is dissipated and all gases are dissolved by surrounding fluids.

Cavitation bubbles have long been studied by a variety of techniques. The high-speed photographic technique was pioneered by Lauterborn (1972). It is very helpful in visualizing the temporal behavior of the cavitation bubble. High-speed photography is performed at framing rates with up to one million frames per second. A typical sequence of growth and collapse of a cavitation bubble is shown in Fig. 3.73. The bubble appears dark in front of a bright background, because it is illuminated from behind and the light was deflected by its wall. Through the center of the bubble, the illuminating light is transmitted without deflection. In the case shown, a Q-switched ruby laser with pulse energies ranging from 100 mJ to 400 mJ was used to induce the cavitation bubble. Its maximum diameter reaches a value of 2.05 mm at a time delay of approximately 300 μs. The end of the first and second collapse is seen in the eighth and thirteenth frames, respectively.

A theory of the collapse of spherical cavitation bubbles was first given by Rayleigh (1917). He derived the relationship

$$r_{max} = \frac{t_c}{0.915 \sqrt{\varrho/(p_{stat} - p_{vap})}} , \tag{3.71}$$

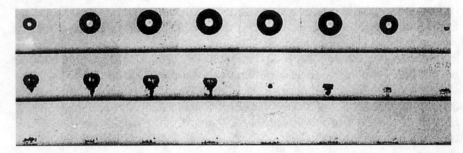

Fig. 3.73. Dynamics of cavitation bubble captured by high-speed photography. Pictures taken at 20 000 frames per second (frame size: 7.3 mm × 5.6 mm). Reproduced from Vogel et al. (1989) by permission. © 1989 Cambridge University Press

where r_{max} is the maximum radius of cavitation, t_c is the duration of the collapse, ϱ is the density of the fluid, p_{stat} is the static pressure, and p_{vap} is the vapor pressure of the fluid. Thus, the time needed for each collapse is proportional to its maximum radius. The latter, on the other hand, is directly related to the bubble energy E_b by means of

$$E_b = \frac{4}{3} \pi \left(p_{stat} - p_{vap}\right) r_{max}^3 , \tag{3.72}$$

according to Rayleigh (1917). This equation states that the bubble energy is given by the product of its maximum volume and the corresponding pressure gradient. The bubble energy is thus readily determined when all kinetic energy has turned into potential energy.

The temporal oscillation of the cavitation bubble can be captured in probe beam experiments as shown in Fig. 3.74. Three complete oscillations are seen. The period of subsequent oscillations decreases in the same manner as their amplitude as already postulated by (3.71). In order to evaluate the dependence of the radius of cavitation on incident pulse energy, probe beam experiments have been performed similar to those discussed for shock wave detection. In Fig. 3.75, some data for picosecond and nanosecond pulses were collected by Zysset et al. (1989). Except for the lowest energy value, their measurements fit well to a straight line with a slope of 1/3. Because of the double-logarithmic scale of the plot, the relation given above by (3.72) is thus experimentally confirmed.

The conversion of incident energy to cavitation bubble energy is summarized in Figs. 3.76a–b. From the slopes, a conversion factor of approximately 19 % is obtained for picosecond pulses, whereas it is roughly 24 % when applying nanosecond pulses. Moreover, it was observed by Vogel et al. (1989) that the average energy loss of the cavitation bubbles during their first cycle is approximately 84 %. The major part of this loss is attributed to the emission of sound. These results were confirmed in theoretical work performed by Ebeling (1978) and Fujikawa and Akamatsu (1980). From (3.72), it can

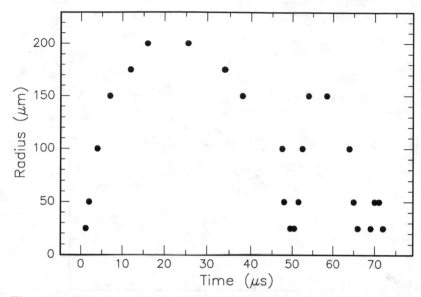

Fig. 3.74. Detection of cavitation bubble by means of a probe beam experiment with a helium–neon laser. Three complete oscillations of a cavitation bubble are observed which was induced in water by a Nd:YLF laser (pulse duration: 30 ps). Unpublished data

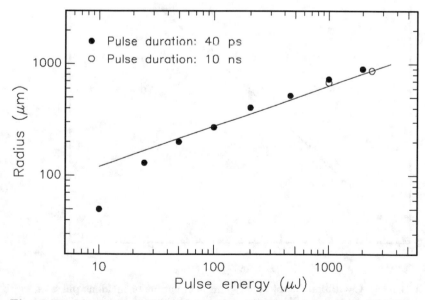

Fig. 3.75. Maximum radius of cavitation bubble as a function of incident pulse energy. Cavitations were induced in water by a Nd:YAG laser (pulse duration: as labeled). Data according to Zysset et al. (1989)

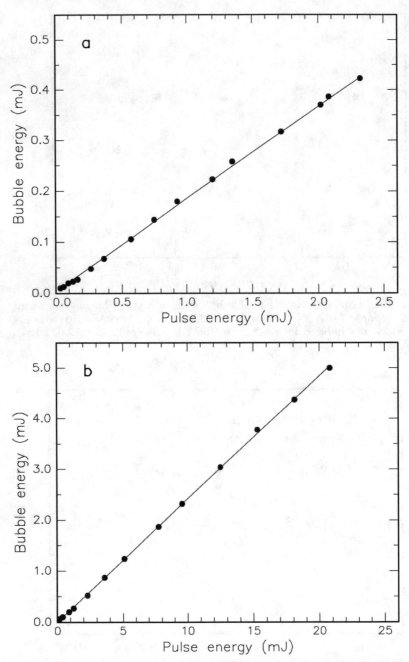

Fig. 3.76. (a) Cavitation bubble energy as a function of incident pulse energy from a Nd:YAG laser (pulse duration: 30 ps). The slope of the fitted line is 0.19. **(b)** Cavitation bubble energy as a function of incident pulse energy from a Nd:YAG laser (pulse duration: 6 ns). The slope of the fitted line is 0.24. Data according to Vogel et al. (1994a)

also be concluded that bubble-induced damage – i.e. the linear extent of the damage zone – scales with the cube root of the contained energy. This is of importance when determining the primary cause of tissue damage. It was observed by Vogel et al. (1990) that tissue damage also scales with the cube root of pulse energy. Thus, cavitations are more likely to induce damage than shock waves, since shock wave related damage should scale with $\sqrt{E}$ as can be derived from (3.70) when inserting $A_\mathrm{s} \sim r^2$.

It has been emphasized above that damage of tissue due to shock waves is limited to a subcellular level due to their short displacement lengths of approximately 1–4 µm. Since the diameter of cavitation bubbles may reach up to a few millimeters, macroscopic photodisruptive effects inside tissues are believed to primarily originate from the combined action of cavitation and jet formation which will be discussed next.

3.5.4 Jet Formation

As already reported by Tomita and Shima (1986), the impingement of a high-speed liquid jet developing during the collapse of a cavitation bubble may lead to severe damage and erosion of solids. Jet formation was first investigated and described by Lauterborn (1974), and Lauterborn and Bolle (1975), when producing single cavitation bubbles by focusing Q-switched laser pulses into fluids. When cavitation bubbles collapse in the vicinity of a solid boundary, a high-speed liquid jet directed toward the wall is produced. If the bubble is in direct contact with the solid boundary during its collapse, the jet can cause high-impact pressure against the wall. Thus, bubbles attached to solids have the largest damage potential.

Jet formation has been thoroughly studied by means of high-speed photography as introduced above. The temporal behavior of cavitation and jet formation is shown in Fig. 3.77. A cavitation bubble was generated near a solid boundary – a brass block located at the bottom of each frame and visible by a dark stripe – and captured by high-speed photography. Jet formation toward the brass block is observed during the collapse of the bubble. Jet velocities of up to 156 m/s were reported by Vogel et al. (1989). The water hammer pressure corresponding to such a velocity is approximately 2 kbar. If the distance between the cavitation bubble and solid boundary is further decreased as seen in the bottom sequence of Fig. 3.77, a counterjet is formed which points away from the solid boundary.

What is the origin of jet formation, and why does it only occur near a solid boundary? To answer these questions, let us take a closer look at the collapse of a cavitation bubble. When the bubble collapses due to external pressure, the surrounding fluid is accelerated toward the center of the bubble. However, at the side pointing to the boundary there is less fluid available. Hence, the collapse takes place more slowly at this side of the bubble. This effect ultimately leads to an asymmetric collapse. At the faster collapsing side, fluid particles gain additional kinetic energy, since the decelerating force – i.e. the

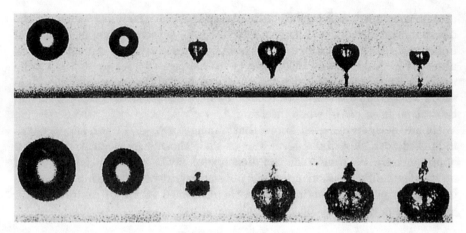

Fig. 3.77. Bubble collapse with jet formation (*top*). Pictures taken at 20 000 frames per second (frame size: 7.3 mm × 5.6 mm). Bubble collapse with counterjet formation (*bottom*). Reproduced from Vogel et al. (1989) by permission. © 1989 Cambridge University Press

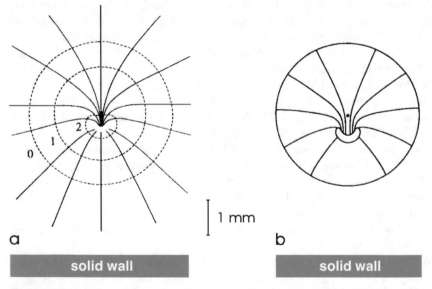

Fig. 3.78. (a) Experimentally obtained pathline portrait of the flow around a collapsing cavitation bubble near a solid wall. Bubble shape 0 represents the bubble at maximum expansion, whereas shapes 1 and 2 correspond to later stages during the collapse. **(b)** Calculated pathline portrait from Kucera and Blake (1988) for several points on the wall of the collapsing bubble. Reproduced from Vogel and Lauterborn (1988) by permission. © 1988 Optical Society of America

force by the slower collapsing opposite wall – is delayed. This explains why jet formation occurs toward the solid boundary. If the jet is relatively slow, the velocity of the central part of the slower collapsing side might even be higher than the jet itself. This is conceivable, because that side of the bubble is accelerated until the very end of the collapse. In this case, a counterjet is formed pointing in the opposite direction. Pathline portraits of the flow around collapsing bubbles have been experimentally and theoretically determined by Vogel and Lauterborn (1988) and Kucera and Blake (1988), respectively. In Figs. 3.78a–b, two of them are shown which offer a good visualization of the fluid flow during the collapse.

The damaging effect of jet formation is extremely enhanced if a gas bubble remaining from an earlier laser pulse is hit by acoustic transients generated by subsequent pulses. According to Vogel et al. (1990), the damage range induced by a 4 mJ pulse can reach diameters of up to 2–3.5 mm if gas bubbles are attached to the corneal tissue. Very small gas bubbles, however, quickly dissolve due to their small volume and strong surface tension. Therefore, these microbubbles should not cause any problem in achieving a certain predictable effect if the repetition rate of the laser pulses is adequately chosen.

3.5.5 Summary of Photodisruption

- *Main idea:*   fragmentation and cutting of tissue by mechanical forces
- *Observable effects:*   plasma sparking, generation of shock waves, cavitation, jet formation
- *Typical lasers:*   Nd:YAG, Nd:YLF, Ti:Sapphire
- *Typical pulse durations:*   100 fs ... 100 ns
- *Typical power densities:* 10^{10} W/cm^2 ... 10^{14} W/cm^2
- *Special applications:*   lens fragmentation, lithotripsy

3.6 Questions to Chapter 3

Q3.1. Which energy density is typical for a laser–tissue interaction?
A: 1 J/m^2.   B: 1 mJ/cm^2.   C: 10 J/cm^2.
Q3.2. Which is toxic to a biological cell?
A: carotenoid.   B: excited singlet oxygen.   C: photosensitizer.
Q3.3. Coagulation occurs at approximately
A: 60°C.   B: 80°C.   C: 100°C.
Q3.4. UV photons have an energy of
A: < 0.3 eV.   B: < 3 eV.   C: > 3 eV.
Q3.5. In a plasma with electron density N, the coefficient of absorption is
A: $\sim \sqrt{N}$.   B: $\sim N$.   C: $\sim N^2$.
Q3.6. Why should there be an appropriate time gap in photodynamic therapy between application of a photosensitizer and laser exposition?

Q3.7. Two different laser pulses from a Nd:YAG laser are used to irradiate living liver tissue: a 100 mJ pulse at a pulse duration of 1 ms and a 100 pJ pulse at a pulse duration of 1 ps. Both laser pulses have an average power of 100 W and are focused to a spot of 1 mm in diameter. How will the tissue react in either case?

Q3.8. Why can a frequency-quadrupoled Nd:YAG laser at a wavelength of 266 nm induce photoablation, while a frequency-doubled Nd:YAG laser at a wavelength of 532 nm cannot?

Q3.9. Which three processes determine the temporal behavior of the free electron density in a plasma?

Q3.10. What is the basic physical mechanism that plasma-induced ablation and photodisruption have in common, and which laser parameter has to be altered to switch from one type of interaction to the other?

4. Medical Applications of Lasers

In this chapter, we will discuss principal applications of lasers in modern medicine. Due to the present boom in developing new lasers, especially diode lasers, and due to the limitations given by the dimensions of this book, not all disciplines and procedures can be taken into account. Our main goal is to focus on the most significant applications and to evoke a basic feeling for using certain techniques. All the examples are chosen to emphasize essential ideas and to assist the reader in grasping the best technical solutions. Potential difficulties and complications arising from either method will be addressed, as well. However, we should always keep in mind that any laser therapy needs to be planned as carefully as any other medical treatment. Every single step should be taken exclusively for the benefit of the patient.

To be in line with the historic sequence, the first section is devoted to applications in *ophthalmology*. Most of the money made with lasers is still earned in this field. *Dentistry* was the second clinical discipline to which lasers were introduced. But, even though considerable research is being done today, there are much less dental applications than applications within the eye. However, photoactivated disinfection (PAD) and laser applications in implantology are very promising. A major effort of today's clinical laser research is focusing on various kinds of tumor treatments such as photodynamic therapy (PDT) and laser-induced interstitial thermotherapy (LITT). These play a significant role in other medical disciplines like *gynecology*, *urology*, and *neurosurgery*. Due to recent advancements in instrumentation for minimally invasive surgery (MIS), for example the development of miniature catheters and endoscopes, novel techniques are under present investigation in *angioplasty* and *cardiology*. Very interesting laser applications can be found in *orthopedics* and *dermatology*. And, last but not least, successful laser treatments have been reported in *gastroenterology*, *otorhinolaryngology*, and *pulmology* as discussed at the end of this chapter.

So, we may conclude that – at the present time – laser medicine is a rapidly growing field of both research and application. This is not at all astonishing, since neither the development of laser systems nor the design of application units have yet come to an end. Moreover, laser medicine is not restricted to a few disciplines. Instead, it is expected that many more clinical applications will be developed in the near future.

© Springer Nature Switzerland AG 2019
M. H. Niemz, *Laser-Tissue Interactions*,
https://doi.org/10.1007/978-3-030-11917-1_4

4.1 Lasers in Ophthalmology

In ophthalmology, various types of lasers are being applied today for either diagnostic or therapeutic purposes. In diagnostics, lasers can be advantageous if conventional light sources fail. One excellent diagnostic tool is confocal laser microscopy which allows the detection of early stages of retinal alterations. By this means, retinal detachment and glaucoma[1] can be recognized in time, thereby increasing the probability of a successful treatment. In this book, however, our interest focuses on therapeutic laser applications. The first indications for laser treatment were given by detachments of the retina. Meanwhile, this kind of surgery has turned into a well-established tool, but only represents a minor part of today's ophthalmic laser procedures. Others are, for instance, treatment of glaucoma and cataract. Yet the most money is made with refractive corneal surgery.

The targets of all therapeutic laser treatments of the eye can be classified into front and rear segments. The front segments consist of *cornea*, *sclera*, *trabeculum*, *iris*, and *lens*. The rear segments are *vitreous body* and *retina*. A schematic illustration of a human eye is shown in Fig. 4.1. In the following paragraphs, we will discuss various treatments of these segments according to the historic sequence, i.e. from the rear to the front.

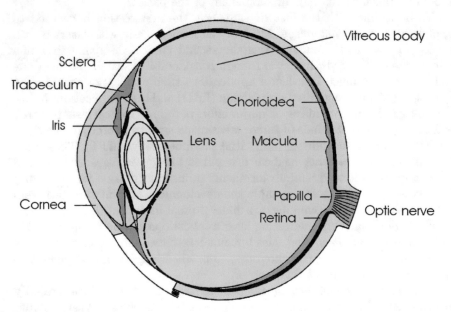

Fig. 4.1. Human eye

---

[1] Since glaucoma is usually associated with a degeneration of the optical nerve fibers, it can be detected by measuring either the thickness of these fibers or alterations of the papilla. Further details are given by Bille et al. (1990).

Retina

The retina is a part of the central nervous system. Its function is to convert an optical image to electrical impulses and send these impulses through the optic nerve to the brain. The retina is a thin and transparent tissue which is permeated with blood vessels. According to Le Grand and El Hage (1980), the thickness of the retina varies from 0.5 mm near the papilla[2] to 0.1 mm at the macula. Anatomically, the retina is subdivided into several different layers, each of them having their own distinct function: pigment epithelium, receptor layer, external limiting membrane, cell layer, nerve fiber layer, and internal limiting membrane. A schematic cross-section of a human retina is shown in Fig. 4.2.

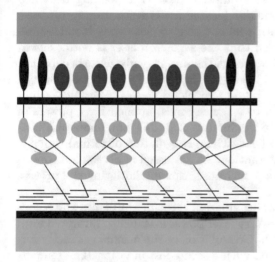

Pigment epithelium

Receptor layer

External limiting membrane

Cell layer

Nerve fiber layer
Internal limiting membrane
Vitreous body

Fig. 4.2. Cross-section of a human retina

The *pigment epithelium* is strongly attached to the chorioidea. The *receptor layer* consists of two types of cells – rods and cones. Rods are used in dim light and are primarily located around the macula. Cones are recepting colors in good light and are very closely packed in the fovea. Obviously, light has to pass through virtually the whole retina beyond the *external limiting membrane*, before it can stimulate any receptor cells. This structural arrangement is known as the "reversed retina" and can be explained by the fact that the retina is an invagination of the embryonic cerebral wall. The *cell layer* is made up of horizontal cells, bipolar cells, amacrine cells, and ganglion cells.

---

[2] The *papilla* is where the optic nerve exits the eye. The *macula* is the site of our best vision. Whenever we read a book, its letters are imaged onto the macula. The central section of the macula is called *fovea* and is the region with the highest density of color receptors.

The main function of these cells is to serve as a first network with corresponding receptive fields. Finally, the *nerve fiber layer* contains the axons of the ganglion cells, whereas the *internal limiting membrane* forms a boundary between retina and vitreous body.

The ophthalmologist Meyer-Schwickerath (1949) was the first to investigate the coagulation of the retina with sunlight for therapeutic purposes. Because of the inconvenient circumstances of this kind of surgery, e.g. the necessity of sunshine, he continued his studies with his famous xenon photocoagulator as he reported in 1956. Shortly after the invention of the laser by Maiman (1960), first experimental studies with the ruby laser were performed by Zaret et al. (1961). The first reports on the treatment of patients were given by Campbell et al. (1963) and Zweng et al. (1964). They discovered that the ruby laser was a very suitable tool when welding detached segments of the retina to the chorioidea located underneath. However, it also became evident that the ruby laser was not able to close open blood vessels or stop bleeding. It was soon found that the argon ion laser is better suited for this aim. Its green and blue wavelengths are strongly absorbed by the hemoglobin of blood – in contrast to the red light from a ruby laser – which finally leads to the coagulation of blood and blood vessels. At typical exposure durations ranging from 0.1 s to a few seconds, applied laser powers of 0.1–1 W, and spot diameters of approximately 200–1000 μm, almost all incident laser energy is converted to heat. Thus, coagulation of retinal tissue is achieved by means of thermal interaction. As discussed in Sect. 3.2, proteins are denatured and enzymes are inactivated, thereby initiating the process of blood clotting.

The surgeon conducts the laser coagulation through a slit lamp and a contact glass. He approaches the necessary laser power from below threshold until the focused area just turns greyish. Coagulation of the macula is strictly forbidden, since it would be associated with a severe loss in vision. The temperatures achieved should generally remain below 80°C to prevent unnecessary vaporization and carbonization. A good localization of blood vessels, i.e. by confocal laser microscopy, and a precise application of the desired energy dose are mandatory when striving for best results.

At the beginning of the 1970s, the krypton ion laser became very significant for ophthalmic applications. Its red and yellow wavelengths at 647 nm and 568 nm, respectively, turned out to be very useful when trying to restrict the interaction zone to either the pigment epithelium or the chorioidea. Detailed histologic studies on this phenomenon were conducted by Marshall and Bird (1979). It was found that the red line is preferably absorbed by the chorioidea, whereas the yellow line is strongly absorbed by the pigment epithelium and also by the xanthophyll contained in the macula. McHugh et al. (1988) proposed the application of diode lasers, since their invisible emission at a wavelength of approximately 800 nm does not dazzle the patient's eye.

There exist six major indications for laser treatment of the retina:

- retinal holes,
- retinal detachment,
- diabetic retinopathy,
- central vein occlusion,
- senile macula degeneration,
- retinoblastoma (retinal tumors).

In the case of *retinal holes*, proper laser treatment prevents their further enlargement which could otherwise lead to retinal detachment. Laser surgery is performed by welding the retina to the underlying chorioidea within a narrow ring-shaped zone around the hole as shown in Fig. 4.3a. The attachment of the coagulated tissue is so strong that further tearing is usually suppressed. If necessary, however, the procedure can be repeated several times without severe complications.

Retinal detachment is often a consequence of undetected retinal holes or tears. It mainly occurs in myopic patients, since the vitreous body then induces an increased tensile stress to the retina. Moderate detachments are treated in a similar mode as retinal holes. In the case of a severe detachment, the treatment aims at saving the fovea or at least a small segment of the macula. This procedure is called panretinal coagulation and is illustrated in Fig. 4.3b. Unfortunately, laser treatment of retinal detachment is often associated with the formation of new membranes in the vitreous body, the retina, or beneath the retina. All these complications are summarized by the clinical term *proliferative retinopathy*. A useful therapeutic technique for the dissection of such membranes was described and discussed by Machemer and Laqua (1978).

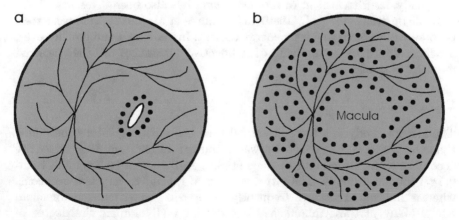

Fig. 4.3. (a) Placement of coagulation spots in the case of retinal holes or moderate detachments. **(b)** Placement of coagulation spots during panretinal coagulation

When a patient is suffering from *diabetic retinopathy*, the concentration of oxygen in the blood is strongly reduced due to disturbances in the body. Because of the lack in oxygen, new blood vessels are formed which is called *neovascularization*. Hemorrhages inside the vitreous body might then lead to severe losses in vision. In order to prevent complete blindness, the whole retina is coagulated except the fovea itself, i.e. a panretinal coagulation is performed as in the case of a severe retinal detachment. By this means, it is assured that the progress of neovascularization is stopped and that at least the fovea does receive enough oxygen. The physiological mechanism of this radical treatment is not completely understood. Most likely, a significant percentage of the receptors, which are consuming most of the oxygen provided to the retina, is turned off. During the treatment, between 1000 and 3000 laser spots should be placed next to each other according to Schulenburg et al. (1979) and Hövener (1980).

Central vein occlusion occurs in the eyes of older patients and is usually restricted to one eye only. As an immediate consequence, retinal veins become dilated and severe edema are formed in the region of the macula. Multiple hemorrhages are associated with a strong decrease in vision. To prevent the occurrence of a secondary glaucoma, the procedure of panretinal coagulation is often performed as stated by Laatikainen et al. (1977).

Senile macula degeneration is increasing among older patients. According to Bird (1974), it is caused by neovascular membranes being formed in the chorioidea. Subretinal fluids emerging from these membranes might lead to severe edema in the region of the macula. Further neovascularization can be prevented by coagulation using the green line of the argon ion laser or the red line of the krypton laser, respectively. These wavelengths are preferably absorbed by subretinal tissues in the pigment epithelium or the chorioidea, but not by xanthophyll contained by the macula itself.

Finally, laser treatment of *retinoblastoma* has also been investigated, for example by Svaasand et al. (1989) and Dimaras et al. (2012). Tumor necrosis is obtained by converting laser energy to heat. In malignant tumors, however, mechanical surgery or implants of radioactive substances are the preferred methods of treatment.

Vitreous Body

The vitreous body is a transparent gel that has a little greater consistency than the white of a raw egg. Its water content varies from 98 % to 99.7 % according to Le Grand and El Hage (1980), and it contains 7 g/l NaCl and 0.5 g/l soluble proteins. The vitreous body of a child is very homogeneous, whereas internal structures frequently appear in the vitreous body of an adult. Many of these inhomogeneities do not really impair the degree of vision. Most of the floating particles can be resorbed by biological mechanisms. Major pathologic alterations, however, are given by the formation of new membranes and neovascularizations extending from the retina into the vitreous body. Their occurrence has already been described when discussing retinal detachment and diabetic retinopathy. It shall be added that only thermally acting lasers should be used for treatment due to the direct vicinity of the retina. Short pulsed lasers evoking photodisruptive effects may only be used for lens surgery and the front segments of the eye.

Lens

The lens grows during the entire human life forming an onion-like structure of adjacent shells. As a result of its continuous development and the associated decrease in water content, the lens interior progressively hardens with age. The bulk of the lens is formed by transparent lens fibers which originate from the anterior lens epithelium. The lens interior is enclosed by a homogeneous elastic membrane called the *capsule*. The capsule is connected to the *ciliary muscle* which is essential for the eye to accommodate. In a cataract, the transparency of the lens is strongly decreasing. The opaqueness is caused by either age, disease, UV radiation, food deficiencies, or trauma. The changes in the lens that lead to the formation of a cataract are not completely understood. They are somehow related to reduced amounts of potassium and soluble lens proteins which go hand in hand with increased amounts of calcium and insoluble lens proteins.

Cataract surgery of the lens is another frequently performed laser treatment in ophthalmology. In order to achieve acceptable vision, the interior of the lens must first be liquified and extracted. Conventional methods rely on fragmentation of the lens by phaco-emulsification followed by aspirating the fragments. Afterwards, either an artificial lens made of silicon is inserted or the patient must wear special cataract glasses. This treatment has already been proposed and documented by Kelman (1967). The posterior lens capsule is retained to prevent a collapse of the vitreous body and subsequent retinal detachment. However, new lens fibers frequently emerge from this posterior capsule forming a scattering membrane. This membrane must be removed during a second invasive surgery.

Posterior capsulotomy with a Nd:YAG laser, on the other hand, is characterized by the advantages of being both non-invasive and feasible during ambulant treatment. It was described in detail by Aron-Rosa et al. (1980) and Terry et al. (1983). Usually, a helium–neon laser is used as an aiming beam. The surgeon first focuses this laser on the posterior capsule and then adds the cutting Nd:YAG laser beam as shown in Fig. 4.4 by pressing a footpedal. Typically, pulse durations of 30 ns, pulse energies of up to 5 mJ, and focus diameters of 50–100 µm are used. With these laser parameters, local power densities exceeding 10^{10} W/cm^2 are achieved, leading to the phenomenon of optical breakdown as described in Sect. 3.4. After having placed several line cuts, the posterior membrane opens like a zipper as illustrated in Fig. 4.5. The whole procedure can be controlled through a slit lamp. The surgeon's eye is protected by a specially coated beamsplitter.

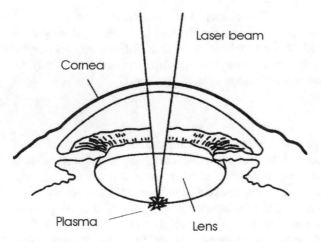

Fig. 4.4. Laser-assisted posterior capsulotomy

Fig. 4.5. Lens before, during, and after posterior capsulotomy

Another laser treatment of the lens is the fragmentation of its interior rather than using ultrasound exclusively[3]. For this kind of treatment, picosecond laser pulses are advantageous, because they are associated with a lower threshold energy for the occurrence of optical breakdown if compared with nanosecond pulses. Thus, more energy can be converted to the ionizing process itself. In Fig. 4.6, the fragmentation of a human lens is shown which was obtained by using a picosecond Nd:YLF laser. The surgeon steadily moves the focus of the laser beam without injuring the capsule. During this treatment, it is important to choose a pulse energy well above the threshold of optical breakdown, because otherwise all laser energy will be absorbed by the retina and other tissues lying underneath.

---

[3] Laser fragmentation can significantly reduce the amount of necessary phaco time.

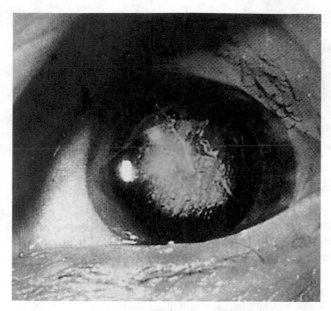

Fig. 4.6. Fragmentation of a human lens using a picosecond Nd:YLF laser (pulse duration: 30 ps, pulse energy: 1 mJ)

Iris

The iris is a tissue which is pierced by a variable circular opening called the *pupil*. Its diameter can vary from 1.5 mm to 8 mm, depending on brightness. In moderate light, the pupil diameter measures approximately 4 mm. The bulk of the iris consists of collagen fibers and pigment cells. The size of the pupil is determined by the action of two smooth muscles – the *sphincter pupillae* and the *dilator pupillae* – which are responsible for contraction and dilation, respectively.

In an acute block glaucoma, the drainage of aqueous humor from the rear to the front chamber is obstructed. Hence, the pressure in the rear chamber increases and shifts the iris forward. This dislocation of the iris induces a closed chamber angle which justifies the clinical term *closed-angle glaucoma*. Thereby, aqueous humor is prevented from entering the trabeculum and the canal of Schlemm. So, the intraocular pressure (IOP) increases to values far above 20 mm Hg, thus inducing strong headaches, severe edema, degeneration of retinal nerve fibers, and a sudden loss in vision. A well-established procedure is called *laser iridotomy*. It provides a high immediate success rate but does not guarantee lasting cure. During this treatment, the iris is perforated as shown in Fig. 4.7 to obtain an additional passage for the aqueous humor to reach the front chamber and the trabeculum.

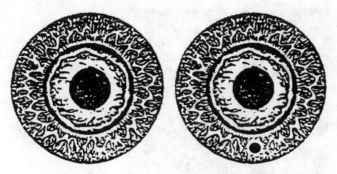

Fig. 4.7. Iris before and after laser treatment

Laser iridotomy can be performed with either argon ion lasers or pulsed neodymium lasers. Prior to laser exposure, the iris is medically narrowed. If applying the argon ion laser, typical exposure durations of 0.1–0.2 s, laser powers ranging from 700–1500 mW, and spot diameters of 50 μm are chosen according to Pollack and Patz (1976) and Schwartz and Spaeth (1980). Iridotomies induced by the argon ion laser are very successful if the iris is dark and strongly pigmented. For bright irises, neodymium lasers with pulse durations in the nanosecond or picosecond range and pulse energies up to a few millijoules are better suited. Detailed data of the procedure are given in the book by Steinert and Puliafito (1985). Perforations around the 12 o'clock position should be avoided because of gas bubbles disturbing the surgeon's vision. Therefore, iridotomies are usually placed between the 3 o'clock and 9 o'clock positions where the gas bubbles can rise upward. The use of a proper contact glass was recommended by Roussel and Fankhauser (1983).

Laser iridotomy is a minor surgical treatment which can be performed ambulantly. Only in rare cases, complications such as severe hemorrhages or infections are induced. Prior to performing iridotomies, however, medicational treatment is provided to set an upper limit for the intraocular pressure.

Trabeculum

Another type of glaucoma is called *open-angle glaucoma*. It is not induced by a dislocation of the iris but by a malfunction of the trabecular meshwork. The drainage of aqueous humor can be improved by a treatment called *laser trabeculotomy* during which the trabeculum is carefully perforated. First results with ruby and argon ion lasers were published by Krasnov (1973) and Worthen and Wickham (1974), respectively. Two years later, it was reported by Ticho and Zauberman (1976) that in some cases a decrease in intraocular pressure was obtained just by shrinking the tissue of the trabeculum instead of perforating it. These observations have been the origin for another treatment technique called *trabeculoplasty* which was described in detail by Wise and Witter (1979).

There are two types of laser trabeculoplasty: argon laser trabeculoplasty (ALT) and selective laser trabeculoplasty (SLT). During ALT, approximately 100 pulses from an argon ion laser are applied to the surface of the trabeculum as described by Wise (1987). Focusing of the laser beam is facilitated by special contact glasses. Typical focus diameters are 50–100 μm. The drop in intraocular pressure arises from thermal interaction, since the heat deposited by the argon ion laser causes a shrinkage of the exposed trabecular meshwork. By this means, tensile forces tend to enlarge intermediate fluid canals located in between exposed tissue areas. SLT is a newer technique that uses a Nd:YAG laser to target melanocytes within the trabecular meshwork. According to Latina and de Leon (2005) and Barkana and Belkin (2007), it creates less thermal damage than ALT.

Trabeculoplasty has meanwhile developed to a standard type of modern laser surgery. Today, clinical data of follow-up periods covering twenty years and more are available. Especially in primarily chronic glaucoma with intraocular pressures below 35 mm Hg, the therapy is very successful. However, ruptures of the trabecular meshwork itself should always be avoided.

Sclera

Among laser treatments of the sclera, external and internal sclerostomies are distinguished. In either case, surgery aims at achieving a continuous channel from the front chamber to the fluid beneath the conjunctiva. Again, this type of filtration treatment is indicated for open-angle glaucomas[4].

External sclerostomies start from the anterior sclera, thereby injuring the conjunctiva. First laser sclerostomies in glaucomateous eyes were performed by Beckman et al. (1971) using a thermally acting CO_2 laser. The main disadvantage of this early technique is the need for dissection—a conjunctival flap frequently causes severe inflammation. Moreover, the very low absorption of scleral tissue at visible and near infrared wavelengths makes it extremely difficult to apply neodymium lasers or other pulsed laser systems. A solution to this problem was offered by L'Esperance (1983) when using exogeneous dyes to artificially increase the absorption coefficient. By dyeing the sclera with sterile india ink at the superior limbus, some fraction of the ink also diffused deeper into the sclera. Then, when focusing an argon ion laser beam on the trabecular meshwork, L'Esperance was able to cut through the trabeculum and the sclera starting from the interior. This was the first *internal sclerostomy* ever performed. Meanwhile, pulsed Nd:YAG lasers are applied to improve filtration in internal sclerostomies. By means of a specially designed goniolens, the incident laser beam is redirected into an acute angle inside the eye. Scleral perforation starts just next to the trabeculum and ends beneath the conjunctiva. A schematic illustration of the surgical procedure is shown in Fig. 4.8.

---

[4] If even this procedure does not stabilize the eye pressure, the ciliary body itself must be coagulated as it is the production site of the aqueous humor.

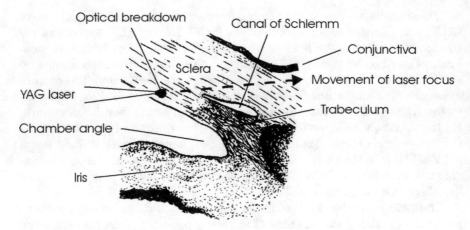

Fig. 4.8. Internal sclerostomy

First experimental and histologic results on Nd:YAG laser sclerostomy were described by March et al. (1984) and March et al. (1985). They found that approximately 250–500 pulses with pulse durations of 12 ns each and energies ranging from 16 mJ to 53 mJ were required to achieve a complete perforation of the sclera. With these laser parameters, thermal damage to adjacent tissue was limited to a few hundred μm. A few years later, visible dye lasers were also applied by again making use of different inks. The diffusion of these inks into the tissue can be accelerated by means of electrophoresis as reported by Latina et al. (1988) and Latina et al. (1990).

Cornea

Cornea and lens together account for the total refraction of the human eye. However, since the anterior surface of the cornea is exposed to air with an index of refraction close to unity, refraction at the anterior surface of the cornea represents the major part. A list of refractive properties of the human eye was first provided by Gullstrand at the beginning of the last century. A theoretical analysis of the refractive properties is found in the book by Le Grand and El Hage (1980). Both sets of data are listed in Table 4.1 together with a third column called the *simplified eye*. In the simplified eye, the same principal planes and focal distances are assumed as for the theoretical eye. However, an integer value of 8 mm is chosen for the radius of curvature of the anterior corneal surface. And, because also assuming the same indices of refraction for both the cornea and aequous humor, refraction at the posterior corneal surface is neglected. From these data, it can be concluded that the power of the cornea is approximately 42 diopters, whereas the total power of the eye is roughly 59 diopters. Therefore, about 70 % of the overall refraction arises from the cornea.

Table 4.1. Parameters of the unaccommodated human eye. Data according to Le Grand and El Hage (1980)

	Gullstrand eye	Theoretical eye	Simplified eye
Index of refraction			
Cornea	1.376	1.377	1.336
Aqueous humor	1.336	1.337	1.336
Lens	1.408	1.420	1.421
Vitreous body	1.336	1.336	1.336
Radius of curvature (mm)			
Cornea (ant. surface)	7.7	7.8	8.0
Cornea (post. surface)	6.8	6.5	–
Lens (ant. surface)	10.0	10.2	10.2
Lens (post. surface)	−6.0	−6.0	−6.0
Power (diopters)			
Cornea	43.05	42.36	42.0
Lens	19.11	21.78	22.44
Total eye	58.64	59.94	59.64

The transparency of corneal tissue in the spectral region from 400 nm to 1200 nm can be attributed to its extremely regular microscopic structure as will be discussed below. The optical zone of the human cornea has typical diameters ranging from 2 mm to 5 mm and is controlled by the iris. The overall thickness of the cornea varies between 500 μm at the center of the optical axis and 700 μm at the periphery. Corneal tissue is avascular and basically consists of five distinct layers: epithelium, Bowman's membrane, stroma, Descemet's membrane, and endothelium. A schematic cross-section of the human cornea is shown in Fig. 4.9.

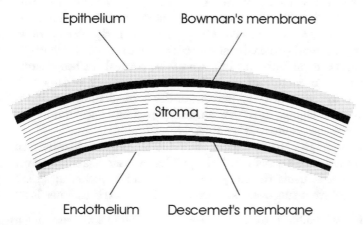

Fig. 4.9. Cross-section of a human cornea

According to Le Grand and El Hage (1980), the *epithelium* is made up of two to three layers of flat cells which – in combination with tear fluid – provide the smooth surface of the cornea. These cells are the only corneal cells capable of regenerating. *Bowman's membrane* consists of densely packed collagen fibers. All these fibers are oriented in planes parallel to the corneal surface, resulting in extremely high transparency. Due to its high density, Bowman's membrane is primarily responsible for the mechanical stability of the cornea. Almost 90 % of the corneal thickness belongs to the *stroma*. It has a structure similar to Bowman's membrane but at a lower density. Since the stroma contributes the major part of the cornea, refractive corneal surgery relies on removing stromal tissue. *Descemet's membrane* protects the cornea from its posterior side. And, finally, the *endothelium* consists of two layers of hexagonally oriented cells. Their main function is to prevent fluid of the front chamber from diffusing into the cornea.

In general, two types of corneal surgeries are distinguished: removal of any pathologic conditions and refractive corneal surgery. The first group includes treatments of irregularly shaped corneas, e.g. keratoconus, externally induced corneal injuries, and corneal transplantations. Prior to laser surgery, all these treatments had to be performed with mechanical scalpels. Today, ophthalmic lasers – among these especially femtosecond lasers and the ArF excimer laser – offer a non-invasive and painless surgery. Successful circular trephinations and smoothing of irregular surfaces were, for instance, reported by Loertscher et al. (1987) and Lang et al. (1989). Very clean corneal excisions can be achieved when using short pulsed neodymium lasers as shown in Figs. 4.10 and 4.11. The dependence of the ablation depth on pulse energy is illustrated in Figs. 4.12 and 4.13.

The other group of corneal surgeries aims at altering its refractive power. Although most cases of wrongsightedness cannot be attributed to a pathologic condition of the cornea[5], the corneal power is the easiest to change. First studies on refractive corneal surgery were reported by Fjodorov and Durnev (1979) using a diamond knife. The procedure is called *radial keratotomy* or *RK* (from Greek: $\kappa\epsilon\rho\alpha\varsigma$ = cornea, $\tau o\mu\alpha\epsilon\iota\nu$ = to be cut). By placing radial incisions into the peripheral cornea, tensile forces are rearranged which causes local steepening and central flattening of the anterior corneal surface, thereby reducing the cornea's refractive power. The major drawback of this early technique was that approximately 95 percent of the corneal thickness had to be incised to achieve a change in refraction. Many patients accidentally lost their vision, because the cornea was completely perforated.

In the beginning of the 1980s, Keates et al. (1981) proposed to perform radial keratotomy not using a scalpel, but applying a laser beam. However, their original concept of using the CO_2 laser failed, and Trokel et al. (1983) were the first to achieve a successful keratotomy with an ArF excimer laser.

---

[5] In most myopias, the bulbus is too long. Presbyopia is due to a decrease in lens accommodation. Only astigmatism is frequently caused by the cornea itself.

Fig. 4.10. Magnification (light microscopy) of a corneal excision achieved with a Nd:YLF laser (pulse duration: 30 ps, bar: 20 µm, original surface: *horizontal*, laser excision: *vertical*)

Fig. 4.11. High magnification (transmission electron microscopy) of a corneal excision achieved with a Nd:YLF laser (pulse duration: 30 ps, bar: 1 µm)

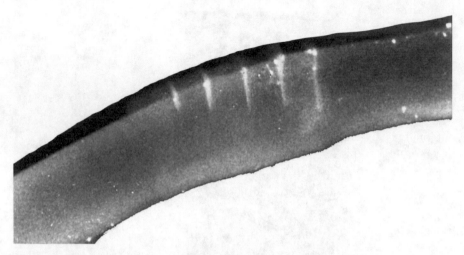

Fig. 4.12. Corneal incisions achieved with a Nd:YLF laser at different pulse energies (from *left* to *right*: 30 μJ, 50 μJ, 70 μJ, 90 μJ, and 110 μJ)

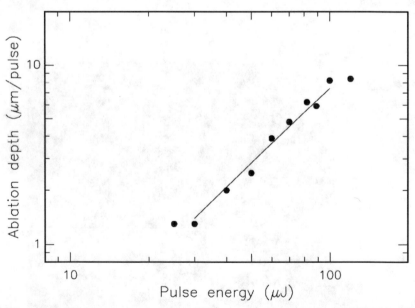

Fig. 4.13. Ablation curve of corneal stroma obtained with a Nd:YLF laser (pulse duration: 30 ps, focal spot size: 15 μm). Data according to Niemz et al. (1991)

Soon after, other groups like Cotlair et al. (1985), Marshall et al. (1985), and Puliafito et al. (1985) also published their results, and they all focused on radial incisions for correcting myopia. Then, Seiler et al. (1988) reported that even astigmatism can be corrected by placing transverse incisions. But again, since 95 percent of the corneal thickness had to be incised to achieve a change in refraction, corneal perforation could not be ruled out completely. A 10-year follow-up study published by Waring et al. (1994) testified radial keratotomy to be unstable and unpredictable in terms of both short-term and long-term outcome. This Prospective Evaluation of Radial Keratotomy (PERK) study, based on 693 eyes, finally dealt a death blow to RK.

Also in the 1980s, a completely different technique called *keratomileusis* (from Greek: $\lambda \upsilon \epsilon \iota \nu$ = to detach) was proposed by Marshall et al. (1986), where the cornea is actually excised rather than being incised. That is why this kind of surgery is also called *photorefractive keratectomy* or *PRK* (from Greek: $\epsilon \kappa \tau o \mu \alpha \epsilon \iota \nu$ = to be cut out). The scientists recognized the superior quality of excimer laser ablations and investigated the direct carving of the cornea to change its optical power. During this treatment, large area ablations are performed around the optical axis rather than a few linear incisions in the periphery. The major advantage of keratomileusis over radial keratotomy is the ability to achieve direct optical correction rather than depending on indirect biomechanical effects of peripheral incisions. PRK turned into a successful treatment, but initial hopes for rapidly achieving a stable refraction and for eliminating scattering effects were never met. The basic ideas of radial keratotomy and photorefractive keratectomy are compared in Fig. 4.14—together with LASIK, today's winner in the race for the best refractive surgery. Before we proceed with all the variations of LASIK, we will calculate how much corneal tissue actually needs to be removed.

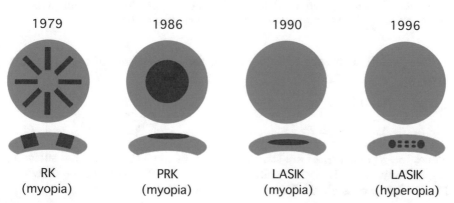

Fig. 4.14. Comparison of RK (radial keratotomy), PRK (photorefractive keratectomy), and LASIK (laser in-situ keramileusis), respectively. Anterior view and cross-section of the cornea are shown. Red color marks incisions (RK) or excisions (PRK and LASIK)

In the following paragraphs, a geometrically derived instruction is given of how the reshaping of a myopic eye has to be done. In Fig. 4.15, the pre- and postoperative anterior surfaces of the cornea are shown together with other geometrical parameters. The optical axis and an arbitrary axis perpendicular to it are labeled x and y, respectively. The curvature of the cornea is given by R, and the indices "i" and "f" refer to the initial and final states of the cornea, respectively.

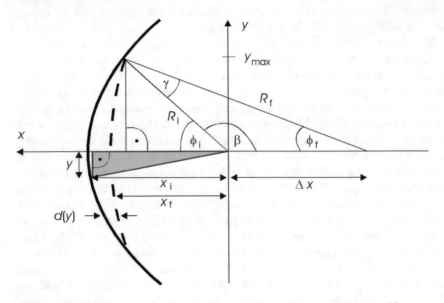

Fig. 4.15. Geometry of performing keratomileusis in a myopic eye. The pre- and postoperative anterior surfaces of the cornea are drawn as a solid curve and a dashed curve, respectively. The alteration in curvature is exaggerated

The equations for the initial and final anterior surfaces of the cornea are given by

$$x_i^2 + y^2 = R_i^2 , \tag{4.1}$$

$$(x_f + \Delta x)^2 + y^2 = R_f^2 , \tag{4.2}$$

where (x_i, y) and (x_f, y) are the coordinates of the initial and final surfaces, respectively, and Δx is the shift in the centers of curvature as shown in Fig. 4.15. We thus obtain for the depth of ablation

$$d(y) = x_i - x_f = \sqrt{R_i^2 - y^2} - \sqrt{R_f^2 - y^2} + \Delta x . \tag{4.3}$$

The shift Δx can be calculated from the sines of the angles β and γ and is expressed by

$$\frac{\Delta x}{\sin \gamma} = \frac{R_f}{\sin \beta} \cdot \tag{4.4}$$

When using the following geometrical relations

$$\sin \beta = \sin \left(180° - \phi_i\right) = \sin \phi_i \ ,$$

$$\gamma = \phi_i - \phi_f \ ,$$

equation (4.4) can be turned into

$$\Delta x = R_f \frac{\sin \left(\phi_i - \phi_f\right)}{\sin \phi_i} \cdot \tag{4.5}$$

Furthermore, we deduce from Fig. 4.15 the two expressions

$$\sin \phi_i = \frac{y_{\max}}{R_i} \ ,$$

$$\sin \phi_f = \frac{y_{\max}}{R_f} \ ,$$

where $y_{\max}$ is the maximum radius of the optical zone to be altered. Hence, substituting ϕ_i and ϕ_f in (4.5) leads to

$$\Delta x = \frac{R_i R_f}{y_{\max}} \sin \left(\arcsin \frac{y_{\max}}{R_i} - \arcsin \frac{y_{\max}}{R_f}\right) . \tag{4.6}$$

From (4.3) and (4.6), we get

$$d(y) = \sqrt{R_i^2 - y^2} - \sqrt{R_f^2 - y^2} + \frac{R_i R_f}{y_{\max}} \sin \left(\arcsin \frac{y_{\max}}{R_i} - \arcsin \frac{y_{\max}}{R_f}\right) .$$

The largest ablation depth must be obtained at $y = 0$ with

$$d(0) = R_i - R_f + \frac{R_i R_f}{y_{\max}} \sin \left(\arcsin \frac{y_{\max}}{R_i} - \arcsin \frac{y_{\max}}{R_f}\right) . \tag{4.7}$$

In (4.7), the depth $d(0)$ is given which the surgeon has to remove at the vertex of the optical axis. The unknown parameter R_f is readily obtained from the basic law of a curved refracting surface

$$\Delta D = (n_c - 1) \left(\frac{1}{R_i} - \frac{1}{R_f}\right) , \tag{4.8}$$

where ΔD is the degree of myopia expressed in units of diopters, and n_c is the refractive index of the cornea. For an attempted correction of ΔD ranging from 1 diopter to 10 diopters, the required values of R_f and $d(0)$ are listed in Table 4.2, assuming $R_i = 7.8\,\text{mm}$, $n_c = 1.377$, and $y_{\max} = 2.5\,\text{mm}$. From these data, it can be concluded that for myopias up to 5 diopters less than one tenth of the corneal thickness needs to be ablated.

Table 4.2. Theoretical values of keratomileusis in the case of myopia. Actual data might slightly differ due to a rearrangement in mechanical stress. Assumed is an optical zone of 5 mm, i.e. $y_{max} = 2.5$ mm

ΔD (diopters)	R_f (mm)	$d(0)$ (μm)
1	7.965	9.0
2	8.137	17.9
3	8.316	26.8
4	8.504	35.7
5	8.700	44.6
6	8.906	53.4
7	9.121	62.2
8	9.347	71.0
9	9.585	79.7
10	9.835	88.4

A similar calculation applies for the correction of hyperopic eyes. However, since a peripheral ring-shaped zone needs to be ablated, the diameter of the optical zone is usually extended to 7–8 mm. It should be added that astigmatism can also be corrected by means of keratomileusis. This is achieved by simply aiming at two different values of R_f in two planes located perpendicularly to the optical axis.

In the 1980s, i.e. during the early stage of performing keratomileusis, this novel technique was further improved and corticosteroids were infrequently used. Early experiences were reported by Aron-Rosa et al. (1987) and Taylor et al. (1989). Normal reepithelialization as well as mild subepithelial haze were observed. McDonald et al. (1990) demonstrated the ability to achieve a measurable myopic refractive correction. However, they also reported on an initial regression and a poor predictability of the refractive effect. Wilson (1990) distinguished between preoperative myopias with less and more than 5.5 diopters. From his clinical observations, he concluded that good predictions can be made in the first case only. Some of his results are shown in Figs. 4.16a–b. Similar statements regarding predictability were published by Seiler and Genth (1994).

Since then, several variations of PRK had been studied all over the world. The ArF excimer laser is well suited for this type of surgery because of its ablation characteristics. As we have already encountered when discussing Fig. 3.39, one pulse from this laser typically ablates 0.1–1 μm of corneal tissue which corresponds to 0.01–0.1 diopters. Usually, energy densities of 1–5 J/cm^2 are applied in order to be less dependent on energy fluctuations, since the ablation curve shown in Fig. 3.39 approaches a saturation limit. Then, the correction of one diopter is achieved with approximately 10 pulses which takes about one second at a repetition rate of 10 Hz. The choice of a proper delivery system was controversially discussed. Most common were the methods of using a scanning slit as described by Hanna et al. (1988) and

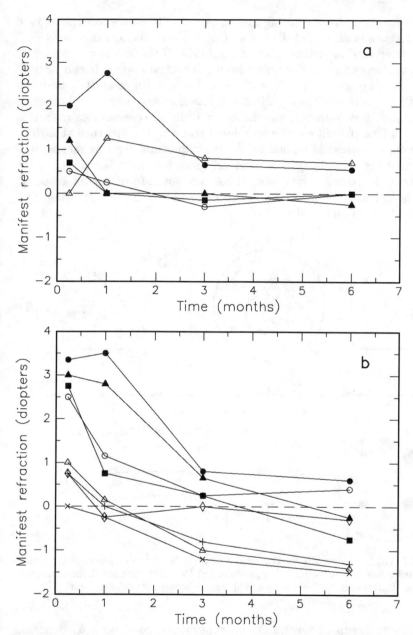

Fig. 4.16. (a) Results of ArF excimer laser keratomileusis performed in five cases of myopia with attempted corrections less than 5.5 diopters. **(b)** Results of ArF excimer laser keratomileusis performed in eight cases of myopia with attempted corrections more than 5.5 diopters. During a follow-up period of six months, eyes with more than 5.5 diopters of preoperative myopia appear to be less stable than eyes with less than 5.5 diopters. Emmetropia is indicated by a dashed line. Data according to Wilson (1990)

Hanna et al. (1989), or a rotating disk mask with different apertures as used by L'Esperance et al. (1989). In the first case, the cornea is exposed to radiation from an ArF laser through a movable slit. If the slit is wider near its center, i.e. if more tissue is removed from the central cornea, it can be used for correcting myopic eyes. If the slit is wider near its ends, it is designed for treating hyperopia. The rotating disk mask, on the other hand, consists of several apertures with different diameters which are concentrically located on a wheel. The patient's eye is irradiated through one aperture at a time. In between, the wheel is turned to the next aperture. By this means, it is assured that the overall exposure gradually decreases from the center to the periphery of the cornea. Thus, more tissue is removed from central areas, i.e. myopia is corrected. Hyperopias cannot be treated using the rotating disk. Either method is illustrated in Figs. 4.17a–c.

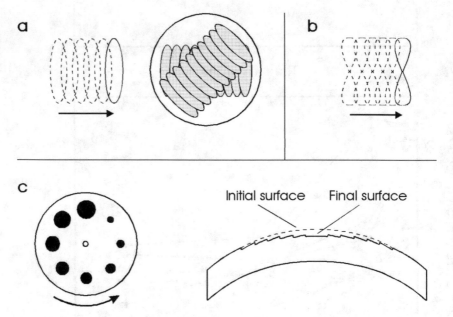

Fig. 4.17. (a) Scanning slit for correction of myopia and anterior view of cornea. **(b)** Scanning slit for correction of hyperopia. **(c)** Rotating disk mask for correction of myopia. Initial and final surfaces are shown in a corneal cross-section

In a large group of keratomileusis treatments, good optical correction was obtained after approximately six months. Despite several improvements, however, three major disadvantages still remained:

– regression of the refractive power initially achieved,
– appearance of a subepithelial haze after the regression period,
– disturbances in night vision due to enhanced scattering.

The existence of a regression effect becomes very obvious when looking at Figs. 4.16a–b. Within the first three months after PRK, a regression of up to three diopters has been observed. It is probably due to the processes of corneal wound healing and reepithelialization. Therefore, the patient's eye is usually transferred to a hyperopic state immediately after surgery. It is hoped that refraction stabilizes at emmetropy after the period of regression. This procedure, of course, requires a lot of patience since the patient has to use several pairs of glasses during the first months after surgery.

Subepithelial haze has not been found in all patients but is a frequent side effect. It disturbs vision especially in darkness when the pupil widens. The primary cause of this haze is yet unknown. It might arise from a rearrangement of collagen fibers inside Bowman's membrane and the stroma. Haze could also be induced by the toxic UV radiation. As stated in Sect. 3.3, cytotoxicity and mutagenicity cannot be excluded for the ArF laser wavelength at 193 nm. Although corneal tumors do not occur, cytotoxic effects might indeed cause a reduction in corneal transparency.

The cause for enhanced scattering, the third drawback of PRK, is well known: the epithelium is removed around the optical axis, so a new epithelial layer must grow after surgery. But since the stromal surface is not perfectly smooth after ArF radiation, the new epithelium will copy these irregularities and will not have the original smoothness anymore. Patients often complain about scattering effects, especially at night.

So, scientists were searching for a new approach to reshape the cornea, keeping the original corneal surface. And that is when the idea of *LASIK* was born. The word LASIK stands for "Laser-Assisted In Situ Keratomileusis" and is a procedure that specifically changes the stromal thickness. Here is how the original LASIK version works (see Fig. 4.18a): in the first step, a mechanical microkeratome (a blade device) is used to cut a flap in the cornea. Typically, the flap has a diameter of 8 – 9.5 mm and a thickness of 100 – 160 µm. A hinge is left at one end of this flap. In the second step, the flap is folded back revealing the stroma. In the third step, pulses from a computer-controlled ArF excimer laser photoablate a portion of the stroma. And finally, in the fourth step, the flap is returned to its original position. So, LASIK's major advantage becomes very obvious: after surgery, the patient's cornea has its original smooth surface. First reports on this original LASIK surgery were published by Pallikaris et al. (1990) and Pallikaris et al. (1991). Further studies were provided by Knorz et al. (1996), Farah et al. (1998), Patel et al. (2000), Lipshitz et al. (2001), and Rosen (2001). LASIK was compared to PRK by Pallikaris and Siganos (1994) and Shortt et al. (2006).

There are several versions of LASIK available today. Let us now see what kind of improvements have since been made to the original version. The mechanical cut performed by the microkeratome is the main cause for complications. In *femto–LASIK* (see Fig. 4.18b), this drawback is overcome by creating the flap using a femtosecond laser. This laser is focused inside the

Fig. 4.18. (a) Original LASIK using microkeratome and UV laser. **(b)** Femto–LASIK using femtosecond and UV laser. **(c)** Smile–LASIK using femtosecond laser only. © Niemz

cornea, generating a plasma and cutting tissue by means of plasma-induced ablation. Femtosecond-created flaps come with a higher accuracy and less variations. The other three steps of the procedure follow as performed in the original LASIK surgery. Femto–LASIK was investigated by several authors, for example by Kurtz et al. (1998), Ratkay-Traub et al. (2001), and Ratkay-Traub et al. (2003). A detailed analysis of the flap creation using femtosecond lasers was published by Salomao and Wilson (2010).

Table 4.3. Three types of LASIK: steps taken and tools needed

Type	Step 1	Step 2	Step 3	Step 4
Original	Create flap **(microkeratome)**	Fold back flap **(forceps)**	Photoablation **(ArF laser)**	Return flap **(forceps)**
Femto	Create flap **(femto laser)**	Fold back flap **(forceps)**	Photoablation **(ArF laser)**	Return flap **(forceps)**
Smile	Detach posterior surf. of lenticule **(femto laser)**	Detach anterior surf. of lenticule **(femto laser)**	Small incision in periphery **(femto laser)**	Extract lenticule **(forceps)**

The newest type of LASIK surgery is called *smile–LASIK* where the word SMILE stands for "SMall Incision Lenticule Extraction". To make it short, the name lives up to its expectations. In smile–LASIK (see Fig. 4.18c), even the need for creating a flap is now overcome. The whole surgery is performed using just one laser—the femtosecond laser which is focused into the stroma. So, there is no mechanical tool and also no UV light needed anymore. In the first step, the posterior surface of a disk-shaped lenticule is detached. In the second step, the anterior surface of this lenticule is detached. In the third step, a small incision is made in the periphery of the cornea. And finally, in the fourth step, the lenticule is extracted using a pair of forceps. Smile–LASIK is the safest of all LASIK surgeries. An excellent review was provided by Reinstein et al. (2014). Comparisons to femto–LASIK were published by Dong et al. (2014) and Lin et al. (2014).

In Fig. 4.19, a high-magnification photograph of a LASIK treatment is shown. Using a Nd:Glass femtosecond laser, a corneal flap was created and folded back. Furthermore, a disk-shaped lenticule of intrastromal tissue was excised and put aside. The fascinating photograph proves the high accuracy associated with this technique. Since smile–LASIK has just recently been developed, no long-term evaluations are available so far. However, as progress with femtosecond laser surgery has demonstrated, research in this field continues to grow rapidly.

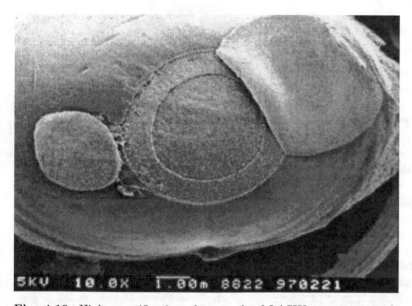

Fig. 4.19. High-magnification photograph of LASIK treatment performed with a Nd:Glass laser (pulse duration: 500 fs, pulse energy: 3 µJ). Photograph kindly provided by Dr. Loesel (Heidelberg)

Actually, smile–LASIK dates back to the beginning of the 1990s. At that time, the idea of *intrastromal ablations* came up as sketched in Fig. 4.20. By means of focusing a laser beam inside the cornea, either a continuous disk-shaped or a ring-shaped cavity is generated, depending on the type of correction needed. When the gaseous vapor inside these cavities has diffused into the surrounding medium, the cavity collapses. Then, the removal of stromal tissue induces a stable change in curvature of the anterior corneal surface. The main advantage of this technique is that the epithelium and Bowman's membrane are not injured. Thus, the stability of the cornea is less affected, and corneal haze is less likely to occur.

Intrastromal ablation
for myopia

Intrastromal ablation
for hyperopia

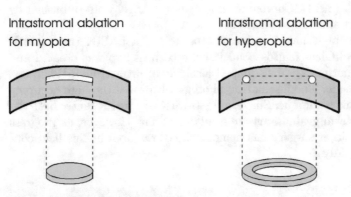

Fig. 4.20. Intrastromal ablations

Intrastromal ablations were first reported by Höh (1990) when focusing a Q-switched Nd:YAG laser beam inside the stroma. A more detailed study followed soon after by Niemz et al. (1993a) using picosecond pulses from a two-stage Nd:YLF laser system. It was shown for the first time that a continuous intrastromal cavity can be achieved which is located approximately 150 μm beneath the epithelium. A typical example of such a laser-induced cavity inside the stroma of a human cornea is shown in Fig. 4.21. This photograph was obtained with a scanning electron microscope. Due to shrinking processes during the preparation, the cavity appears closer to the epithelium than 150 μm. The collapse of an intrastromal cavity is captured in Fig. 4.22. Two vacuoles are still visible which have not yet collapsed.

Using the algorithm of *finite elements*, predictions can be made concerning any changes in refraction. The method of finite elements is a very powerful tool of modern engineering science. Reshaping of the cornea by mechanical alterations is a typical problem of a special field called *biomechanics* or *bioengineering*. The governing equations of biological tissues have already been discussed by Fung (1981). Mechanical properties of the cornea have been reported by Jue and Maurice (1989). A detailed review of corneal biomechanics after LASIK and smile–LASIK has been published by Yang et al. (2016).

Fig. 4.21. Intrastromal cavity achieved with a Nd:YLF laser (pulse duration: 30 ps, pulse energy: 140 μJ)

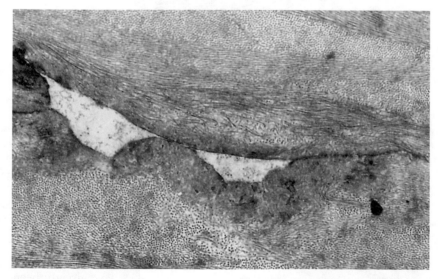

Fig. 4.22. Collapse of an intrastromal cavity achieved with a Nd:YLF laser (pulse duration: 30 ps, pulse energy: 100 μJ). The two vacuoles which have not yet collapsed measure about 20 μm

When applying the method of finite elements, the object, e.g. the cornea, is subdivided into a certain number of finite elements. The size of these elements is chosen such that each of them can be characterized by constant physical parameters. The advantage of this procedure is that all governing equations are readily solved within each element. Then, by implementing proper boundary conditions, each element is related to its adjacent elements which finally leads to a physical description of the whole object. Any alteration of a single element consequently influences other elements, as well. In our example shown in Fig. 4.23, two intrastromal elements within the optical zone are removed. Using iterative calculations on fast processing computers, the deformed postoperative state of the cornea is simulated. In order to obtain reliable results, such a model should comprise the following properties of the cornea:

– radial dependence of corneal thickness,
– anisotropy of corneal tissue,
– incompressibility of corneal tissue,
– non-linear stress–strain behavior[6],
– non-uniform initial stress distribution.

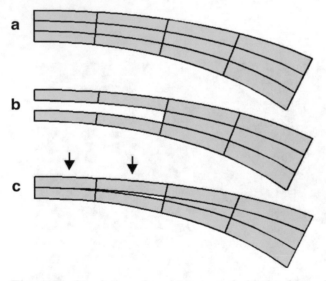

Fig. 4.23. Simulation of an intrastromal ablation by twelve finite elements. One half of the corneal cross-section is shown. (**a**) Preoperative state. (**b**) Intrastromal ablation (immediate postoperative state). (**c**) Steady postoperative state

---

[6] Most biological tissues do not obey *Hooke's law* of elasticity.

4.2 Lasers in Dentistry

Historically, dentistry was the second medical discipline where lasers were applied, but it took three decades until scientists finally succeeded in finding useful applications. The main reason was that early research focused on treating hard tissues, although only thermally acting lasers were available in the 1960s and 1970s. These lasers caused heating and severe cracking of teeth. Especially in caries therapy – the most frequent dental surgery – conventional mechanical drills are still superior compared to most types of lasers. Only laser systems capable of providing ultrashort pulses are an alternative to mechanical drills as shown by Niemz et al. (1993b) and Pioch et al. (1994). However, many clinical studies and extensive engineering effort still remain to be done. As always with new technology, that kind of treatment will finally win the race which provides the best benefit to the patient. Other topics of interest in dentistry include photoactivated disinfection, laser treatment of soft tissue, laser-welding of dental bridges and dentures, and implantology. In all of these areas, research has meanwhile been very successful.

The Human Tooth

Before going into the details of laser dentistry, a brief summary of the anatomy of the human tooth as well as its physiology and pathology shall be given. In principle, the human tooth consists of mainly three distinct segments called enamel, dentin, and pulp. A schematic cross-section of a human tooth is shown in Fig. 4.24.

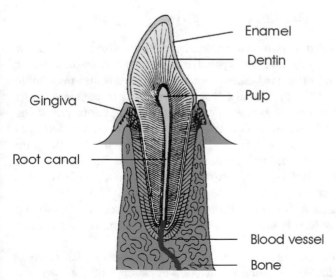

Fig. 4.24. Human tooth

The *enamel* is the hardest substance of the human body. It is made of approximately 95 % (by weight) hydroxyapatite, 4 % water, and 1 % organic matter. Hydroxyapatite is a mineralized compound with the chemical formula $Ca_{10}(PO_4)_6(OH)_2$. Its substructure consists of tiny crystallites which form so-called *enamel prisms* with diameters ranging from 4 μm to 6 μm. The crystal lattice itself is intruded by several impurities, especially Cl^-, F^-, Na^+, K^+, and Mg^{2+}.

The *dentin*, on the other hand, is much softer. Only 70 % of its volume consists of hydroxyapatite, whereas 20 % is organic matter – mainly collagen fibers – and 10 % is water. The internal structure of dentin is characterized by small tubuli which measure up to a few millimeters in length, and between 100 nm and 3 μm in diameter. These tubuli are essential for the growth of the tooth.

The *pulp*, finally, is not mineralized at all. It contains the supplying blood vessels, nerve fibers, and different types of cells, particularly odontoblasts and fibroblasts. Odontoblasts are in charge of producing the dentin, whereas fibroblasts contribute to both stability and regulation mechanisms. The pulp is connected to peripheral blood vessels by a small channel called *root canal*. The tooth itself is embedded into soft tissue called *gingiva* which keeps the tooth in place and prevents bacteria from attacking the root.

The most frequent pathologic condition of teeth is called *decay* or *caries*. It originates from both a cariogeneous nourishment and insufficient oral hygiene. Microorganisms multiply at the tooth surface and form a layer of *plaque*. These microorganisms produce lactic and acetic acid, thereby reducing the pH down to values of approximately 3.5. The pH and the solubility of hydroxyapatite are strongly related by

$$Ca_{10}(PO_4)_6(OH)_2 + 8\,H^+ \longleftrightarrow 10\,Ca^{2+} + 6\,HPO_4^{2-} + 2\,H_2O\ .$$

By means of this chemical reaction, the enamel is demineralized within a few days only. Calcium bound to the hydroxyapatite is ionized and washed out by saliva. This process turns the hard enamel into a very porous and permeable structure as shown in Fig. 4.25. Usually, this kind of decay is associated with a darkening in color. Sometimes, however, carious lesions appear bright at the surface and are thus difficult to detect. At an advanced stage, the dentin is demineralized, as well. In this case, microorganisms can even infect the pulp and its interior which often induces severe pain. Then, at the latest, must the dentist remove all infected substance and refill the tooth with suitable alloys, gold, ceramics, or composites. Among alloys, amalgam has been a very popular choice of the past. Recently, though, a new controversy has arisen concerning the toxicity of this filling material, since it contains a significant amount of mercury.

The removal of infected substance is usually accomplished with conventional mechanical drills. These drills do evoke additional pain for two reasons. First, tooth nerves are very sensitive to induced vibrations. Second,

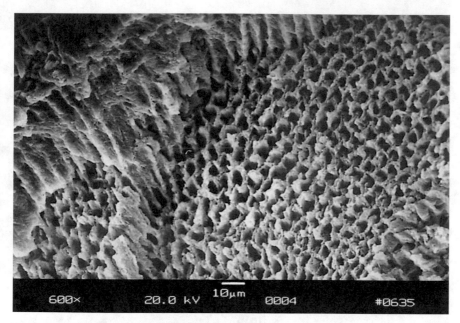

Fig. 4.25. High-magnification photograph of caries taken with a scanning electron microscope. Reproduced from Niemz (1994a). © 1994 Springer-Verlag

tooth nerves also respond to temperatures higher than 45°C which are easily induced by friction during the drilling process. Pain relief without injection of an anaesthetic was the ulterior motive when looking for laser applications in caries therapy. However, it turned out that not all types of lasers fulfill this task. Although vibrations are avoided due to the contactless technique, thermal side effects are not always eliminated when using lasers. CW and long-pulse lasers, in particular, induce extremely high temperatures in the pulp as shown in Figs. 4.26a–b. Even air cooling does not reduce this temperature to a tolerable value. Thermal damage is negligible only when using ultrashort pulses according to Sect. 3.2.

Meanwhile, other advantages are being discussed which could even be more significant than just pain relief[7]. Very important among these are the so-called conditioning of dental substance and a possibly more precise procedure of caries removal. Conditioning provides additional protection of the tooth by means of sealing its surface. Thereby, the occurrence of caries can be significantly delayed. Improved control of caries removal, e.g. by a spectroscopic analysis of laser-induced plasmas as shown in Figs. 3.54a–b, could minimize the amount of healthy substance to be excised. Then, indications for expensive dental crowns or bridges are effectively reduced.

---

[7] It should be kept in mind that pain relief alone would probably not justify the application of more expensive machines.

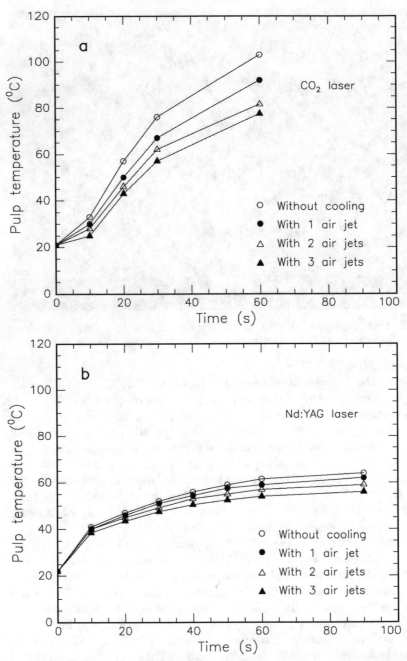

Fig. 4.26. (a) Mean temperatures in the pulp during exposure to a CW CO_2 laser (power: 5 W) without and with air cooling, respectively. **(b)** Mean temperatures in the pulp during exposure to a CW Nd:YAG laser (power: 4 W) without and with air cooling, respectively. Data according to Frentzen and Koort (1992)

Laser Treatment of Hard Tooth Substance

First experiments with teeth using the laser as a surgical tool were performed by Goldman et al. (1964) and Stern and Sognnaes (1964). Both of these groups used a pulsed ruby laser at a wavelength of 694 µm. This laser induced severe thermal side effects such as irreversible injury of nerve fibers and tooth cracking. Thus, it is not very surprising that these initial studies never gained clinical relevance. A few years later, a CO_2 laser system was investigated by Stern et al. (1972). However, the results did not improve very much with respect to the ruby laser. These observations are due to the fact that both the ruby and CO_2 lasers are typical representatives of thermally acting lasers. Thus, it was straightforward to conclude with Stern (1974) that without being able to eliminate these thermal effects, lasers would never turn into a suitable tool for the preparation of teeth.

Meanwhile, several experiments have been conducted using alternative laser systems. At the end of the 1980s, the Er:YAG laser was introduced to dental applications by Hibst and Keller (1989), Keller and Hibst (1989), and Kayano et al. (1989). The wavelength of the Er:YAG laser at 2.94 µm matches the resonance frequency of the vibrational oscillations of water molecules contained in the teeth as described in Sect. 3.2. Thereby, the absorption of the Er:YAG radiation is strongly enhanced, resulting in a high efficiency. However, the sudden vaporization of water is associated with a pressure gradient. Small microexplosions are responsible for the break-up of the hydroxyapatite structure. High-magnification photographs of a human tooth after Er:YAG laser exposure were shown in Sect. 3.2 in Figs. 3.14a–b. The coincidence of thermal (e.g. vaporization) and mechanical (e.g. pressure gradient) ablation effects has led to the term "thermomechanical interaction" as used by Frentzen and Koort (1991).

Initially, Er:YAG lasers seemed to be very promising because of their high efficiency in ablating dental substances. Meanwhile, though, some indication has been given that microcracking is induced by Er:YAG laser radiation. It was observed by Niemz et al. (1993b) and Frentzen et al. (1994) – using scanning electron microscopy and dye penetration tests – that these fissures can extend up to 300 µm in depth. They could thus easily serve as an origin for the development of new decay. External cooling of the tooth might help to reduce the occurrence of cracking.

Even worse results were found with the Ho:YAG laser as reported by Niemz et al. (1993b). High-magnification photographs of a human tooth after laser exposure were shown in Figs. 3.16a–b. Severe thermal effects including melting of tooth substance were observed. Cracks up to 3 mm in depth were measured when performing dye penetration tests.

Dye penetration tests are suitable experiments for the detection of laser-induced tooth fissures. After laser exposure, the tooth is stained with a dye, e.g. neofuchsine solution, for several hours. Afterwards, the tooth is sliced using a microtome, and the maximum penetration depth of the dye is determined. The results of some representative measurements are summarized in Fig. 4.27. Obviously, tooth fissures induced by Ho:YAG and Er:YAG lasers must be considered as a severe side effect.

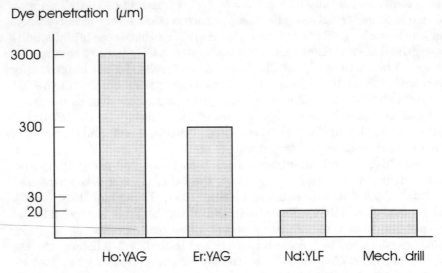

Fig. 4.27. Results of dye penetration tests for three different solid-state lasers and the mechanical drill. Listed are the maximum penetration depths inside the enamel of human teeth. Pulse durations: 250 μs (Ho:YAG), 90 μs (Er:YAG), and 30 ps (Nd:YLF). Data according to Niemz (1993b)

Another laser type – the ArF excimer laser – was investigated by Frentzen et al. (1989) and Liesenhoff et al. (1989) regarding its usefulness in dentistry. Indeed, initial experiments proved that only very slight thermal effects were induced which was attributed to the shorter pulse duration of approximately 15 ns and the gentle interaction mechanism of photoablation. However, the ablation rate achieved with this laser, i.e. the ablated volume per unit time, is too low for clinical applications. Although very successful in refractive corneal surgery because of its high precision, it is exactly this accuracy with ablation depths less than 1 μm per pulse and the rather moderate repetition rates which pull the ablation rate down. This ineffectiveness and the general risks of UV radiation are the major disadvantages concerning the use of the ArF laser in dentistry.

A second UV laser – the frequency-doubled Alexandrite laser at 377 nm – was studied by Steiger et al. (1993) and Rechmann et al. (1993). They observed that this laser offers a better selectivity for carious dentin than the Er:YAG laser, i.e. the fluence required for the ablation threshold of healthy dentin is higher when using the Alexandrite laser, whereas the thresholds for ablating carious dentin are about the same.

The most promising approach to laser caries therapy was probably made by Niemz et al. (1993b) and Pioch et al. (1994) when using picosecond pulses from a Nd:YLF laser system. Although, at the early stage of experiments, uncertainty predominated concerning potential shock wave effects, it has meanwhile been verified by five independent tests that mechanical impacts are negligible. These consist of *scanning electron microscopy*, *dye penetration tests*, *hardness tests*, *histology*, and *polarized microscopy*, respectively. The results were published by Niemz (1995b).

Scanning Electron Microscopy. In Figs. 4.28a–b, two SEM are shown demonstrating the ability of a picosecond Nd:YLF laser to produce extremely precise tetragonal cavities in teeth. The cavities are located in healthy and carious enamel, respectively. Both of them have lateral dimensions of $1 \times 1\,mm^2$ and a depth of approximately $400\,\mu m$. They were created by distributing $1\,mJ$ laser pulses onto 40 lines over the tooth surface with 400 lasered spots per line, and repeating this procedure ten times for the cavity shown in Fig. 4.28a and only once for the cavity in Fig. 4.28b. Thus, a total number of 160 000 laser shots was necessary for the cavity in healthy enamel, and only one tenth of this number was needed to achieve a similar depth in carious enamel. This observation already indicates that the ablation rate of demineralized enamel is about ten times higher than that of healthy enamel, i.e. the Nd:YLF laser provides a caries-selective ablation.

In Figs. 4.29a–b, the cavity wall and bottom are shown, respectively. The cavity wall is extremely steep and is characterized by a sealed glass-like structure. This is of great significance for the prevention of further decay. The roughness of the cavity bottom is of the order of $10\text{--}20\,\mu m$ and thus facilitates the adhesion of most filling materials. The scanning ablation becomes more evident when using fewer pulses to produce a shallow cavity as shown in Fig. 4.30a. In this case, a circular ablation pattern with $2\,500$ pulses was selected. In Fig. 4.30b, the effect of a conventional drill on the cavity wall is demonstrated. Deep grooves and crumbled edges are clearly visible.

Dye Penetration Tests. The results of dye penetration tests after exposure to a picosecond Nd:YLF laser have already been presented in Fig. 4.27. Laser-induced fissures typically remained below $20\,\mu m$. This value is of the same order as fissure depths obtained with the mechanical drill. One potential cause for the extremely small dye penetration might be the sealing effect demonstrated in Fig. 4.29a.

Hardness Tests. One obvious test for the potential influence of shock waves is the measurement of hardness of a tooth before and after laser exposure. In hardness tests according to Vickers, the impact of a diamond tip into a tooth surface is determined. Softer material is characterized by a deeper impact of the diamond tip – and thus a larger impact diameter. The hardness itself is defined as

$$H_V = 1.8544 \, \frac{K}{D^2} \,,$$

Fig. 4.28. (a) Cavity in healthy enamel achieved with 160 000 pulses from a Nd:YLF laser (pulse duration: 30 ps, pulse energy: 1 mJ). **(b)** Cavity in carious enamel achieved with 16 000 pulses from a Nd:YLF laser (pulse duration: 30 ps, pulse energy: 1 mJ)

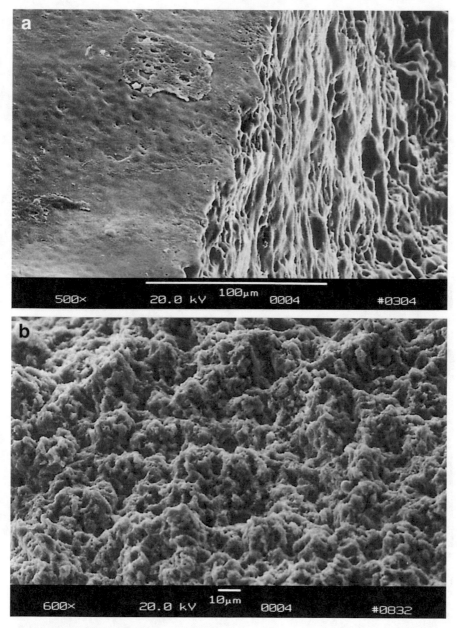

Fig. 4.29. (a) Cavity wall in healthy enamel achieved with a Nd:YLF laser (pulse duration: 30 ps, pulse energy: 1 mJ). (b) Cavity bottom in healthy enamel achieved with a Nd:YLF laser (pulse duration: 30 ps, pulse energy: 1 mJ)

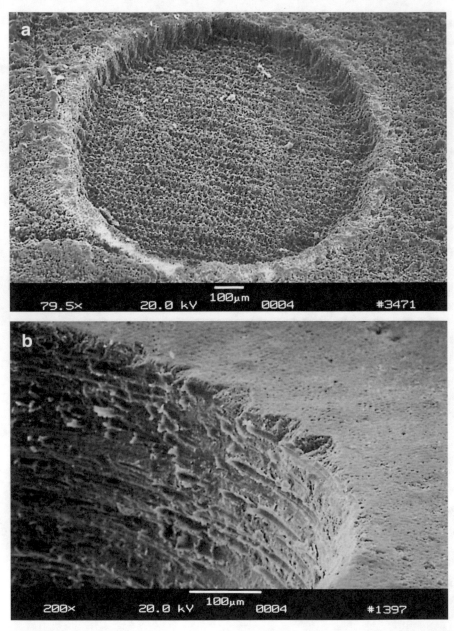

Fig. 4.30. (a) Cavity in carious enamel achieved with 2500 pulses from a Nd:YLF laser (pulse duration: 30 ps, pulse energy: 1 mJ). **(b)** Cavity wall achieved with a conventional diamond drill

where $K = 5.0 \times 10^4$ N, and D is the impact of a small diamond tip (angle: 136°) expressed in millimeters. The results of hardness tests after exposure to picosecond Nd:YLF pulses are presented in Table 4.4. According to Niemz (1995c), no significant alteration in hardness is observed in exposed and unexposed enamel. As expected, though, dentin appears much softer due to its lower content of hydroxyapatite.

Table 4.4. Mean hardness values of teeth before and after exposure to a Nd:YLF laser (pulse duration: 30 ps, pulse energy: 1 mJ)

	D (mm)	H_V (N/mm^2)
Exposed enamel	5.9	2660
Unexposed enamel	5.8	2760
Unexposed dentin	11.5	700

Histology. The most important touchstone for the introduction of a new therapeutic technique is the biological response of the tissue, i.e. the survival of cells. Histologic sections enable specific statements concerning the condition of cells due to highly sophisticated staining techniques. In Fig. 4.31a, the dentin–pulp junction of a human tooth is shown. It was located underneath a 1×1 mm^2 area exposed to 16 000 pulses from a Nd:YLF laser. Along the junction, several odontoblasts are clearly visible. They have not intruded into the dentin and have a similar appearance as in unexposed teeth. Thus, potential shock waves do not have a detectable impact on the pulp – not even on a cellular level.

Polarized Microscopy. Polarized microscopy is an efficient tool for detecting alterations in optical density which might also arise from the exposure to shock waves. If these shock waves are reflected, e.g. at the enamel–dentin junction, such alterations might even be enhanced and should thus become visible. For polarized microscopy, exposed teeth are dehydrated in an upgraded series of ethanol. Afterwards, they are kept in fluid methacrylate for at least three days. Within the following period of seven days, polymerization takes place in a heat chamber set to 43°C. Then, the embedded samples are cut into 100 μm thick slices using a saw microtome. Finally, the slices are polished and examined with a polarized light microscope. In Fig. 4.31b, the enamel–dentin junction of a human tooth is shown. It was located underneath a 1×1 mm^2 area exposed to 16 000 pulses from a Nd:YLF laser. In the top and bottom parts of the picture, dentin tubuli and enamel prisms are found, respectively. Due to the different optical densities of dentin and enamel, these two structures appear blue and yellow in the corresponding color photograph. However, no substantial alteration in color is observed within either the dentin or the enamel. Hence, no evidence of an altered optical density due to laser-induced shock waves is given.

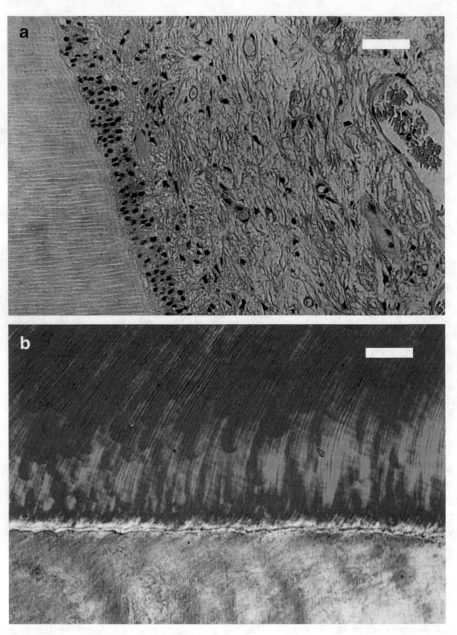

Fig. 4.31. (a) Histologic section of a human tooth after exposure to 16 000 pulses from a Nd:YLF laser (pulse duration: 30 ps, pulse energy: 500 μJ, bar: 50 μm). The junction of dentin (*left*) and pulp (*right*) is shown which was located next to the application site. (b) Polarized microscopy of a human tooth slice after exposure to 16 000 pulses from the same laser (bar: 50 μm). The junction of dentin (*top*) and enamel (*bottom*) is shown which was located next to the application site

From the results of the above tests, i.e. the negligibility of mechanical effects, it can be concluded that the cavities shown in Figs. 4.28a–b were produced by means of plasma-induced ablation as discussed in Sect. 3.4. The main reason for these observations is that the applied pulse energies were just slightly above the threshold energy of optical breakdown. In Fig. 4.32, the ablation curves of healthy enamel, healthy dentin, and carious enamel are given, respectively. In healthy enamel, plasma sparking was already visible at approximately 0.2 mJ. Then, when taking the corresponding focal spot size of 30 μm into account, the ablation threshold is determined to be about 30 J/cm^2. For carious enamel, plasma generation started at roughly 0.1 mJ, i.e. at a threshold density of 15 J/cm^2. In the range of pulse energies investigated, all three ablation curves are mainly linear. Linear regression analysis yields that the corresponding slopes in Fig. 4.32 are 1 μm/0.2 mJ, 3 μm/0.2 mJ, and 8 μm/0.2 mJ, respectively. Thus, the ablation efficiency increases from healthy enamel and healthy dentin to carious enamel. From the ablation volumes, we derive that – at the given laser parameters – approximately 1.5 mm^3 of carious enamel can be ablated per minute. To cope with conventional mechanical drills, a ten times higher ablation efficiency would be desirable. This has meanwhile been achieved by increasing both the pulse energy and repetition rate. So, the Nd:YLF picosecond laser is indeed a considerable alternative to mechanical drills. The potential realization of a clinical laser system is being evaluated.

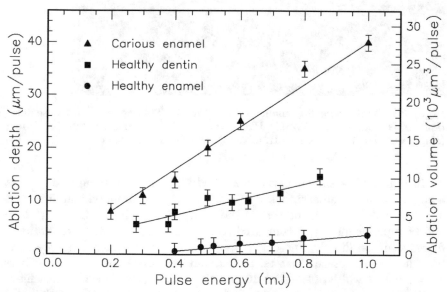

Fig. 4.32. Ablation curves of carious enamel, healthy dentin, and healthy enamel, respectively, obtained with a Nd:YLF laser (pulse duration: 30 ps, focal spot size: 30 μm). Data according to Niemz (1994a) and unpublished data

The results with the Nd:YLF picosecond laser described above have proven that ultrashort laser pulses are a considerable alternative to the mechanical drill for the removal of caries. Due to the tremendous progress in the generation of even shorter laser pulses, femtosecond lasers have become very promising tools, as well. First experiments with these ultrashort pulse durations have been reported by Niemz (1998). The cavity shown in Fig. 4.33 was achieved with 660 000 pulses from a Ti:Sapphire femtosecond laser at a pulse duration of 700 fs. The geometrical accuracy of the cavity and its steep walls are fascinating.

Fig. 4.33. Cavity in healthy enamel achieved with a Ti:Sapphire laser (pulse duration: 700 fs, pulse energy: 100 µJ). Photograph kindly provided by Dipl.-Ing. Bauer (Hannover), Dr. Kasenbacher (Traunstein), and Dr. Nolte (Jena)

In Fig. 4.34, a similar cavity was produced with the same laser but at a larger spacing of adjacent laser pulses. The impacts of individual line scans are clearly visible at the bottom of the cavity. The cavity itself is very clean and of superior quality, if compared to results achievable with conventional diamond drills. Finally, Fig. 4.35 demonstrates the extremely high precision provided by femtosecond lasers. In between exposed areas, unexposed bars remain with a width of less than 10 µm. No mechanical drill is able to achieve similar results. Furthermore, Fig. 4.35 provides the ultimate proof that mechanical shock waves are negligible when applying femtosecond laser pulses at a suitable energy.

Fig. 4.34. Cavity in healthy enamel achieved with a Ti:Sapphire laser (pulse duration: 700 fs, pulse energy: 100 μJ). Photograph kindly provided by Dipl.-Ing. Bauer (Hannover), Dr. Kasenbacher (Traunstein), and Dr. Nolte (Jena)

Fig. 4.35. Cavities in healthy enamel achieved with a Ti:Sapphire laser (pulse duration: 700 fs, pulse energy: 100 μJ). Photograph kindly provided by Dipl.-Ing. Bauer (Hannover), Dr. Kasenbacher (Traunstein), and Dr. Nolte (Jena)

One very important issue associated with dental laser systems is the temperature increase inside the pulp where odontoblasts, blood vessels, and tooth nerves are located. Only increments below 5°C are tolerable, otherwise thermal side effects might occur as discussed in Sect. 3.2. Moreover, the feeling of pain is enhanced at pulp temperatures exceeding approximately 45°C. It is thus very important to remain below these temperatures when striving for clinical applicability. In Fig. 4.36, the temperature increments induced by a picosecond Nd:YLF laser at a repetition rate of 1 kHz are summarized. For this experiment, human teeth were cut into 1 mm thick slices. On one surface of these slices, the laser beam was scanned over a $1 \times 1\,\mathrm{mm}^2$ area, while the temperature was measured at the opposite surface by means of a thermocouple. The observed temperature increments depend on the number of consecutive pulses as well as on the total duration of exposure. We have already stated in Sect. 3.2 that high repetition rates can also induce an increase in temperature even when using picosecond pulses. Hence, a higher temperature is obtained when applying 30 instead of only 10 consecutive pulses before moving the focal spot to the next position. The total duration of exposure also affects the final temperature, although the increase during the first minute is most significant. From these results, we can conclude that up to approximately 10 consecutive pulses may be applied to a tooth at a repetition rate of 1 kHz if the temperature in the pulp shall not increase by more than 5°C.

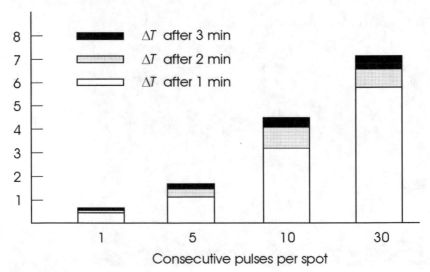

Fig. 4.36. Increase in temperature in a distance of 1 mm from cavities achieved with a Nd:YLF laser (pulse duration: 30 ps, pulse energy: 1 mJ, repetition rate: 1 kHz). Unpublished data

Another useful dental application of lasers has been proposed by Niemz et al. (1995d). Using a confocal laser scanning microscope, the space to be occupied by dental fillings can be very precisely determined. The confocal principle requires that only light reflected within the focal pane is detected. Thereby, very high axial resolutions can be obtained compared to conventional microscopy. With the confocal microscope, several layers of the cavity to be filled are scanned. From the reflected intensities, a three-dimensional plot of the cavity is calculated as shown in Fig. 4.37a. These data can be inverted to form a direct pattern for the milling of inlays or crowns with a CNC-machine as demonstrated in Fig. 4.37b.

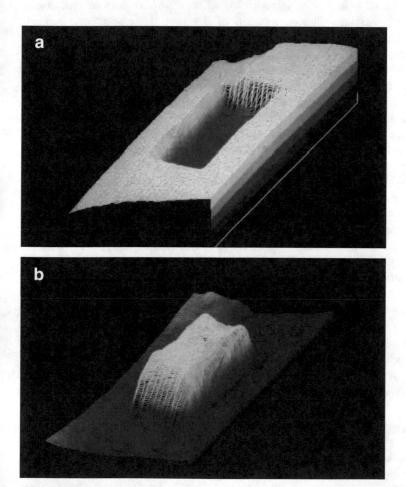

Fig. 4.37. (a) Three-dimensional plot of a laser-induced cavity captured with a confocal laser scanning microscope. The cavity was achieved with a Nd:YLF laser (pulse duration: 30 ps). (b) Inverted image of the same cavity as calculated with a computer. Reproduced from Niemz et al. (1995d). © 1995 Springer-Verlag

Lasers in Endodontics

Endodontics is concerned with the treatment of infections of the root canal. These arise from either a breakthrough of decay into the pulp or from plaque accumulation beneath the gingiva and subsequent bacterial attacks of the root. In either case, once the pulp or the root canal are infected by bacteria, the only treatment is to sterilize both pulp and root, thereby taking into account the associated death of the tooth. However, even a dead tooth may reside in place for years.

The mechanical removal of bacteria, plaque, infected root cementum, and inflammated soft tissues is regarded as an essential part of systematic periodontal treatment. The excavation of the root itself is a very complicated and time-consuming procedure, since roots are very thin and special tools are required. The procedure can be supported by adding antimicrobial chemicals to ensure sterility which is mandatory for a successful treatment. Lasers can improve conventional techniques of endodontics, especially in the removal of plaques and sterilization. First experimental results using CO_2 and Nd:YAG lasers were published by Weichmann and Johnson (1971) and Weichmann et al. (1972). By melting the dentin next to the root, the canal walls appeared to be sealed and thus less permeable for bacteria. Melcer et al. (1987) and Frentzen and Koort (1990) were able to confirm that lasers indeed have a sterilizing effect. Sievers et al. (1993) observed very clean surfaces of the root canal after application of an ArF excimer laser. However, both the CO_2 laser and ArF laser will not gain clinical relevance in endodontics, since their radiation cannot be applied through flexible fibers.

The breakthough in laser endodontics was achieved when Moritz et al. (1997a) suggested to use high power diode lasers for disinfecting the root canal. According to their microbiological examinations, extensive bacterial reduction was achieved in all cases by repeating laser treatment only once. The authors conclude that, compared to conventional techniques, high power diode lasers are highly suitable for killing bacteria in infected root canals. Since then, several groups reported on applications of diode lasers in endodontics. Detailed reviews were published by Kimura et al. (2000) and Stabholz et al. (2004). Lee et al. (2004) and Williams et al. (2006) reported on photo-activated disinfection (PAD) using tolonium chloride and diode lasers.

Laser Treatment of Soft Dental Tissue

Several studies have been reported on the use of a CO_2 laser in the management of malignant, premalignant, and benign lesions of the oral mucosa, e.g. by Strong et al. (1979), Horch and Gerlach (1982), Frame et al. (1984), and Frame (1985). Since the oral environment is very moist, radiation from the CO_2 laser is predestined for such purposes because of its high absorption. When treating soft tissue lesions inside the mouth, the surgeon has a choice of two techniques—either excision or vaporization. Usually, it is preferable to excise the lesion because it enables histologic evidence of its complete removal and a correct diagnosis. During vaporization, a risk always remains that not all altered tissue was eliminated. Hence, if a dentist decides to vaporize a pathologic lesion, a biopsy should be obtained from the adjacent tissue, after the treatment is completed.

The CO_2 laser is particularly useful for very small mucosal lesions. Most of them can be vaporized at a power of 5–10 W in pulsed or CW mode. After laser treatment, the wound is sterile and only minimal inflammatory reactions of the surrounding tissue occur. One major advantage is that there is no need to suture the wound, since small blood vessels are coagulated and bleeding is thus stopped. The wound edges can even be smoothed with a defocused beam. Wound healing usually occurs within a period of two weeks, and the process of reepithelialization is complete after about 4–6 weeks. Frame (1985) states that patients complain of little pain only.

Cases of leukoplakia are difficult to treat by conventional surgery, since they are frequently widespread inside the mouth. The lesion is usually outlined with a focused CO_2 laser beam for easy visualization. Afterwards, it is vaporized with a defocused beam at a power of about 15–20 W. According to Horch (1992), laser-treated leukoplakias heal very well, and there is only little evidence of recurrence. Even leukoplakias on the tongue and lips can be treated without losses in performance of these organs.

Malignant lesions require a higher laser power of approximately 20–30 W to deal with the bulk of the tumor. Lanzafame et al. (1986) state that the recurrence of local tumors is reduced when using the CO_2 laser rather than a mechanical scalpel. The authors claim that the thermal effect of the laser radiation is responsible for this observation. Other treatments of oral cancer include hemangioma and granuloma. Genovese et al. (2010) applied a diode laser at a wavelength of 980 nm to remove oral hemangioma. And Azma and Safavi (2013) used different diode lasers at 810 nm, 940 nm, and 980 nm for excisional biopsy of oral granuloma.

One of the most effective applications of lasers in soft dental tissue is the bacterial reduction in periodontal pockets using high power diode lasers. Moritz et al. (1997b) irradiated a total of 37 patients at a wavelength of 805 nm, an output power of 2.5 W, a pulse duration of 10 ms, and a pulse rate of 50 Hz. The final bacterial counts revealed that irradiation with the diode laser facilitates considerable bacterial elimination, especially of *Actinobacillus actinomycetemcomitans*.

Laser Treatment of Dental Fillings and Alloys

In dental practice, not only tooth substance needs to be ablated but also old fillings have to be removed, for example when a secondary decay is located underneath. For the removal of metallic fillings, infrared lasers cannot be used, since the reflectivity of these materials is too high in that spectral range. Amalgam should never be ablated with lasers at all. In Figs. 4.38a–b, two samples of amalgam are shown which were exposed to a Nd:YLF laser and an Er:YAG laser, respectively. During irradiation, the amalgam has melted and a significant amount of mercury has been released which is extremely toxic for both patient and dentist. For other filling materials, e.g. composites, little data are available. Hibst and Keller (1991) have shown that the Er:YAG laser removes certain kinds of composites very efficiently.

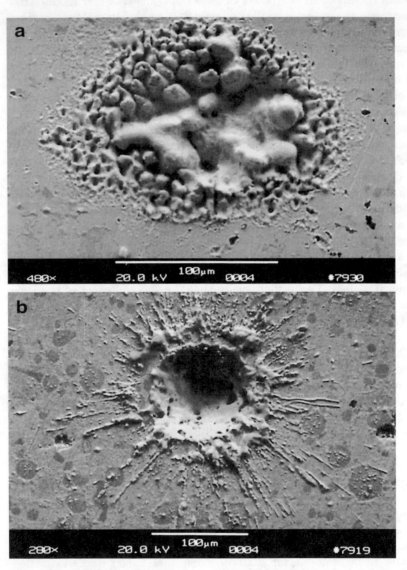

Fig. 4.38. (a) Removal of amalgam with a Nd:YLF laser (pulse duration: 30 ps, pulse energy: 0.5 mJ). (b) Removal of amalgam with an Er:YAG laser (pulse duration: 90 µs, pulse energy: 100 mJ)

Another very interesting topic in dental technology is laser-welding of dental bridges and dentures. It can be regarded as an alternative to conventional soldering. During soldering, the parts to be joined are not melted themselves but are attached by melting an additional substance which, in general, is meant to form a link between them. Laser-welding, on the other hand, attaches two parts to each other by means of transferring them to a plastic or fluid state. According to van Benthem (1992), CO_2 lasers and Nd:YAG lasers are preferably used. Since the reflectivity of metals is very high in the infrared spectrum, it must be assured that either a laser plasma is induced at the surface of the target or that the target is coated with a highly absorbing layer prior to laser exposure.

Dobberstein et al. (1991) state that some laser-welded alloys are characterized by a higher tear threshold than soldered samples as shown in Fig. 4.39. However, van Benthem (1992) argues that such behavior cannot be observed in all alloys, but tear thresholds in laser-welded alloys can definitely reach the same values as in the original cast. According to his studies, the major advantages of laser-welding are:

– higher resistance against corrosion,
– the ability to weld different metals,
– the ability to weld coated alloys,
– lower heat load.

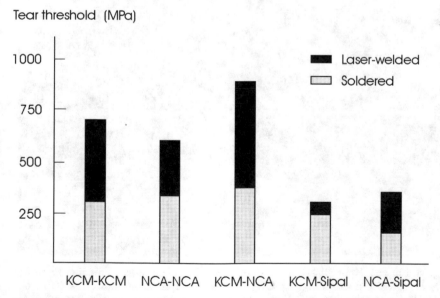

Fig. 4.39. Tear thresholds of laser-welded and soldered dental alloys (KCM: cobalt-based alloy, NCA: nickel-based alloy, Sipal: silver-palladium-based alloy). Data according to Dobberstein et al. (1991)

Lasers in Implantology

Advancements in the development of biomaterials have steadily increased the demand for replacing missing teeth with dental implants. Two of the most common biological complications are called *peri-implant hyperplasia* and *peri-implantitis*. Hyperplasia (from Greek: $\upsilon\pi\epsilon\rho$ = over, $\pi\lambda\alpha\sigma\iota\varsigma$ = formation) is a pathologic increase in organic tissue that results from cell proliferation. It is a common neoplastic response to a stimulus (here: to the implant). Peri-implantitis is a destructive inflammatory process affecting the soft and hard tissues surrounding dental implants. It is often associated with the loss of supporting bone. According to Heitz-Mayfield (2008), peri-implantitis results from an imbalance between bacterial load and host defense. In their review, Zitzmann and Berglundh (2008) reported that peri-implantitis occurs in 12–43 percent of implant sites after 5 years.

Fig. 4.40a–d show the sequence of a laser treatment of a peri-implant hyperplasia. Romanos et al. (2009) applied a CW CO_2 laser at a wavelength of 10.6 μm and a power of 4 W to excise, coagulate, and sufficiently carbonize peri-implant soft tissue. The final result one year after the laser treatment is a very homogeneous, natural tissue surface.

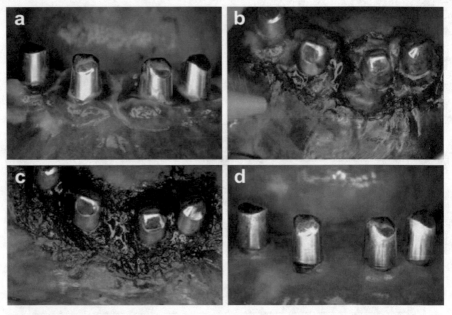

Fig. 4.40. (a) Hyperplastic soft tissue around an implant immediately before CO_2 laser-assisted excision. **(b)** CO_2 laser-assisted excision (power: 4 W) of the hyperplastic peri-implant soft tissue. **(c)** CO_2 laser-assisted coagulation and sufficient carbonization (power: 4 W) of the peri-implant soft tissue. **(d)** Final result one year after laser treatment. Reproduced from Romanos et al. (2009) by permission. © 2009 Springer-Verlag

Management of peri-implantitis involves decontamination of the implant surface, using chemical treatment, mechanical cleaning, or ultrasonic scalers. Unfortunately, all these protocols can damage the implant surface. Romanos et al. (2000) investigated alternatives such as a Nd:YAG laser at a wavelength of 1064 nm and a high power diode laser at 980 nm. They observed that the Nd:YAG laser easily melts titanium discs, whereas the diode laser does not modify the titanium disc surface (see Figs. 4.41a–b).

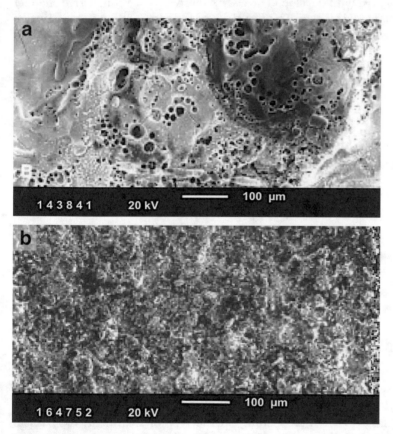

Fig. 4.41. (a) Titanium disc exposed to a pulsed Nd:YAG laser (wavelength: 1064 nm, power: 4 W). **(b)** Titanium disc exposed to a CW diode laser (wavelength: 980 nm, power: 15 W). Photographs kindly provided by Dr. Romanos (Stony Brook)

This result was a breakthrough for high power diode lasers in implantology, since Bach et al. (2000) reported at the same time that these lasers can indeed successfully decontaminate implant surfaces. Yet one important question remains: There are several diode lasers with different wavelengths available—which of these is best suited? Further research will eventually help us find an answer to this challenging question.

4.3 Lasers in Gynecology

Beside ophthalmology, gynecology is another multifaceted discipline for laser applications. This is primarily due to the high success rate of about 93–97 % in treating *cervical intraepithelial neoplasia (CIN)*, i.e. uncommon growth of new cervical tissue, using the CO_2 laser. CIN is the most frequent alteration of the *cervix* and should be treated as soon as possible. Otherwise, cervical cancer is very likely to develop. The cervix represents the connective channel between *vagina* and *uterus*. The locations of the cervix and adjacent tissues are illustrated in Fig. 4.42.

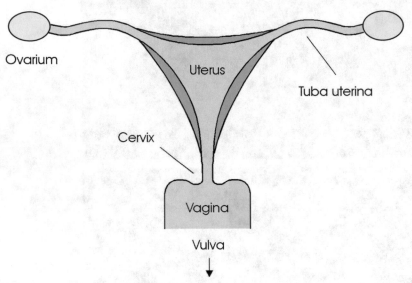

Fig. 4.42. Female reproduction organs

The CO_2 laser is the standard laser in gynecology. Beside treating CIN, it is applied in vulvar intraepithelial neoplasia (VIN) and vaginal intraepithelial neoplasia (VAIN). Depending on the type of treatment, CO_2 lasers can be operated in three different modes – CW radiation, chopped pulse, and superpulse – as shown in Fig. 4.43. Chopped pulses with durations in the millisecond range are obtained from CW lasers when using rotating apertures. Superpulses are achieved by modulating the high voltage discharge. Thereby, pulse durations less than 1 ms can be generated. The peak power is inversely related to the pulse duration. The mean powers of CW radiation and chopped pulses are nearly the same, whereas it decreases in the case of superpulses. As discussed in Sect. 3.2, shorter pulse durations are associated with a reduction of thermal effects. Hence, best surgical results can be obtained by adjusting the mode in which the laser operates.

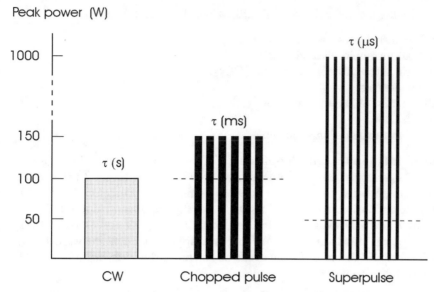

Fig. 4.43. CW, chopped pulse, and superpulse modes of a CO_2 laser. Dashed lines denote mean powers

Beside selecting the temporal mode, the surgeon has to decide whether he applies a focused or defocused mode as shown in Fig. 4.44. Only in tightly focused mode are deep excisions achieved. In partially focused mode, less depth but a larger surface is vaporized. In defocused mode, finally, the power density decreases below the threshold of vaporization, and irradiated tissue is coagulated only.

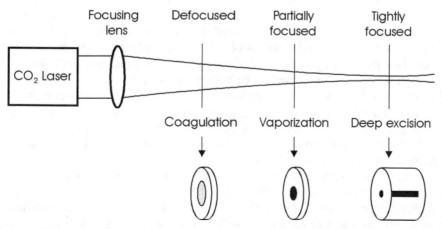

Fig. 4.44. Coagulation, vaporization, and excision modes of a CO_2 laser, depending on a defocused, partially focused, or tightly focused beam

In gynecology, there exist several indications for laser treatment:

- vulvar intraepithelial neoplasia (VIN),
- vaginal intraepithelial neoplasia (VAIN),
- cervical intraepithelial neoplasia (CIN),
- endometriosis,
- obstruction of the uterine tube,
- sterilization,
- twin-twin transfusion syndrome.

Vulvar Intraepithelial Neoplasia (VIN). As already mentioned above, neoplasia generally describes uncommon growth of new tissue. In the case of VIN, intraepithelial tissue of the vulva is significantly proliferated. After histologic examination, the altered tissue is usually vaporized using a CW CO_2 laser in a partially focused mode. According to Baggish and Dorsey (1981), typical power densities of $100\,W/cm^2$ are applied to achieve vaporization depths of about 3–4 mm. In the event of bleeding, the laser beam is immediately switched to a defocused mode.

Vaginal Intraepithelial Neoplasia (VAIN). VAIN is a similar diagnosis as VIN, except that it occurs inside the vagina. Initial trials using the CO_2 laser were reported by Stafl et al. (1977). A more detailed study was provided by Wang et al. (2014). Due to the thinner and more sensitive vaginal tissue, slightly lower power densities are applied. The use of a proper surgical microscope is indicated.

Cervical Intraepithelial Neoplasia (CIN). The tissue at risk for the development of cervical cancer is the columnar epithelium which is located in the transformation zone. This type of epithelium can migrate up and down the endocervical channel. Therefore, it is very important to determine its exact location prior to any treatment. This is usually achieved with a colposcope which essentially is an endoscope specially designed for gynecologic purposes. The extent of columnar epithelium is relatively constant. Thus, the more it is exposed at the ektocervix, the less likely is the existence of diseased tissue inside the channel. In order to exclude any potential inflammation, a biopsy specimen is obtained and a second or third control examination is performed after 3–6 months. According to the histologic evaluation of the biopsy, three grades (CIN I – CIN III) and cervical carcinoma are distinguished, depending on the progress of neoplasia. If the biopsy reveals the presence of cervical carcinoma, a complete resection of the cervix is indicated. At a late stage of cancer, adjacent tissues might have to be removed, as well.

In the case of CIN I, the columnar epithelium is usually located at the ektocervix as shown in Fig. 4.45a. The ablated epithelium is vaporized in a similar fashion as in VIN or VAIN. According to Wright et al. (1983), the procedure should aim at a treated depth of approximately 6 mm. Fast movements of the laser beam cause a more homogeneous distribution of heat and thus reduce the probability of carbonization. Scanning mirror devices are available to assist the surgeon in steadily moving the beam. In Fig. 3.15b, a vaporization (including coagulation and some carbonization) of cervical tissue by a CO_2 laser is shown. Escape of smoke is usually inevitable during surgery, but can be managed with specially designed suction tubes.

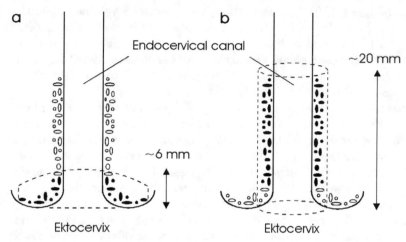

Fig. 4.45. (a) CO_2 laser treatment in the case of CIN I. Diseased epithelium is illustrated by filled ovals. Dashed lines indicate the dome-shaped volume to be vaporized. **(b)** CO_2 laser treatment in the case of CIN II or CIN III. Dashed lines indicate the cylindrical volume to be excised

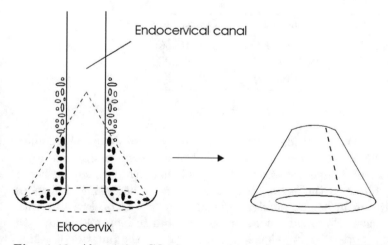

Fig. 4.46. Alternative CO_2 laser treatment in the case of CIN II or CIN III. Diseased epithelium is illustrated by filled ovals. Dashed lines indicate the cone-shaped volume to be excised (*left*). The excised cone is unzipped for histologic examination (*right*)

If CIN II or CIN III is diagnosed, parts of the cervix must be removed to reduce the probability of recurrence. Two different treatment techniques were proposed by Dorsey and Diggs (1979) and Wright et al. (1983). Either surgery is performed only after careful evaluation of the obtained biopsy and determination of the cervical length with a micrometer probe. Dorsey and Diggs (1979) suggest excising a cone-shaped volume out of the cervix with a focused CO_2 laser as demonstrated in Fig. 4.46. The angle of the cone can be adjusted to individual extents of neoplasia. The procedure itself is called *laser conization*. The excised cone is then unzipped for histologic examination as shown in Fig. 4.46. By this means, it can be determined whether all altered tissue has been removed.

An alternative method was proposed by Wright et al. (1983) which is illustrated in Fig. 4.45b. Instead of a cone-shaped volume, a cylindrical volume is removed. The vertical cylindrical excisions are achieved with a focused CW CO_2 laser, whereas the horizontal excision inside the cervix is made using either a mechanical scalpel or the same CO_2 laser but in superpulse mode. After excision of the complete cylinder, the remaining surface is coagulated by defocusing the CW laser beam to achieve local hemostasis. According to Heckmann (1992), cylindrical excisions are better adjusted to individual cases than cone-shaped excisions. In either case, most of the cervical tissue regenerates after being removed. A ten-year review on the treatment of CIN using CO_2 lasers was published by Baggish et al. (1989). In Table 4.5, typical cure rates after one laser treatment are summarized.

Table 4.5. Results of CIN treatment using the CO_2 laser. Data according to Baggish et al. (1989)

	Vaporization	Conization
Number of patients	3070 (100 %)	954 (100 %)
Cured	2881 (93.8 %)	925 (97.0 %)
Persistent	189 (6.2 %)	29 (3.0 %)

According to Wright et al. (1983), laser surgery is an excellent modality for treating CIN if compared to conventional techniques, for example cryotherapy. However, Baggish et al. (1992) argued that high-frequency electric currents can often do the same job as CO_2 lasers. When using thin loops of only 10–15 mm in height, similar thermal damage was observed as with a 40 W CO_2 laser. Profound studies using low voltage loop diathermy had already been reported by Prendeville et al. (1986) for the purpose of taking cervical biopsies. Since electrically induced excisions require less expensive equipment, they might often be the better choice. This is one of those applications where we must balance the pros and cons of lasers.

Endometriosis. The cyclic growth of uterine-like mucosa outside the uterus is called *endometriosis*. It appears as dark "burns", deep nodules, or vesicles. Endometriosis can be either coagulated, vaporized, or excised. Deeply located endometriosis is usually excised rather than vaporized. A scalpel is still necessary to cut off the distal end. Sutton and Hill (1990) reported on the successful use of a CO_2 laser in 228 patients. Meuleman et al. (2009) found that the cumulative recurrence rate of endometriosis was only 2% (or 7%) at 1 year (or 4 years) after surgery.

Obstruction of the Uterine Tube. Tubal obstructions are caused by either adhesions, proliferated growth of tissue, or tubal pregnancies. Adhesions are vaporized using CO_2 or Nd:YAG lasers to obtain a free tubal lumen. In the presence of proliferated growth of tissue, additional openings of the tube are generated in a treatment called *salpingostomy*. Tubal pregnancies are usually managed by either salpingostomy or *salpingectomy*, where salpingectomy denotes the complete removal of one tube. Successful laser treatment of tubal pregnancies was reported by Huber et al. (1989).

Sterilization. A sure way to achieve sterilization is to artifically occlude both uterine tubes. This is performed by either suturing the tube or by coagulating it using a Nd:YAG laser. According to Bailer (1983), safe sterilization is obtained when coagulating both tubes on a length of about 1 cm.

Twin-Twin Transfusion Syndrome. This syndrome is caused by a misplaced shunt vessel between the twins leading to an unbalanced blood supply (see Fig. 4.47). The situation is often lethal to both twins because one of them is overnourished while the other one is undernourished. de Lia et al. (1995), Ville et al. (1995), and Sohn et al. (1996) demonstrated the successful occlusion of the shunt by coagulating it using a Nd:YAG laser. For this indication, there is no alternative to lasers. Only laser radiation can penetrate into the transparent amniotic sac without causing it to burst.

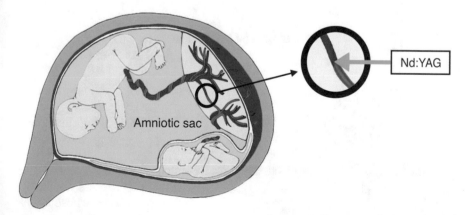

Fig. 4.47. Nd:YAG laser treatment of the twin-twin transfusion syndrome

4.4 Lasers in Urology

The workhorse lasers of urology are CO_2, argon ion, Nd:YAG, and Ho:YAG, but the initial euphoria has meanwhile calmed down—due to the fact that coagulation-based techniques for treating benign prostatic hyperplasia have been abandoned. CO_2 lasers are best in precise cutting of biological tissues. Argon ion lasers and Nd:YAG lasers are used for the coagulation of highly vascularized tumors or malformations. Among these two lasers, the Nd:YAG laser is preferably applied for coagulating large volumes because its radiation deeply penetrates into most of the tissues. The photodisruptive effect of Q-switched Nd:YAG lasers has been investigated in lithotripsy as an alternative to ultrasound fragmentation. Finally, Ho:YAG lasers found very promising applications in both lithotripsy and benign prostatic hyperplasia.

After the development of the first fiberoptic endoscope by Nath et al. (1973), Staehler et al. (1976) performed initial experimental studies with the argon ion laser in urology. Meanwhile, the indications for urologic laser treatments have significantly increased. They extend from the *external genital*, the *lower urinary tract (urethra)*, the *bladder*, the *upper urinary tract (ureter)*, all the way up to the *kidneys* as illustrated in Fig. 4.48. In addition, lasers have been applied to treat benign hyperplasia of the *prostate* which completely embraces the urethra. Various laser therapies for all of these different organs require specific strategies and parameters. They shall now be discussed in the above order.

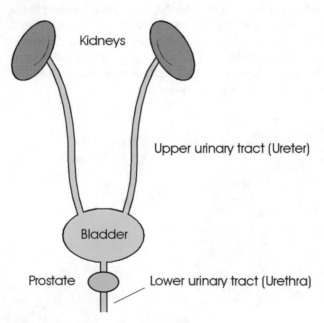

Fig. 4.48. Male urinary tract

The most frequent malformations of the external genital are called *condy-lomata acuminata*. These benign warts must be treated as early as possible, because they tend to be very infectious and degenerating. After circumcision and application of 4% acetic acid to suspected areas, they are coagulated using either Nd:YAG or CO_2 lasers. Occasional recurrences cannot be excluded, especially in the treatment of intraurethral condylomata. With both laser types, however, the rate of recurrence is less than 10% as reported by Baggish (1980) and Rosemberg (1983). Hemangiomas of the external genital should be treated by radiation from Nd:YAG lasers because of their higher penetration depth. Hofstetter and Pensel (1993) stated that additional cooling of the tissue surface may even improve the procedure. Carcinoma of the external genital are best treated using a Nd:YAG laser if they are at an early stage. This significantly reduces the risk of having to perform a partial amputation. According to Eichler and Seiler (1991), powers of 40 W and focal spot sizes of 600 μm are usually applied. At an advanced stage, the tumor is mechanically excised and the remaining tissue surface is coagulated.

Frequent diseases of the lower urinary tract are stenoses induced by either inflammation, tumor growth, or unknown origins. In these cases, *urethrotomy* by endoscopic control is usually performed as proposed by Sachse (1974). During this conventional technique, stenotic material is removed using a cold scalpel. Unfortunately, restenoses often occur due to scarring of the tissue. Further urethrotomies are not of any great help, since they only enhance additional scarring. The first recanalizations of urethral stenoses using an argon ion laser were performed by Rothauge (1980). However, the results obtained were not as promising as initially expected. Then, no further progress was made until Wieland et al. (1993) published first results using a Ho:YAG laser. Follow-up periods of up to 20 months after Ho:YAG laser treatment were reported by Nicolai et al. (1995). They concluded that this technique is a considerable alternative to mechanical urethrotomy in virgin stenoses and restenoses. The probability for the occurrence of laser-induced restenoses is approximately 10% only. In Figs. 4.49a–b, the effects of the Ho:YAG laser on the urethra and ureter are shown, respectively. In either of these samples, thirty pulses with an energy of 370 mJ and an approximate duration of 1 ms were applied.

Tumors of the bladder are very difficult to treat, since they tend to recur after therapy. It is yet unknown whether this is due to metastasis induced either prior to or by the treatment. Unfortunately, bladder tumors also easily break through the bladder wall. Thus, any treatment is successful only if it completely removes the tumor, does not perforate the bladder wall, and does not damage the adjacent intestine. Frank et al. (1982) have compared the effects of CO_2, Nd:YAG, and argon ion lasers on bladder tissue. Among these, the Nd:YAG laser has proven to be best suited in coagulating bladder tumors. Argon ion lasers are applicable only in superficial bladder tumors. According to Hofstetter et al. (1980), the rate of recurrence after laser treat-

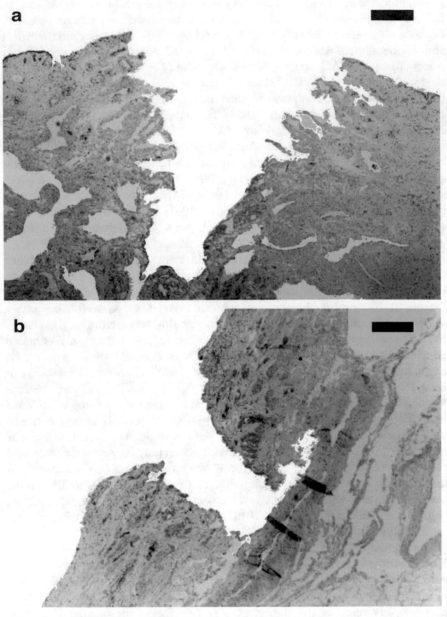

Fig. 4.49. (a) Effect of thirty pulses from a Ho:YAG laser (pulse duration: 1 ms, pulse energy: 370 mJ, bar: 250 μm) on the urethra. **(b)** Effect of thirty pulses from a Ho:YAG laser (pulse duration: 1 ms, pulse energy: 370 mJ, bar: 250 μm) on the ureter. Photographs kindly provided by Dr. Nicolai (Regensburg)

ment is approximately 1–5 %, whereas it ranges from 40–60 % if conventional transurethral resection (TUR) is performed. Even advanced tumors can be efficiently removed by Nd:YAG lasers, since the hemostatic treatment guarantees best vision. Pensel (1986) suggests the application of 30–40 W of laser power and a working distance between 1 mm and 2 mm. The tumor should be irradiated until it visibly pales. Afterwards, coagulated necrotic tissue is mechanically removed. For safety reasons, the remaining tissue surface should be coagulated, as well.

It was emphasized by Hofstetter and Pensel (1993) that tumors can still be graded and staged by biopsies after coagulation. Usually, control biopsies should be obtained within the next 3–6 months. The laser treatment itself is extremely safe, since perforations of the bladder wall are very unlikely and the function of the bladder remains unaffected. All transurethral treatments are performed using a rigid cytoscope and a flexible fiber. In most cases, local anesthetization is sufficient.

Moreover, photodynamic therapy (PDT) has gained increasing significance in the treatment of bladder tumors. First endoscopic applications of HpD had already been investigated by Kelly and Snell (1976). Several clinical reports on PDT were published by Benson (1985), Nseyo et al. (1985), and Shumaker and Hetzel (1987). A complete treatment system including in vivo monitoring and dose control was described by Marynissen et al. (1989). A list of potential complications arising when using dihematoporphyrin ether was provided by Harty et al. (1989). Today, photodynamic therapy is considered as a useful supplement to other techniques, since it enables the resection of tumors which are not visible otherwise. The ability to simultaneously diagnose – by means of laser-induced fluorescence – and treat tumors surely is one of the key advantages of photodynamic therapy. Senior (2005) showed in a phase I study of 24 patients with recurrent superficial bladder cancer that a combination therapy with sequential mitomycin and 5-aminolaevulinic acid (ALA) is safe, well tolerated, and effective. Bader et al. (2013) reported on a therapy of bladder cancer using hexaminolevulinate (HAL) as photosensitizer. An excellent review on treating bladder cancer using 5-aminolaevulinic acid was published by Inoue (2017).

Lithotripsy of urinary calculi is often based on ultrasound techniques. However, not all calculi are equally indicated for such an external therapy. In particular, those calculi which are stuck inside the ureter are in an extremely inconvenient location. In these cases, laser-induced lithotripsy offers the advantage of directly applying energy to the vicinity of the calculus by means of a flexible fiber. First experiments regarding laser lithotripsy had already been performed by Mulvaney and Beck (1968) using a ruby laser. From today's perspective, though, it is quite obvious that these initial studies had to be restricted to basic research, since they were associated with severe thermal side effects. Watson et al. (1983) first proposed the application of a Q-switched Nd:YAG laser. Shortly after, pulsed dye lasers were investigated by Watson et al. (1987). With the decrease in pulse durations, additional complications arose concerning induced damage of the fiber. Extensive calculations on the limits of fiber transmission were published by Hering (1987). The advantages of different approaches like bare fibers or focusing fiber tips were discussed by Dörschel et al. (1987) and Hofmann and Hartung (1987). In a more recent study, even the feasibility of a femtosecond laser lithotripsy was investigated by Qiu et al. (2010).

Currently, best results are achieved with Ho:YAG lithotripters. In their review, Kronenberg and Somani (2018) conclude that complications are rare, making laser lithotripsy one of today's safest tools in urology. A detailed description of the optimal power settings was provided by Sea et al. (2012). At low pulse energies (0.2 J), less fragmentation occurs and small fragments are produced. At high pulse energies (2.0 J), more fragmentation occurs with larger fragments. Optimal lithotripsy laser dosimetry depends on the desired outcome. The diameter of the optical fiber varies between 200 μm and 600 μm. Interestingly, Vassar et al. (1999) showed that Ho:YAG lithotripsy is based on photothermal interaction, since it does not create significant shock waves. Peak pressures were reported to be less than 2 bars.

Since the 1980s, research in urology has increasingly focused on various treatments of the prostate. This very sensitive organ embraces the urethra. Diseases of the prostate, e.g. benign hyperplasias or carcinoma, thus often tend to handicap the discharge of urine. A profound analysis of the development of benign prostatic hyperplasia (BPH) was given by Berry et al. (1984). Several conventional therapies are available, e.g. the initial application of phytopharmaka or transurethral resection in severe cases. Other techniques such as cryotherapy or photodynamic therapy have also been investigated, e.g. by Bonney et al. (1982) and Camps et al. (1985). A complete list of potential treatment methods was provided by Mebust (1993). During the first few years, research was restricted to the treatment of prostatic carcinoma. Böwering et al. (1979) were the first to investigate the effect of Nd:YAG laser radiation on tumors of the prostate. Shortly after, several detailed reports followed, e.g. by Sander et al. (1982) and Beisland and Stranden (1984). The latter study pointed out the extreme importance of a continuous temperature monitoring of the adjacent rectum. Extensive clinical results were reported by McNicholas et al. (1988).

At the beginning of the 1990s, patients' demand for minimally invasive techniques significantly increased. In the treatment of BPH, two milestones were achieved with the development of improved surgical techniques called *transurethral ultrasound-guided laser-induced prostatectomy (TULIP)* and *laser-induced interstitial thermotherapy (LITT)*. The idea of TULIP was proposed by Roth and Aretz (1991) and Johnson et al. (1992). Detailed clinical results were published by McCullough et al. (1993). The key element of TULIP is to position a 90° prism inside the urethra by ultrasound control. Thereby, the precision in aiming at the target is strongly enhanced.

In other studies, Siegel et al. (1991) have shown that hyperthermia alone, i.e. temperatures up to 45°C, is not sufficient for treating BPH. This finding led to the idea of LITT as previously described in Sect. 3.2. During LITT, biological tissue is coagulated, which means that temperatures above 60°C are obtained. The technical realization of suitable ITT fibers was discussed by Hessel and Frank (1990). In urology, initial experimental results with LITT were published by McNicholas et al. (1991) and Muschter et al. (1992). With typical laser powers of 1–5 W, coagulation volumes with diameters of up to 40 mm were achieved.

But, as already mentioned before, the initial euphoria regarding LITT did not last very long. Hoffman et al. (2003) and Laguna et al. (2003) observed the occurrence of severe adverse events such as early postoperative dysuria, urgency, and the need for prolonged catheterization. To make it even worse, Daehlin and Frugard (2007) reported in a study with a 46–61 month follow-up that half of the coagulated patients needed retreatment. Because of the unacceptable adverse events and the high retreatment rate, coagulation-based techniques for treating BPH have meanwhile been abandoned. Instead, it is again the Ho:YAG laser that is being used—not for coagulating, but for ablating or for enucleating part of the prostate. The two procedures are called *holmium laser ablation of the prostate (HoLAP)* and *holmium laser enucleation of the prostate (HoLEP)*, respectively.

HoLAP was introduced by Gilling et al. (1995) and Gilling et al. (1996). The first 7-year follow-up was published by Tan et al. (2003). HoLAP surgery is performed under local anesthesia. A Ho:YAG laser is used to ablate prostate tissue that blocks urine flow. Severe complications were not observed so far. First results with HoLEP were published by Gilling and Fraundorfer (1998) and Mackey et al. (1998). Later on, a technical update on HoLEP was provided by Kuo et al. (2003). From today's perspective, both HoLAP and HoLEP are equally effective and safe. HoLEP requires greater experience and training to perform. It also allows treatment of larger sized prostates.

But how do HoLAP and HoLEP compare with *transurethral resection of the prostate (TURP)*—a mechanical procedure that most urologists still consider to be the gold standard for treating BPH? Mottet et al. (1999) showed in a randomized prospective study that the efficacy of both HoLAP and TURP is about the same. When comparing HoLEP with TURP, Westenberg et al. (2004) observed no differences in terms of peak flow, potency, and continence. Gravas et al. (2011) found that HoLAP and TURP show a similar improvement regarding peak flow and the international prostate symptom score (IPSS). Michalak et al. (2015) stated that patients undergoing HoLEP benefit from a shorter catheterization time, shorter hospitalization, and fewer complications. And in a network meta-analysis involving 36 studies with 3831 patients, Zhang et al. (2016) concluded that holmium lasers generally achieve a better peak urinary flow and less postvoid residual volume than any other lasers or even TURP.

4.5 Lasers in Neurosurgery

Neurosurgery deals with diseases of the central nervous system (CNS), i.e. the brain and the spine. Surgery of brain tumors is very difficult, since extremely localized operations are necessary due to the complicated structure and fragility of the brain. Moreover, the tumor itself is often not easily accessible, and very important vital centers are situated beside it. Therefore, it is not surprising that a considerable amount of research funds is currently being spent in this field, especially since any kind of brain tumor – even benign tumors – are extremely life-threatening. This is because space inside the skull is very limited. Hence, growth of new tissue increases the pressure inside the brain which leads to mechanical damage of other neurons. A schematic cross-section of the human brain is shown in Fig. 4.50.

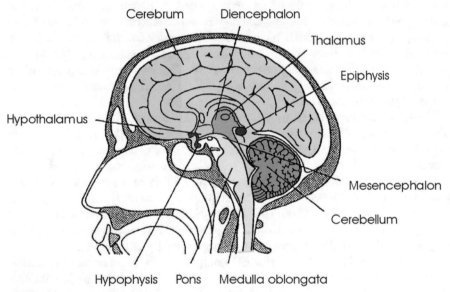

Fig. 4.50. Human brain

The major parts of the brain are the cerebrum, diencephalon, cerebellum, and brainstem. The diencephalon can be further divided into the hypothalamus, hypophysis, thalamus, and epiphysis, whereas the brainstem consists of the mesencephalon, pons, and medulla oblongata. Usually, tumors of the brainstem are highly malignant and, unfortunately, they reside in an inaccessible location. In general, brain tissue can be divided into *gray matter* and *white matter* which are made up of cell nuclei and axons, respectively. Blood perfusion of gray and white matter differs remarkably. The corresponding ratio is about five to one.

The application of lasers in neurosurgery has been extremely slow compared with other medical fields, e.g. ophthalmology. This was mainly due to two reasons. First, studies by Rosomoff and Caroll (1966) revealed that the ruby laser was not of great help in neurosurgery. Second, initial experiments with the CO_2 laser were performed at too high energy levels, e.g. by Stellar et al. (1970), which was dangerous and completely unnecessary. It then took some time until Ascher (1979), Beck (1980), and Jain (1983) reawakened interest in neurosurgical lasers, especially moderate CO_2 lasers and Nd:YAG lasers. The principal advantages of lasers in neurosurgery are evident. Lasers are able to cut, vaporize, and coagulate tissue without mechanical contact. This is of great significance when dealing with very sensitive tissues. Simultaneous coagulation of blood vessels eliminates dangerous hemorrhages which are extremely life-threatening when occurring inside the brain. Moreover, the area of operation is sterilized as lasing takes place, thereby reducing the probability of potential infections.

According to Ascher and Heppner (1984) and Stellar (1984), the main advantage of the CO_2 laser is that its radiation at a wavelength of 10.6 μm is strongly absorbed by brain tissue. By this means, very precise cuts can be performed. However, CO_2 lasers are not appropriate for the coagulation of all blood vessels. In particular, arteries and veins with diameters > 0.5 mm tend to bleed after being hit by the laser beam. Nd:YAG lasers, on the other hand, are effective in coagulating blood vessels as stated by Fasano et al. (1982) and Wharen and Anderson (1984b). Ulrich et al. (1986) even observed very good results on both ablation and coagulation when combining a Nd:YAG laser emitting at 1.319 μm and a 200 μm fiber. The biological response of brain tissue to radiation from Nd:YAG lasers was extensively studied by Wharen and Anderson (1984a). A preliminary report on the clinical use of a Nd:YAG laser was given by Ascher et al. (1991). Moreover, neurosurgical applications of argon ion lasers had been investigated by Fasano (1981) and Boggan et al. (1982), but they seem to be rather limited, since radiation from these lasers is strongly scattered inside brain tissue.

The main problem with CW lasers is that they do not remove brain tumors but only coagulate them. Necrotic tissue remains inside the brain and can thus lead to the occurrence of severe edema. Moreover, adjacent healthy tissue might be damaged due to heat diffusion, as well. In the 1990s, two alternative lasers have been investigated concerning their applicability to neurosurgery: Er:YAG and Nd:YLF lasers. Cubeddu et al. (1994) and Fischer et al. (1994) have studied the ablation of brain tissue using free-running and Q-switched Er:YAG lasers. They observed limited thermal alterations of adjacent tissue. However, mechanical damage was very pronounced. Since the Er:YAG laser emits at a wavelength of 2.94 μm, its radiation is strongly absorbed in water as already discussed in Sect. 3.2. Thus, soft brain tissue – having a high water content – is suddenly vaporized which leads to vacuoles inside the tissue with diameters ranging up to a few millimeters. In Fig. 4.51a, mechanical damage

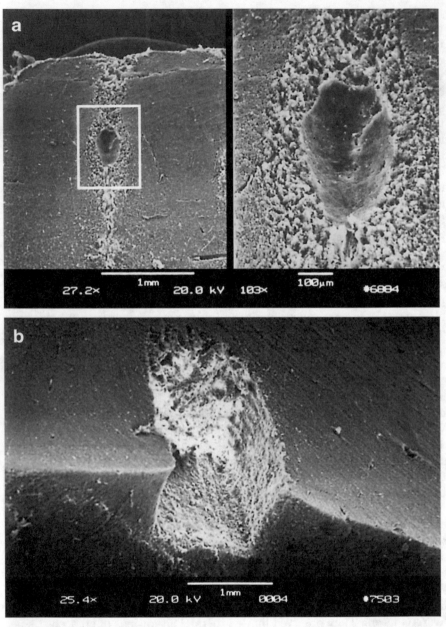

Fig. 4.51. (a) Brain tissue after exposure to an Er:YAG laser (pulse duration: 90 μs, pulse energy: 60 mJ). Mechanical damage is evident. (b) Voluminous ablation of brain tissue achieved with the same laser. Reproduced from Fischer et al. (1994) by permission. © 1994 Springer-Verlag

up to a depth of at least 1.5 mm is clearly visible. Therefore, it is not very helpful that even large volumes of brain tissue can be ablated using Er:YAG lasers as shown in Fig. 4.51b.

The ablation of brain tissue using a picosecond Nd:YLF laser system was investigated by Fischer et al. (1994). In Fig. 4.52, the ablation depths of white and gray brain matter are given, respectively. Obviously, there is no significant difference in ablating either substance. It is interesting to observe, though, that there is no saturation in ablation depth even at energy densities as high as $125 \, \text{J/cm}^2$. Thus, higher laser powers will probably enable ablation depths $> 200 \, \mu\text{m}$. Fischer et al. (1994) state that the corresponding ablation threshold is at approximately $20 \, \text{J/cm}^2$.

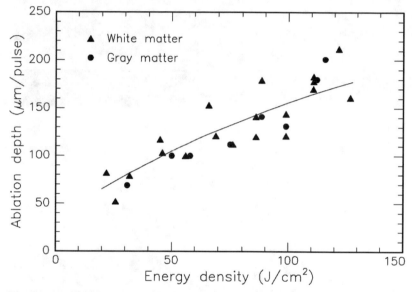

Fig. 4.52. Ablation curve of white and gray brain matter obtained with a Nd:YLF laser (pulse duration: 30 ps, focal spot size: 30 μm). Data according to Fischer et al. (1994)

Two samples of brain tissue which were exposed to the Nd:YLF laser are shown in Figs. 4.53a–b. A rectangular ablation geometry was achieved by scanning the laser beam. The lesion in Fig. 4.53a is characterized by steep walls and is approximately 600 μm deep. In Fig. 4.53b, a histologic section of an ablation edge is shown as obtained with the Nd:YLF laser. The tissue was stained with cluever barrera to visualize any thermal effects. There is no evidence of either thermal or mechanical damage to adjacent tissue. Hence, removal of tissue can be attributed to the process of plasma-induced ablation as described in Sect. 3.4. Non-thermal ablation of tissue is a mandatory requirement for precise functional surgery of the brain.

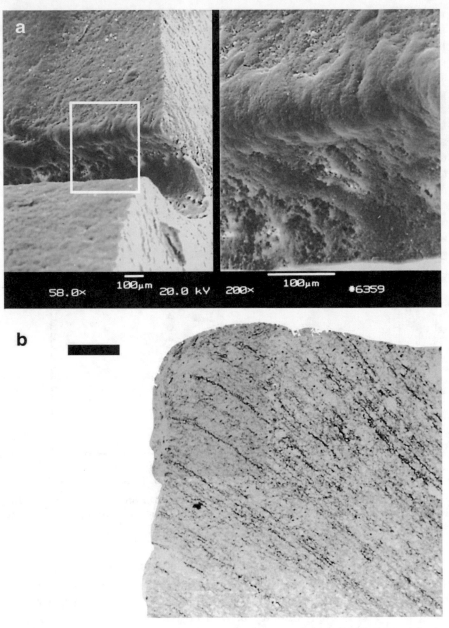

Fig. 4.53. (a) Brain tissue after exposure to a picosecond Nd:YLF laser (pulse duration: 30 ps, pulse energy: 0.5 mJ). **(b)** Histologic section of brain tissue after exposure to the same laser (bar: 50 μm). Reproduced from Fischer et al. (1994) by permission. © 1994 Springer-Verlag

A very precise technique is called *stereotactic neurosurgery* which was described in detail by Kelly et al. (1982). It requires a so-called stereotactic head ring made of steel or carbon fibers which is tightly fastened to the patient's skull by several screws. This ring defines a coordinate system which serves as a valuable means of orientation during surgery. The head ring appears on CT or magnetic resonance images (MRI) and thus determines the coordinates of the tumor. Various kinds of aiming devices can be mounted to the ring allowing for precise operation in all three dimensions. The main goal of stereotactic neurosurgery is to plan a suitable penetration channel in advance of surgery, set its coordinates with respect to the head ring, and then keep this channel during surgery. By this procedure, the risk of hitting a vital center within the brain can be significantly reduced, and the success of a treatment becomes more predictable.

The concept of stereotactic laser–neurosurgery according to Bille et al. (1993) is illustrated in Fig. 4.54. By means of a stereotactic head ring, a laser probe is inserted into the brain. CT and NMR data are used to correctly position the distal end of the probe inside the tumor. A schematic drawing of the laser probe is given in Fig. 4.55. It basically consists of a conical tube which contains a rotating mirror at its distal end, a movable focusing lens, and additional channels for aspiration and rinsing. Aspiration is necessary to maintain a constant pressure at the site of operation. The laser probe is rinsed to remove debris from the rotating mirror and to increase the efficiency of the ablation process. The rotating mirror deflects the laser beam perpendicularly to the axis of rotation. Tissue is thus ablated in cylindrical layers as shown in Fig. 4.55. If a confocal laser scanning microscope is integrated into this system, blood vessels can automatically be detected during surgery as illustrated in Fig. 4.56.

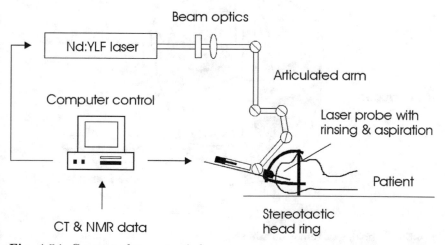

Fig. 4.54. Concept of stereotactic laser–neurosurgery

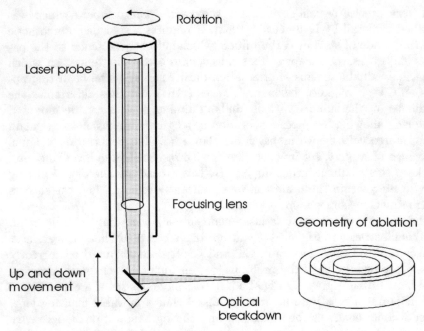

Fig. 4.55. Schematic drawing of laser probe

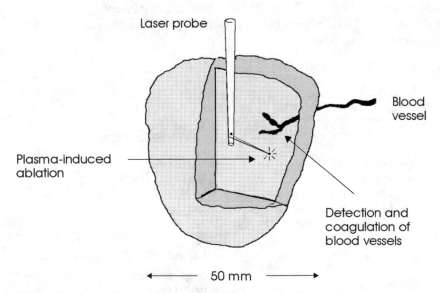

Fig. 4.56. Topology of tumor ablation

It should be mentioned that stereotactic neurosurgery is already a well-established clinical discipline. Stereotactic techniques are not only applied in combination with lasers but with alternative therapies, e.g. insertion of radioactive seeds (^{60}Co or ^{125}J) and high-frequency coagulation, as well. In general, any of these procedures alone might not lead to a complete necrosis of all tumor cells. In these cases, however, the stereotactic concept provides a useful combination of several treatment techniques by simply exchanging the surgical equipment mounted to the head ring of the patient. The principal advantages of stereotactic surgery, of course, are its high precision – within tenths of a millimeter – in aiming at the tumor and the ability to manage surgery with a tiny hole in the skull of less than 1 cm in diameter[8]. Stereotactic surgery thus certainly belongs to the favored treatments of minimally invasive surgery (MIS).

An exciting technique of sutureless microvascular anastomosis using the Nd:YAG laser was developed by Jain (1980). It is performed in some cases of cerebrovascular occlusive disease. During anastomosis, a branch of the superficial temporal artery is connected to a cortical branch of the middle cerebral artery. Typical laser parameters are powers of 18 W, focal spot sizes of 0.3 mm, and single exposure durations of 0.1 s. This method is considerably faster than conventional suture techniques, it does not induce damage to the endothelium of the vessel, and it can be performed on relatively small and/or deeply located blood vessels, as well. The mechanism of vessel welding is not completely understood but is believed to rely on heat-induced alterations in collagen of the vessel. First clinical results had already been reported by Jain (1984a), but a high rate of associated complications soon slowed down the initial euphoria. Later, Neblett et al. (1986) and Ulrich et al. (1988) combined the application of a Nd:YAG laser with conventional techniques of anastomosis, and they achieved more promising results. In blood vessels with diameters of 0.8–1.2 mm, neither short-term nor long-term complications occurred.

Spinal surgery is the other principal field of neurosurgical treatments. According to Jain (1984b), the CO_2 laser has proven to be useful in treating tumors of the spinal cord. Such tumors can be coagulated without severe complications. Ascher and Heppner (1984) have reported on the successful dissection of intramedullary gliomas of the spinal cord with a pulsed CO_2 laser. Moreover, some basic procedures concerning pain relief of the spinal cord can be performed using this laser. Spinal laser surgery is still in its infancy, and considerable progress is expected within the next few years when miniaturized surgical instruments become available, e.g. in the technique of laser-assisted nerve repair as already proposed by Bailes et al. (1989). The combination of highly sophisticated endoscopes and appropriate laser systems might then turn into a powerful joint venture.

---

[8] Conventional craniotomies usually require openings in the skull of at least 5 cm in diameter.

4.6 Lasers in Angioplasty and Cardiology

Angioplasty is concerned with the treatment of blood vessels which are narrowed by atherosclerosis[9]. The obstructions stem from the formation of an anorganic plaque inside the vessels which reduces or even completely suppresses blood flow. The degree of a so-called *stenosis* is determined by

$$\% \text{ stenosis} = 100 \; \frac{\text{intimal area}}{\text{intimal area} + \text{lumen area}} \; ,$$

where the areas are obtained from a cross-sectional view of the vessel, the intima is the interior wall of the vessel, and the lumen is the space available for blood flow.

The promotion factors of plaques are not completely understood. The formation of a plaque might be favored at sites of a local vessel injury where cells capable of repairing the vessel wall tend to gather. If some of the secreted products of these cells are not carried away, a plaque is formed. After the cells have died, primarily anorganic concrements with a high content of calcium are left behind.

The non-surgical treatment of atherosclerosis was introduced by Dotter and Judkins (1964) when performing angioplasty of femoral arterial stenoses with a specially designed dilation catheter. In the 1970s, Grüntzig (1978) and Grüntzig et al. (1979) modified this technique to enable its application in coronary arteries, i.e. the blood vessels supplying the heart itself. Atherosclerotic plaques inside these vessels are extremely life-threatening, since their obstruction necessarily induces myocardial infarction. *Percutaneous transluminal coronary angioplasty (PTCA)* has since been used in many patients with angina or acute myocardial infarction, and a large variety of balloon catheters is available today. In general, the method of PTCA has been widely accepted, and several profound reviews have been published, e.g. by Grüntzig and Meier (1983) and Landau et al. (1994).

The main mechanisms by which PTCA increases the size of the vessel lumen are cracking, splitting, and disruption of the atherosclerotic plaque. Resorption of plaque material is also initiated by simply pushing it into the vessel wall. All these effects are evoked by inflation of a balloon placed inside the blood vessel. According to Waller (1983), balloon inflation may be deleterious, however, causing plaque hemorrhage, extensive dissection of the vessel wall, and thrombus formation. Therefore, the treatment needs to be performed extremely carefully and under X-ray control. Aspirin and heparin are usually administered to reduce the incidence of thrombosis.

Although PTCA is generally safe, some short-term and long-term complications do occur. Among the first, arterial dissection and intracoronary thrombosis are most severe. On the other hand, a recurrence of the original stenosis may take place even months after the treatment. This process of

---

[9] The term *arteriosclerosis* applies in arteries only.

so-called *restenosis* has been extensively studied by Lange et al. (1993) who also considered it as the "Achilles' heel" of coronary angioplasty. Serruys et al. (1988) reported on the occurrence of restenoses in 30 % of patients treated with PTCA. Restenoses are believed to be initiated by accidental injury of the vessel wall, resulting in the subsequent release of thrombogenic, vasoactive, and mitogenic factors. Endothelial damage, in particular, leads to the activation of macrophages and smooth muscle cells as stated by Austin et al. (1985). Thereby, growth factors are released which in turn may promote their own synthesis. Thus, a self-perpetuating process is initiated which is associated with a thickening of the intima, i.e. the interior part of the vessel wall. Finally, the vessel is obstructed again. Since the occurrence of these restenoses is not predictable, extensive follow-up studies are usually performed. According to Hombach et al. (1995), there is a slight decrease in the probability of restenoses when implanting specially designed mechanical stents inside the vessel wall immediately after balloon dilation.

Beside PTCA, other surgical treatments are available in cases of coronary arteriosclerosis. These are *bypass surgery, atherectomy*, and *high-frequency rotational coronary angioplasty (HFRCA)*. Among these, only bypass surgery is performed during complete anesthetization. It is a very complicated type of surgery, since the chest must be opened and the heart beat is interrupted. Atherectomy is a more rigorous version of PTCA, where the plaque is additionally planed away by means of mechanical abrasion. Finally, in HFRCA, a miniaturized mechanical drill called a rotablator is used for vessel recanalization as described by Tierstein et al. (1991).

First experiments regarding laser angioplasty were performed by Macruz et al. (1980), Lee et al. (1981), Abela et al. (1982), and Choy et al. (1982). While these in vitro studies left no doubt that laser light could ablate atherosclerotic plaque, it was quite uncertain whether such a treatment could be transferred to in vivo surgery. Choy et al. (1984) and Ginsburg et al. (1985) were the first to try clinical laser angioplasties. Laser light was applied to the plaque by means of optical fibers. However, only thermally acting lasers – i.e. argon ion, CO_2, and Nd:YAG lasers – were investigated at that time which induced severe thermal injuries such as extensive coagulation, necrosis of vascular tissue, and perforation of the vessel wall. In addition, mechanical perforations often occurred due to the bare distal end of the optical fibers. All these complications turned out to be extremely critical when applying laser angioplasty to coronary arteries as initially suggested by Selzer et al. (1985) and Sanborn et al. (1986).

It was Hussein (1986) who developed a novel tip design, the so-called *hot tip*. It consists of a simple metal cap which completely encloses the distal end of the fiber, thereby converting all laser energy to heat by means of absorption. Instead of using a tightly focused laser beam, plaques are removed by homogeneously distributed heat as shown in Fig. 4.57. Usually, CW argon ion lasers and Nd:YAG lasers are applied, although any kind of laser radiation could be used which is absorbed by the metal cap. Detailed measurements of the temperature distribution associated with various parameters were performed by Labs et al. (1987). According to Cumberland et al. (1986), the concept of using a hot tip diminishes the incidence of thermal perforations to a degree that is acceptable for peripheral angioplasty. However, the method of thermal angioplasty has led to considerable controversy, as well. Other groups, for example Diethrich et al. (1988), observed the occurrence of so-called vasospasm – i.e. thermally induced shrinkage of the vessel wall – when using a hot tip applicator. In general, these vasospasms are not predictable, and they can induce severe secondary obstructions.

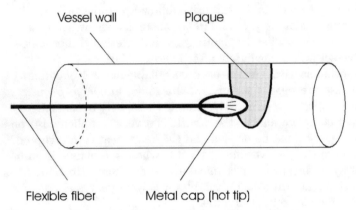

Fig. 4.57. Laser-driven hot tip for vessel recanalization

The common denominator of the above concerns is that laser-induced thermal injury is a virtually unavoidable by-product of CW lasers. Basic improvement could not be achieved until moving from one laser–tissue interaction to another. As discussed in Sect. 3.2, only pulsed lasers with pulse durations shorter than 1 µs provide non-thermal ablation.

Grundfest et al. (1985) first demonstrated that pulsed XeCl excimer lasers are capable of performing efficient plaque ablations with only minimal thermal injury of adjacent tissue. These studies were done shortly after the description of the photoablative interaction mechanism by Srinivasan and Mayne-Banton (1982). Thus, it was straightforward that researchers also focused on other applications for excimer laser radiation. Unfortunately, though, an unpredictable type of complication occurred as discussed below which soon slowed down initial enthusiasm. Karsch et al. (1989) were the first to report on clinical results of percutaneous coronary excimer laser angioplasty. In this study, thirty patients were treated using a 1.3 mm laser catheter consisting of twenty 100 µm quartz fibers. These fibers were located concentrically around a 0.35 mm thick flexible guide wire as shown in Fig. 4.58.

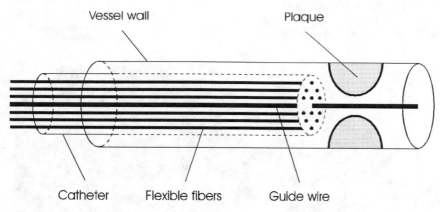

Fig. 4.58. Laser angioplasty for vessel recanalization

After moving the guide wire into the coronary artery, the catheter was guided into the correct position. The fibers were coupled to a XeCl excimer laser emitting at a wavelength of 308 nm and pulse durations of 60 ns. Typical energy densities of up to $5\,J/cm^2$ were applied. The mean percentage of stenosis fell from 85 % initially to 41 % immediately after laser treatment, and the primary success rate was as high as 90 %. In twenty patients, subsequent balloon dilation was additionally performed. Perforations of the vessel wall did not occur in any of the patients. However, it was only shortly after when Karsch et al. (1991) published a second report admitting that one patient suddenly died two months after laser treatment. Postmortem histologic examination proved that a severe restenosis had occurred which had led to an acute myocardial infarction.

In Figs. 4.59a–b, two photographs are shown demonstrating the removal of an atherosclerotic plaque using a XeCl excimer laser. For this ablation, Hanke et al. (1991) have applied pulse durations of 60 ns at a repetition rate of 20 Hz. An enlargement of the plaque itself is captured in Fig. 4.59a. On the right half of the picture shown in Fig. 4.59b, parts of the plaque have already been removed without injuring the vessel wall.

Today, it is well accepted that restenoses are extremely pronounced following excimer laser angioplasty. Their occurrence can be attributed to an enhanced proliferation of smooth muscle cells as has been demonstrated by Hanke et al. (1991). Most of these cells undergo DNA synthesis during two weeks after laser treatment, resulting in intimal thickening within the first four weeks. Obviously, the mechanism of photoablation is more stimulating than only mechanical cracking or abrasion. Thus, even though photoablation is a rather gentle technique for removing plaques, its long-term effects forbid its use for the purpose of vessel recanalization. Therefore, excimer laser angioplasty is generally being rejected today, and it is rather doubtful whether it will ever gain clinical relevance.

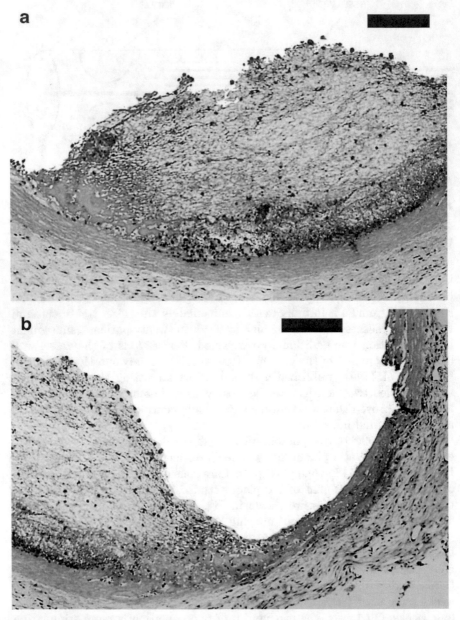

Fig. 4.59. (a) Histologic section of atherosclerotic plaque inside a blood vessel (bar: 150 µm). The vessel wall is located at the *bottom* of the picture. **(b)** Ablation of atherosclerotic plaque with a XeCl excimer laser (pulse duration: 60 ns, repetition rate: 20 Hz, bar: 150 µm, plaque: *left*, ablation: *right*). Photographs kindly provided by Dr. Hanke (Tübingen)

Meanwhile, other laser types have also been investigated concerning their application in angioplasty and cardiology. One of them is the Ho:YAG laser which has been studied in detail by Hassenstein et al. (1992). However, an extensive increase in intimal thickening was observed within the first six weeks after laser treatment. And, again, the proliferation of smooth muscle cells seems to be responsible for this effect. Thus, the clinical use of holmium laser angioplasty appears to be extremely limited.

More promising are CO_2 laser systems which can be used to create additional channels for the blood supply of the heart. These channels originate from the epicardium, i.e. the periphery of the heart, and remain open after laser treatment. This technique is called *transmyocardial laser revascularization (TMLR)* and was initially proposed by Mirohseini et al. (1982). Shortly after, first clinical experiences were reported by Mirohseini et al. (1986). Yano et al. (1993) have confirmed the effect of revascularization. Other investigators, though, could not verify their results, e.g. Whittaker et al. (1993). Horvath et al. (1995) were able to judge treatment effects by measuring the local contractility of the heart muscle. They concluded that acute infarcts treated by TMLR show improved contractility both in the short- and long-term. Cooley (1996) achieved TMLR applying the CO_2 laser at an extremely high power of 800 W. Horvath (2008) provided a profound review of various TMLR studies.

Even if laser treatment of blood vessels should fail to become a safe procedure, laser diagnostics will always play a significant role in angioplasty and cardiology. Apart from X-ray and ultrasound angiography, Doppler angiography and laser endoscopy are very sensitive techniques. A typical example of a laser endoscope is shown in Fig. 4.60. Visible laser radiation is emitted from the distal end of a flexible fiber and illuminates the area of interest. Modern engineering science has meanwhile enabled the design of extremely miniaturized and highly sophisticated laser endoscopes.

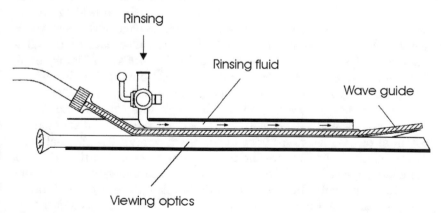

Fig. 4.60. Laser endoscope for angioplasty

4.7 Lasers in Orthopedics

Progress in surgical medicine is often related to an improved technique of performing *osteotomies*, i.e. bone excisions. Standard tools in orthopedics are saws, milling-machines, and mechanical drills. All of them operate in contact mode and possibly induce severe mechanical vibrations and hemorrhage. It is thus straightforward to ask whether lasers might represent a considerable alternative in orthopedic surgery.

Bone fulfills three major functions: mechanical support of the body, protection of soft tissues, and supply of minerals and blood cells. The hardness of bone results from a complex structure of hydroxyapatite, water, soluble agents, collagen, and proteins. The chemical composition of human bone is listed in Table 4.6. The high water content is responsible for strong absorption of infrared radiation. Therefore, CO_2, Er:YAG, and Ho:YAG lasers are predestined for the efficient treatment of bone.

Table 4.6. Mean composition of human bone

Matter	Percentage	Constituent
Anorganic	50 – 60 %	Hydroxyapatite
	15 – 20 %	Water
	5 %	Carbonates
	1 %	Phosphates
Organic	20 %	Collagen
	1 – 2 %	Proteins

In the 1970s, Moore (1973), Verschueren and Oldhoff (1975), and Clayman et al. (1978) reported on osteotomies performed with CO_2 lasers. Extensive studies on bone healing were published by Gertzbein et al. (1981) and Pao-Chang et al. (1981). All researchers agreed on a delayed healing process compared with conventional osteotomies. Thermal damage of the bone rim is exclusively made responsible for this time delay. Detailed data on the ablation characteristics were given by Kaplan and Giler (1984) and Forrer et al. (1993). The ablation curves of fresh and dried bone obtained with the CO_2 laser are illustrated in Fig. 4.61a. From the above, we could conclude that CO_2 lasers always evoke severe thermal side effects in bone. This statement, however, is not generally true. Forrer et al. (1993) have also demonstrated the potential of CO_2 lasers for bone ablation with very little thermal damage. When selecting the laser transition at 9.6 μm, a pulse duration of 1.8 μs, and an energy density of 15 J/cm^2, they found thermally altered damage zones of 10–15 μm only. In this case, both wavelength and pulse duration play a significant role. First, the absorbance of bone at 9.6 μm is higher than at 10.6 μm. Second, shorter pulse durations tend to be associated with less thermal damage as already discussed in Sect. 3.2.

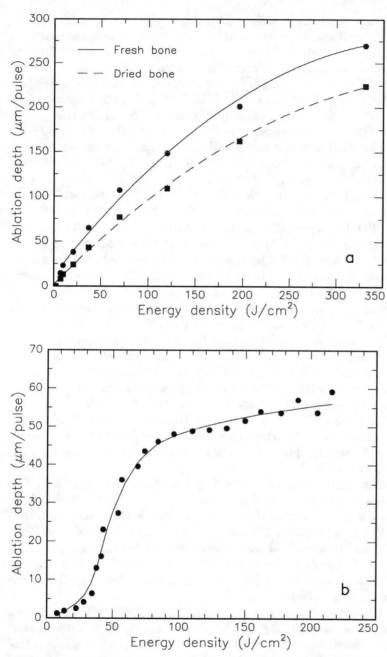

Fig. 4.61. (a) Ablation curves of fresh and dried bone obtained with a CO_2 laser (pulse duration: 250 µs, wavelength: 10.6 µm). Due to its higher water content, fresh bone is ablated more efficiently. Data according to Forrer et al. (1993). **(b)** Ablation curve of bone obtained with an Er:YAG laser (pulse duration: 180 µs, wavelength: 2.94 µm). Data according to Scholz and Grothves-Spork (1992)

In the 1980s, most research focused on laser radiation at a wavelength of approximately $3\,\mu m$ which is strongly absorbed by water. For instance, Wolbarsht (1984) compared the effects induced by CO_2 lasers at $10.6\,\mu m$ and HF* lasers[10] at $2.9\,\mu m$ with each other. From his observations, he concluded that the latter wavelength is better suited for orthopedic applications. Similar results were published by Izatt et al. (1990). Unfortunately, though, HF* lasers are very unwieldy machines. Walsh and Deutsch (1989), Nelson et al. (1989a), and Gonzales et al. (1990) reported on the application of compact Er:YAG lasers at a wavelength of $2.94\,\mu m$. They stated that this radiation efficiently ablates both bone and cartilage. The ablation curve of bone obtained with the Er:YAG laser is illustrated in Fig. 4.61b.

Another promising laser in orthopedics is the Ho:YAG laser which emits at a wavelength of $2.12\,\mu m$. Nuss et al. (1988), Charlton et al. (1990), and Stein et al. (1990) have investigated short-term and long-term effects of bone ablation using this laser. Its major advantage is that its radiation can be efficiently transmitted through flexible fibers. However, thermal effects are significantly enhanced compared to those induced by Er:YAG lasers at a wavelength of $2.94\,\mu m$ as observed by Romano et al. (1994). They found that thermal damage is extremely pronounced when applying $250\,\mu s$ pulses from a Ho:YAG laser. At an incident energy density of $120\,J/cm^2$, a thermal damage zone of roughly $300\,\mu m$ is determined. On the other hand, pulses from an Er:YAG laser are associated with very little thermal damage. At an energy density of $35\,J/cm^2$, a damage zone of only $12\,\mu m$ is estimated. The corresponding histologic sections are shown in Figs. 4.62a–b. In the case of the Er:YAG laser, a lower energy density was chosen to obtain a similar ablation depth as with the Ho:YAG laser. One potential application field of erbium lasers is microsurgery of the stapes footplate in the inner ear. This treatment belongs to the discipline of otorhinolaryngology, and it will therefore be addressed in Sect. 4.10.

Due to their high precision in removing tissues, excimer lasers have also been proposed for the ablation of bone material, e.g. by Yow et al. (1989). However, it was soon observed that their efficiency is much too low to justify their clinical application. Moreover, osteotomies performed with XeCl lasers at $308\,nm$ are associated with severe thermal damage as reported by Nelson et al. (1989b). As in the case of CO_2 laser radiation, these thermal effects are believed to be responsible for the manifest delay in healing of the laser-induced bone excisions.

A very interesting approach to determine laser effects on bone has been reported by Barton et al. (1995) and Christ et al. (1995). By using a confocal laser scanning microscope, they were able to analyze both the ablation rate and morphology as a function of incident pulse energy from a Ho:YAG laser. They concluded that scattering is a dominant factor in the interaction of Ho:YAG laser radiation and bone.

---

[10] Hydrogen fluoride lasers.

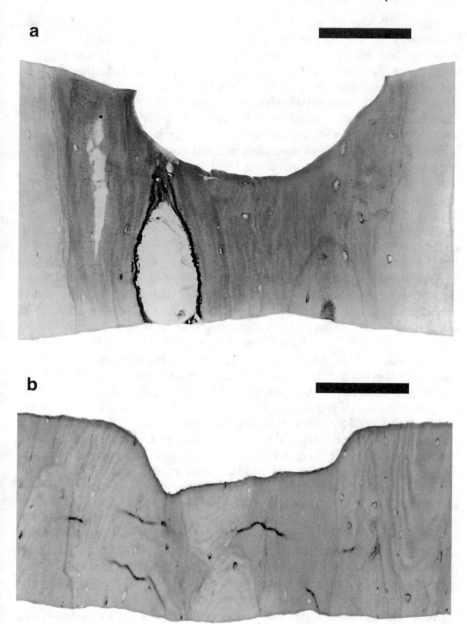

Fig. 4.62. (a) Histologic section of bone after exposure to a Ho:YAG laser (pulse duration: 250 µs, energy density: 120 J/cm^2, bar: 100 µm). (b) Histologic section of bone after exposure to an Er:YAG laser (pulse duration: 250 µs, energy density: 35 J/cm^2, bar: 200 µm). Photographs kindly provided by Dr. Romano (Bern)

Another discipline for laser applications within orthopedics is arthroscopy. Preliminary results regarding laser *meniscectomy*, i.e. the treatment of the meniscus, had already been reported by Glick (1981) and Whipple (1981) when using Nd:YAG and CO_2 lasers, respectively. At that time, though, suitable delivery systems were not available. Moreover, the CW mode of these lasers led to unacceptable thermal damage. Katsuyuki et al. (1983) and Bradrick et al. (1989) applied Nd:YAG lasers in arthroscopic treatment of the jaw joint. Significant improvements were not achieved until O'Brien and Miller (1990) made use of specially designed contact probes consisting of ceramics. Limbird (1990) pointed out the necessity of blood perfusion measurements after surgery. Major limitations for all infrared lasers in arthroscopic surgery arise from the optical delivery system. Transmission through flexible fibers can be regarded as a mandatory requirement for an efficient surgical procedure. Therefore, CO_2 lasers will never gain clinical relevance in arthroscopic treatments.

A new era of laser arthroscopy began with the application of holmium and erbium lasers. Trauner et a. (1990) reported promising results when using the Ho:YAG laser for the ablation of cartilage. A few years later, Ith et al. (1994) investigated the application of a fiber-delivered Er:YSGG laser emitting at a wavelength of 2.79 μm. The transmittance of zirconium fluoride (ZrF_4) fibers at this specific wavelength is satisfactory. Ith et al. (1994) used fresh human meniscus from the knee joint which was obtained during surgery. They observed a thermally damaged zone of 60 μm when exposing the tissue in air to five laser pulses at a pulse energy of 53 mJ and a pulse duration of 250 μs. On the other hand, when exposing the tissue through water at a slightly higher energy of 65 mJ, thermal damage extended to only 40 μm close to the surface and was even negligible elsewhere. In either case – whether exposed in air or through water – a crater depth of approximately 1 mm was achieved. The surprising result of this study is that laser radiation at 2.79 μm can be effectively used for tissue ablation, although it should be strongly absorbed by surrounding water. Thus, Ith et al. (1994) concluded that light – after exiting the fiber – is guided through a water-vapor channel created by the leading part of the laser pulse. The period during which this channel is open was found to be dependent on the duration of the laser pulse. For pulse durations of 250–350 μs, most of the laser energy is transmitted through the water-vapor channel to the target.

Meanwhile, even femtosecond lasers have been investigated for orthopedic applications. Liu and Niemz (2007) used a thin disk Yb:KYW femtosecond laser to ablate femoral bone. The study proved that up to 9 mm^3 of bone can be removed per minute when adjusting the pulse energy to 100 μJ and the repetition rate to 40 kHz. At these parameters, the cutting of bone in a knee replacement surgery would take about twenty minutes. However, further investigations still need to be performed to evaluate both thermal and mechanical side effects at this high repetition rate.

4.8 Lasers in Dermatology and Cosmetics

In dermatology, photothermal effects of laser radiation are frequently used, especially coagulation and vaporization. Since both the absorption and scattering coefficients are strongly wavelength-dependent, various kinds of tissue reactions can be evoked by different laser systems. In clinical practice, mainly four types of lasers are currently being applied: argon ion lasers, dye lasers, CO_2 lasers, and Nd:YAG lasers.

A schematic cross-section of human skin is sketched in Fig. 4.63. The skin grossly consists of three layers: *epidermis*, *dermis*, and *subcutis*. The outer two layers – epidermis and dermis – together form the *cutis*. The epidermis contains so-called keratocytes and melanocytes which produce keratin and melanin, respectively. Both keratin and melanin are important protective proteins of the skin. Most of the dermis is a semi-solid mixture of collagen fibers, water, and a highly viscous gel called ground substance. The complex nature of the skin creates a remarkable tissue with a very high tensile strength which can resist external compression but remains pliable at the same time. Blood vessels, nerves, and receptors are primarily located within the subcutis and the dermis.

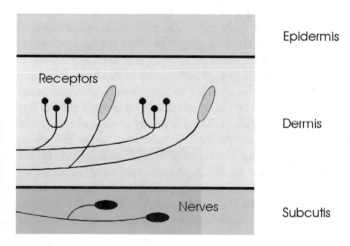

Epidermis

Receptors

Dermis

Nerves

Subcutis

Fig. 4.63. Human skin

On a microscopic scale, the air–skin interface is quite rough and therefore scatters incident radiation. Absorption of light by chromophores such as hemoglobin or melanin give skin its unique color. Optical scattering by collagen fibers in the ground substance largely determines which fraction of incident light penetrates into connective tissues. Detailed studies regarding the optical parameters of skin were performed by Graaff et al. (1993b).

Radiation from the argon ion laser is strongly absorbed by hemoglobin and melanin as already illustrated in Fig. 2.4. This laser is thus predestined for superficial treatments of highly vascularized skin. Apfelberg et al. (1978) and Apfelberg et al. (1979a) investigated laser-induced effects on various abnormalities of the skin. The most frequent indications for the application of argon ion lasers are given by port wine stains (*naevi flammei*). Earlier methods of treating these malformations – e.g. cryotherapy, X-ray, or chemical treatment – had failed, and patients were advised to accept their misery. The idea of removing port wine stains with argon ion lasers has led to the most significant progress of lasers in dermatology so far. The treatment itself requires a lot of patience, since several sessions are necessary over a period of up to a few years. The faster the treatment is to come to an end, the higher is the probability for the occurrence of scarring. However, "patient" patients are usually rewarded with an acceptable outcome. In Figs. 4.64a–b, two photographs of the pre- and postoperative states of a laser-treated port wine stain are shown.

Treatment of port wine stains with argon ion lasers is usually performed in several sessions. First, a small test area of approximately $4\,mm^2$ is irradiated. During this test, a suitable laser power is determined by gradually increasing it until the skin visibly pales. According to Dixon and Gilbertson (1986) and Philipp et al. (1992), laser powers of 2–5 W are applied during an exposure time of 0.02–0.1s. Immediately after laser exposure, inflammation of the skin frequently occurs. After four weeks, the test area is checked for recanalization and scarring. And after another four weeks, a second test area is treated. If both tests lead to acceptable results, the whole stain is exposed. Multiple exposures of the same area should be avoided in any case. Laser treatment may be repeated after a few years, but it is advisable to choose pulsed dye lasers for the second time. Haina et al. (1988) did not recommend treatment of patients up to 16 years of age, since otherwise severe scarring might occur. Laser radiation is usually applied by means of a flexible handpiece. In the treatment of facial stains, the eyes of both patient and surgeon must be properly protected. One disadvantage of treating port wine stains with argon ion lasers is that it is rather painful to the patient. Depending on the location and spatial extent of the stain, treatment is performed during either local or complete anesthetization.

Less painful and probably even more efficient is the treatment of port wine stains with dye lasers. Although quite expensive and bulky, these machines have gained increasing significance in dermatology, especially in the treatment of port wine stains and capillary hemangiomas. Detailed studies were published by Morelli et al. (1986), Garden et al. (1988), and Tan et al. (1989). Frequently, Rhodamine dye lasers are used which emit radiation at wavelengths in the range 570–590 nm. Typical pulse durations of 0.5 ms and energy densities of 4–$10\,J/cm^2$ have been recommended. About 20–60 s after laser exposure, the color of the treated skin turns red, and after another few

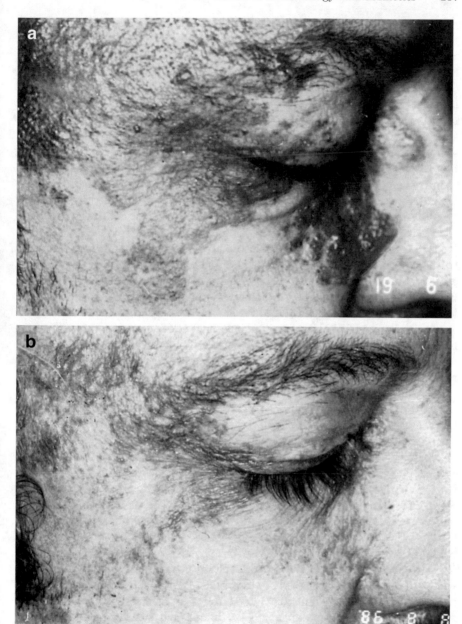

Fig. 4.64. (a) Preoperative state of a port wine stain. **(b)** Postoperative state of the same stain after five treatments with an argon ion laser (pulse duration: 0.3 s, power: 2.5 W, focal spot size: 2 mm). Photographs kindly provided by Dr. Seipp (Darmstadt)

minutes livid blue. Although pain is less pronounced as with argon ion lasers, patients frequently talk of triple pain perception: mechanical impact during the light flash, stabbing pain shortly afterwards, and finally a longer lasting heat wave within the skin. The irradiated area itself might be irritating for several days. One major advantage of treating port wine stains with dye lasers is that this procedure can be successfully performed among children as reported by Tan et al. (1989).

The basic mechanism by which pulsed laser radiation can cause selective damage to pigmented structures in vivo has been termed *selective photothermolysis* and was thoroughly described by Anderson and Parrish (1983). It requires the presence of highly absorbing particles, e.g. pigments of the skin. Extensive experimental and theoretical studies were performed by Kimel et al. (1994) and van Gemert et al. (1995). Their results significantly improved the treatment of port wine stains.

In dermatology, the CO_2 laser is used for tissue vaporization. Compared to the conventional scalpel, it offers the possibility of precise tissue removal without touching the tissue. Thus, feeling of pain is significantly reduced. External ulcers and refractory warts are common indications. In warts, deep lesions should be performed to reduce the probability of recurrence.

Argon ion and CO_2 lasers have also gained a lot of attention in efficiently removing tattoos. Clinical studies were reported by Apfelberg et al. (1979b) and Reid and Muller (1980). Even ruby lasers are commonly used for tattoo removal as stated by Scheibner et al. (1990) and Taylor et al. (1990). Indeed, good results can be obtained, although they do depend on the dyes used in the tattoo. It is extremely important that all dye particles are removed during the same session. In Figs. 4.65a–b, two photographs are shown which prove the efficiency of laser-induced tattoo removal.

Radiation from the Nd:YAG laser is significantly less scattered and absorbed in skin than radiation from the argon ion laser. The optical penetration depth of Nd:YAG laser radiation is thus much larger. According to Seipp et al. (1989), major indications for Nd:YAG laser treatments in dermatology are given by deeply located hemangiomas or semimalignant skin tumors. However, argon ion and CO_2 lasers should never be replaced by Nd:YAG lasers when treating skin surfaces.

Dermatology is one of the few disciplines in medicine where stimulating effects of laser radiation have been reported. Positive stimulation on wound healing is one of the current topics of controversy as discussed in Sect. 3.1.3. A considerable number of papers has been published, but not all of the results could be reproduced by others, and so not all of the initial claims could be verified. In general, dermatologists should be very careful when using laser radiation for such purposes, especially when applying so-called "soft lasers" with extremely low output powers of only a few mW. According to Alora and Anderson (2000), these lasers most likely do not evoke any effect at all—other than additional expenses.

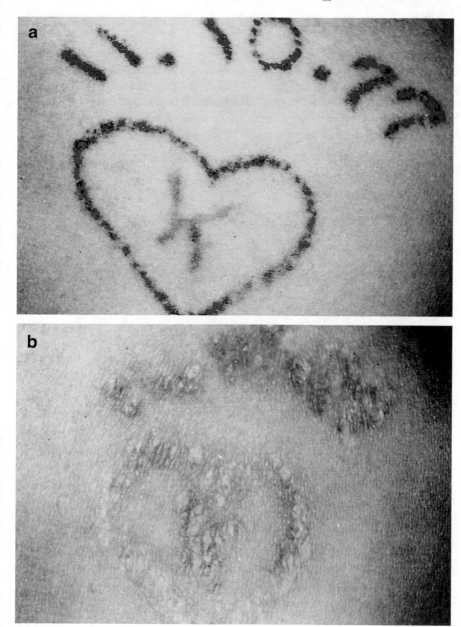

Fig. 4.65. (a) Preoperative state of a tattoo. (b) Postoperative state of the same tattoo after six complete treatments with an argon ion laser (pulse duration: 0.3 s, power: 3 W, focal spot size: 0.5 mm). Photographs kindly provided by Dr. Seipp (Darmstadt)

4.9 Lasers in Gastroenterology

Gastrointestinal diseases primarily include ulcers and tumors of the *esophagus*, *stomach*, *liver*, *gallbladder*, and *intestine*. The intestine further consists of the *jejunum*, *ileum*, *colon*, and *rectum*. According to the location of these organs, the gastrointestinal tract is subdivided into an upper and lower tract. Both tracts are schematically illustrated in Fig. 4.66. Most intestinal tumors are reported to occur inside the colon or the rectum.

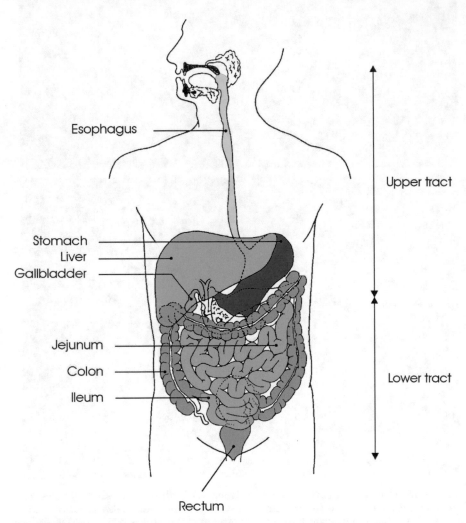

Fig. 4.66. Cross-section of upper and lower gastrointestinal tracts. The upper tract includes esophagus, stomach, liver, and gallbladder, whereas the lower tract consists of jejunum, ileum, colon, and rectum

In general, any kind of ulcer or tumor can be treated with lasers if it is accessible with endoscopic surgery. Ulcers and tumors tend to occupy additional space and are thus likely to induce severe stenoses. According to Sander et al. (1990), short and scarred stenoses of the lower tract are better suited for laser treatment than long and inflamed stenoses of the upper tract. If a tumor itself is no longer completely resectable due to its rather late detection – which unfortunately is quite often the case – laser application is restricted to a palliative treatment. The major concern of a palliative treatment is to provide an improved quality of the remaining life which also includes pain relief.

Gastroenterology is one of the major domains of the CW Nd:YAG laser. Only in photodynamic therapy are dye lasers applied. There exist mainly two indications for laser therapy: gastrointestinal hemorrhages and benign, malignant, or non-neoplastic[11] stenoses. Since the CW Nd:YAG laser is acting thermally, it can stop bleeding by means of coagulation. At higher power levels, i.e. in the vaporizing mode, it may serve in recanalization of stenoses. The application of other lasers was also investigated, e.g. the argon ion laser by Prauser et al. (1991), but was not associated with any significant advantages so far. The CO_2 laser is not suitable for clinical gastroenterology, since it is not transmitted through optical fibers which belong to the mandatory equipment of successful endoscopic surgery.

Stenoses of the esophagus are a common indication for laser treatment, since lasers can assist the surgeon in opening the stenosis. If even endoscopes cannot pass the stenosis, it must first be mechanically widened by specially designed dilators. Afterwards, the stenotic tissue may be coagulated using a Nd:YAG laser and a flexible fiber. Frequently, quartz fibers with a diameter of 600 µm are used. The fiber is protected by a Teflon tube with diameters of 1.8–2.5 mm. CO_2 gas is provided to cool the fiber tip and to keep debris away from the tissue. During coagulation, the tissue significantly shrinks, thereby reducing the occupied lumen inside the esophagus. If the stenosis was induced by a tumor, the fiber should be placed inside the tumor – by endoscopic control – and the tumor is coagulated starting from its interior. Induced bleeding can be stopped by a temporary increase in laser power. Dilation of the stenosis and coagulation of the tumor are usually performed during the same session. Up to 30 kJ may be necessary during one treatment as stated by Semler (1989). In most cases, necrotic tissue is repelled during the next few days, resulting in a further widening of the available lumen inside the esophagus.

Unfortunately, restenoses of the esophagus frequently occur after a few weeks. According to Bader et al. (1986), they can be efficiently prevented by a second treatment called *afterloading*. This treatment involves a radioactive source, e.g. ^{192}Ir, which is placed inside the esophagus for a few minutes by means of a computer-controlled probe. Between three and five of these

---

[11] Stenoses which are not related to tumor formation are called *non-neoplastic*.

afterloading treatments are normally performed starting approximately two weeks after laser coagulation. In malignant esophagus tumors, the mean survival rate is extremely low, since most of them are diagnosed at a very late stage. Treatment can then only be of a palliative character according to Fleischer and Sivak (1985), and *esophagectomy*, i.e. the complete or partial removal of the esophagus, must be performed. Afterwards, an artificial tube can be implanted.

Laser treatments of tumors belonging to the lower gastrointestinal tract were described by Hohenberger et al. (1986) and Kiefhaber et al. (1987). The conventional technique is called *cryotherapy*, since it induces tumor necrosis by freezing the tissue to temperatures of approximately $-180°$ C. In contrast to laser and afterloading therapy, it can be performed only during a complete anesthetization of the patient. Especially among older patients, this is one of the major disadvantages of cryotherapy. On the other hand, laser therapy is associated with an enhanced formation of edema. However, such edema can be treated with proper medicamentation.

Non-neoplastic stenoses of the lower gastrointestinal tract are treated by applying 80–100 W of laser power to an optical quartz fiber and slowly moving this fiber backwards out of the stenosis. Thereby, 1–2 mm deep grooves along the stenosis are induced according to Sander et al. (1990). After 3–5 days the grooves have dilated to a permeable path, and endoscopic passage is enabled without any mechanical pressure. In extended and inflamed stenoses, several treatments may be necessary.

The first laser therapy in gastroenterology was performed in the case of a massive hemorrhage by Kiefhaber et al. (1977). Since then, several extensive studies have been published, e.g. by Rutgeerts et al. (1982), Macleod et al. (1983), and Swain et al. (1986). From these studies, we may conclude that all localized and acute hemorrhages are suitable for laser coagulation. Inside the rectum and stomach, powers of 50–70 W and 70–100 W, respectively, are applied. After a complete clearance of the bleeding source, the tissue is coagulated from a distance of 5–10 mm by performing circularly shaped movements of the laser beam. There is no time limit for this procedure. The operation is stopped by releasing a footpedal if the desired effect is achieved. Patients are normally supervised by intensive care for at least three days following laser treatment.

An improved technique for the laser treatment of ulcers or hemorrhages was developed by Sander et al. (1988). It is based on a combination of laser beam and water jet. The laser beam is guided through a water jet to the site of application. First results reported by Sander et al. (1989) have shown that – using this technique – the percentage of successful hemostatic treatments could be raised from 82 % to 93 %. Moreover, fewer emergency surgeries needed to be performed and less mortality was observed. Sander et al. (1990) added that this technique has proven to be useful for other kinds of tissue coagulation, as well.

Despite early expectations concerning the potential of lasers in gastroenterology, e.g. by Fleischer et al. (1982), they could only partially be fulfilled
so far. It is not obvious that the Nd:YAG laser at a wavelength of 1.064 μm
provides the optimum radiation for gastrointestinal diseases. An extensive
analysis of potential complications arising from the use of this laser was given
by Mathus-Vliegen and Tytgat (1990). In the near future, alternative lasers
with different wavelengths will certainly be investigated, as well. The surgical
results obtained with these lasers must be compared to those achieved with
the Nd:YAG laser. Most probably, not just one single laser will then prove
to be best in the treatment of all diseases. There will rather be a variety of
different lasers which – in combination with alternative treatments – should
be used in specific cases.

One important field of modern gastroenterology – which has not been
addressed yet – is based on photodynamic therapy (PDT). The procedure of
PDT has already been described in detail in Sect. 3.1. After injection of an appropriate photosensitizer and a time delay of approximately 48–72 hours, tumors of the gastrointestinal tract are exposed to a dye laser, e.g. a Rhodamine
dye laser pumped by an argon ion laser. The clearance of the photosensitizer
leads to a concentration gradient among benign and malignant tissue ranging
from about 1:2 to 1:4. Meanwhile, several reports on PDT in gastroenterology have been published, e.g. by Barr et al. (1989), Barr et al. (1990), and
Karanov et al. (1991). The success is inversely related to the tumor size at the
time of treatment. According to Gossner and Ell (1993), tumors are curable
only if their infiltration depths remain below 5–10 mm. Overholt et al. (1993)
have shown that normal epithelium might then cover the interior of the
esophagus again. The extent of a tumor is usually determined by ultrasound
techniques. If a tumor is diagnosed at an early stage, more than 75 % of cases
can be completely cured. In advanced cancer, the corresponding rate is less
than 30 %. Tumors of the stomach are often more difficult to access for PDT
due to wrinkles of the mucosa. Thus, treatments of the stomach are frequently
associated with the application of higher energy doses. One major advantage
of PDT is that fewer endoscopic sessions are usually required compared to
treatments based on the Nd:YAG laser. Hence, the overall duration of a PDT
treatment is significantly shorter and easier to tolerate, as well.

In general, PDT is applied at early stages of cancer and in otherwise
inoperable patients, e.g. if alternative methods are associated with a high risk
for the patient. At advanced stages of esophagus cancer, PDT usually cannot
provide a complete cure. However, it might significantly facilitate the act of
swallowing. With the development of novel photosensitizers having a higher
efficiency in the red and near infrared spectrum, additional applications of
PDT might be indicated in the near future. Then, after careful evaluation of
clinical studies, improved treatments characterized by even higher cure rates
might be achievable.

4.10 Lasers in Otorhinolaryngology and Pulmology

Otorhinolaryngology is concerned with diseases of the *ear*, the *nose*, and the *throat*. Application of lasers in otorhinolaryngology primarily aims at micro-surgery of the *larynx*, for example laryngeal stenoses or laryngeal carcinoma. Stenoses of the larynx can be inherent or acquired. In either case, they are often associated with a severe impairment of the airway and should be treated immediately. Laryngeal carcinoma are the most frequent malignant tumors of the throat. Diagnosis and treatment of laryngeal carcinoma are performed by means of a laryngoscope which consists of a rigid tube being connected to a surgical microscope. During treatment, the patient must be intubated as demonstrated in Fig. 4.67.

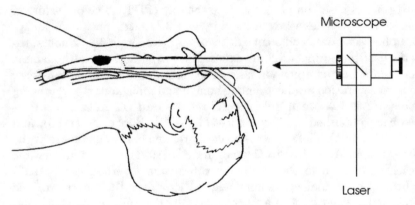

Fig. 4.67. Direct laryngoscopy with simultaneous intubation. Reproduced from Boenninghaus (1980) by permission. © 1980 Springer-Verlag

In the 1970s, Jako (1972) and Strong et al. (1976) had already shown that benign and malignant lesions of the glottis, i.e. the vocal cords, can be more safely excised by means of CO_2 lasers rather than mechanically. In laryngeal carcinoma, complete removal is confirmed once histopathology proves the absence of tumor cells in adjacent tissue. If the tumor is already at an advanced stage, however, laser therapy aims at a palliative treatment only. Complications arise when the laser treatment requires complete anesthetization of the patient, since the associated gases are inflammable. For a proper operation of the intubation tube, different materials are being investigated. Among these, metal tubes probably provide highest safety.

In the 1980s, Holinger (1982) and Duncavage et al. (1985) demonstrated that CO_2 lasers are superior to conventional therapy when treating laryngeal or subglottic stenoses. However, Steiner (1989) emphasized that not all kinds of stenoses are equally indicated for laser treatment. Large stenoses, in particular those extending several centimeters in size, should always be assigned to conventional surgery.

Major indications for laser treatment of the nose are highly vascular-ized tumors such as hemangiomas or premalignant alterations of the mucosa. The principal advantage of lasers is again their ability to simultaneously perform surgery and coagulate blood vessels. Lenz and Eichler (1984) and Parkin and Dixon (1985) have reported on argon ion laser treatments in vas-cular surgery of the nose. In proximal nose segments, Steiner (1989) suggests the application of a CO_2 laser which is connected to a surgical microscope. By means of tissue coagulation, even chronic nose-bleeding can be efficiently and safely treated.

Another useful laser application has been developed in the treatment of otosclerosis which is caused by an abnormal growth of bone in the middle ear. Otosclerosis usually affects the stapes shown in Fig. 4.68 and ultimately leads to its fixation. It is often associated with hearing impairment, because a movable stapes is necessary for the transport of sound to the cochlea. Ac-tually, the auditory ossicles in our middle ear (malleus, incus, and stapes) are a very clever impedance-matching device, since they match low-impedance airborne sounds to the higher-impedance fluid of the inner ear. Two poten-tial treatments of otosclerosis are called *stapedectomy* and *stapedotomy*. In stapedectomy, the stapes footplate in the middle ear is mechanically removed and replaced by an artificial implant. In stapedotomy, a hole is drilled into the stapes to improve the propagation of sound to the oval window. Perforation of the stapes is achieved with either miniaturized mechanical drills or with suitable lasers and application units. Laser stapedotomy can be considered as a typical method of minimally invasive surgery (MIS).

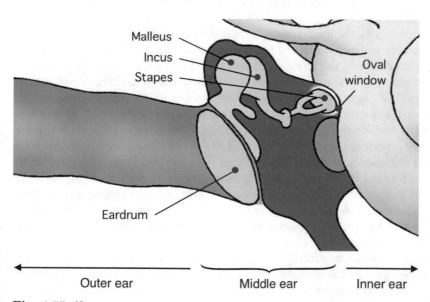

Fig. 4.68. Human ear

Stapedotomy seems to be a predestined treatment for laser radiation, since it requires least mechanical damage. It is generally accepted that a contactless method is certainly the best choice in preventing the inner ear structures from externally induced compression. Perkins (1980) and Gantz et al. (1982) performed the first stapedotomies using an argon ion laser. And just a few years later, Coker et al. (1986) investigated the application of a CO_2 laser for the same purpose. Finally, Lesinski and Palmer (1989) compared surgical results achieved with the CO_2 laser, argon ion laser, and a frequency-doubled Nd:YAG laser. The main disadvantage of radiation from the CO_2 laser is that it cannot be guided through optical fibers. On the other hand, radiation from argon ion or frequency-doubled Nd:YAG lasers is strongly absorbed only in highly pigmented tissue. The efficiency of these lasers in ablating cortical bone is thus rather weak. Due to the low absorption, their radiation might even induce severe lesions in adjacent tissue.

In the 1990s, Frenz et al. (1994) and Pratisto et al. (1996) performed stapedotomies in patients using Er:YAG and Er:YSGG lasers, respectively. Radiation from erbium lasers combines the advantages of being strongly absorbed in bone and of being transmitted through zirconium fluoride (ZrF_4) fibers. In Figs. 4.69a–b, a perforation of the stapes is shown which was induced by only five pulses from an Er:YAG laser at a rather moderate fluence of only $10\,J/cm^2$. The edge of the perforation is very precise and does not indicate any mechanical damage. In Fig. 4.70, the ablation curve of cortical bone obtained with the Er:YSGG laser is shown. Due to the high ablation depths achieved with erbium lasers, only a few pulses are necessary to perforate the stapes. According to Pratisto et al. (1996), the ablation threshold was less than $5\,J/cm^2$.

Potential risks in laser-assisted stapedotomy evolve from either an excessive increase in temperature of the perilymph or too high pressures induced inside the cochlea of the inner ear. It was found by Frenz et al. (1994) that the temperature at the stapes increases by only less than $5°C$ if the power output of the Er:YAG laser is limited to $10\,J/cm^2$. In a specially designed ear model, the temperature increase in the perilymph was even negligible. The pressure inside the cochlea during the laser treatment is very important, since the ear is a very sensitive organ. Frenz et al. (1994) have measured pressure signals in their ear model using a PVDF foil as described in Sect. 3.5. The PVDF foil was located $3\,mm$ below the exposed area. The corresponding pressure data are presented in Fig. 4.71. Frenz et al. (1994) have compared these data with maximum tolerable sound pressures as published by Pfander (1975). They stated that erbium lasers permit a safe pressure level if their fluence is limited to $10\,J/cm^2$.

From today's perspective, we conclude that laser-assisted stapedotomy has turned into a considerable alternative to mechanical treatments. Erbium lasers are capable of performing safe and very precise stapedotomies with negligible thermal damage.

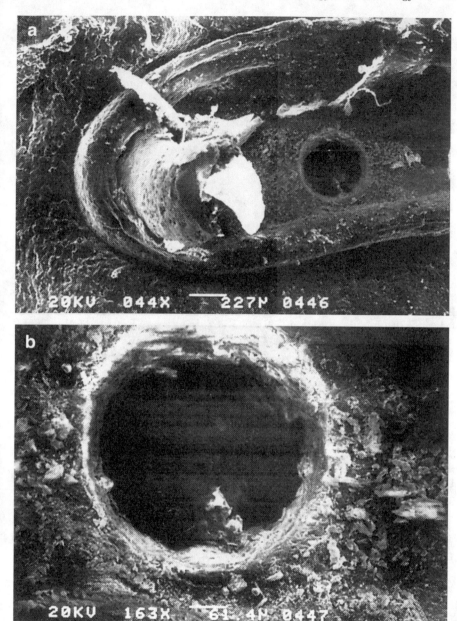

Fig. 4.69. (a) Stapedotomy performed with five pulses from an Er:YAG laser (pulse duration: 200 μs, fluence: 10 J/cm^2). (b) Enlargement of the perforation. Reproduced from Frenz et al. (1994) by permission. © 1994 Verlag Huber AG

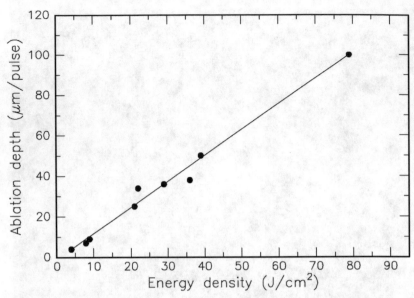

Fig. 4.70. Ablation curve of cortical bone obtained with an Er:YSGG laser (wavelength: 2.79 µm, pulse duration: 200 µs). Data according to Pratisto et al. (1996)

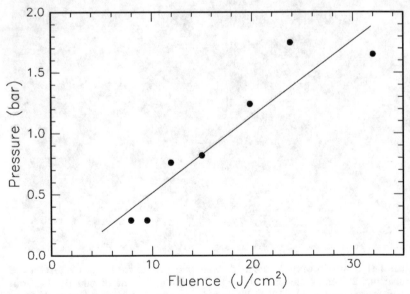

Fig. 4.71. Pressure in the perilymph of the inner ear during Er:YAG laser (wavelength: 2.94 µm, pulse duration: 200 µs) irradiation of the stapes footplate. Data according to Frenz et al. (1994)

Pulmology is concerned with diseases of the lung. In the western civilized world, tracheobronchial tumors are the primary cause of death due to cancer. According to Macha (1991), only less than 6–8 % of patients survive the next five years after diagnosis, because the tumor is often diagnosed at a rather late stage. The resection of tracheobronchial tumors is conventionally performed with a rigid bronchoscope. Severe and life-threatening hemorrhage is often inevitable. Beside mechanical removal, electrocoagulation and cryotherapy are performed. Dumon et al. (1982) and Hetzel et al. (1985) have investigated the application of a Nd:YAG laser and a flexible fiber in the treatment of tracheobronchial lesions. Since the Nd:YAG laser provides the surgeon with the ability of immediate coagulation, the occurrence of severe hemorrhage can be significantly reduced. Macha et al. (1987) proposed a combined therapy of a radioactive source, e.g. [192]Ir, and laser radiation to improve the surgical outcome. Gao et al. (2011) provided a detailed review on the endoscopic management of tracheobronchial tumors.

In summary, the Nd:YAG laser is a valuable supplement in the therapy of tracheobronchial tumors, although it is usually limited to a palliative treatment. However, patients soon perceive pain relief during breathing. Further clinical studies with alternative lasers and sophisticated application units are needed. Preliminary results concerning photodynamic therapy in the treatment of lung cancer had been published by Lam (1994). More recent reviews were provided by Simone et al. (2012) and Shafirstein et al. (2016).

4.11 Questions to Chapter 4

Q4.1. During the treatment of hyperopia with PRK, the curvature of the anterior corneal surface must
A: be flattened.   B: remain unchanged.   C: be steepened.
Q4.2. Er:YAG lasers are not suitable for the treatment of caries because of
A: cytotoxicity.   B: low efficiency.   C: thermal effects.
Q4.3. Angioplasty with excimer lasers is likely to induce
A: atherosclerosis.   B: restenoses.   C: thermal injury.
Q4.4. In stapedotomy with Er:YAG lasers, the fluence should be limited to
A: $1\,J/cm^2$.   B: $10\,J/cm^2$.   C: $100\,J/cm^2$.
Q4.5. The deepest layer of the skin is called
A: dermis.   B: epidermis.   C: subcutis.
Q4.6. Why is LASIK inducing less scattering than PRK?
Q4.7. Why is laser dentistry with picosecond or femtosecond pulses painfree?
Q4.8. Which is the workhorse laser in gynecology?
Q4.9. How can BPH be treated optically?
Q4.10. What is the purpose of a stereotactic ring?

5. Laser Safety

Most parts of this chapter are adapted from the booklet "Laser Safety Guide" (Editor: D.H.Sliney, 9th edition, 1993) published by the *Laser Institute of America, Orlando, Florida, USA*. The permission obtained for reproduction is gratefully acknowledged.

5.1 Introduction

The increasingly widespread use of lasers requires more people to become familiar with the potential hazards associated with the misuse of this valuable new product of modern science. Applications exist in many technologies, including material processing, construction, medicine, communications, energy production, and national defense. Of recent importance from a safety consideration, however, is the introduction of laser devices into more consumer-oriented retail products, such as the laser scanning devices, office copy and printing machines, and audio/visual recording devices. Most devices in these markets emit relatively low energy levels and, consequently, are easily engineered for safe use.

5.2 Laser Hazards

The basic hazards from laser equipment can be categorized as follows:

Laser Radiation Hazards

Lasers emit beams of optical radiation. Optical radiation (ultraviolet, visible, and infrared) is termed *non-ionizing* radiation to distinguish it from *ionizing* radiation such as X-rays and gamma rays which are known to cause different biological effects.

© Springer Nature Switzerland AG 2019
M. H. Niemz, *Laser-Tissue Interactions*,
https://doi.org/10.1007/978-3-030-11917-1_5

- Eye hazards: Corneal and retinal burns (or both), depending upon laser wavelength, are possible from acute exposure; and corneal or lenticular opacities (cataracts), or retinal injury may be possible from chronic exposure to excessive levels.
- Skin hazards: Skin burns are possible from acute exposure to high levels of optical radiation. At some specific ultraviolet wavelengths, skin carcinogenesis may occur.

Chemical Hazards

Some materials used in lasers (i.e. excimer, dye, and chemical lasers) may be hazardous and/or contain toxic substances. In addition, laser-induced reactions can release hazardous particulate and gaseous products.

Electrical Hazards

Lethal electrical hazards may be present in all lasers, particularly in high-power laser systems.

Other Secondary Hazards

These include:

- cryogenic coolant hazards,
- excessive noise from very high energy lasers,
- X radiation from faulty high-voltage ($> 15\,\mathrm{kV}$) power supplies,
- explosions from faulty optical pumps and lamps,
- fire hazards.

5.3 Eye Hazards

The ocular hazards represent a potential for injury to several different structures of the eye. This is generally dependent on which structure absorbs the most radiant energy per volume of tissue. Retinal effects are possible when the laser emission wavelength occurs in the visible and near infrared spectral regions ($0.4\,\mu\mathrm{m}$ to $1.4\,\mu\mathrm{m}$). Light directly from the laser or from a specular (mirror-like) reflection entering the eye at these wavelengths can be focused to an extremely small image on the retina. The incidental corneal irradiance and radiant exposure will be increased approximately $100\,000$ times at the retina due to the focusing effect of the cornea and lens.

Laser emissions in the ultraviolet and far infrared spectral regions (outside $0.4\,\mu\mathrm{m}$ to $1.4\,\mu\mathrm{m}$) produce ocular effects primarily at the cornea. However, laser radiation at certain wavelengths may reach the lens and cause damage to that structure.

Optical Radiation Hazards

Effects of optical radiation at various wavelengths on various structures of the eye are shown in Figs. 5.1a–c. Actinic-ultraviolet, at wavelengths of 180 nm to 315 nm, is absorbed at the cornea. These wavelengths are responsible for "welder's flash" or *photokeratitis*. Near ultraviolet (UV-A) radiation between 315 nm and 400 nm is absorbed in the lens and may contribute to certain forms of *cataracts*.

Radiation at visible wavelengths, 400 nm to 780 nm, and near infrared wavelengths, 780 nm to 1400 nm, is transmitted through the ocular media with little loss of intensity and is focused to a spot on the retina 10 μm to 20 μm in diameter. Such focusing can cause intensities high enough to damage the retina. For this very reason, laser radiation in the 400 nm to 1400 nm range is termed the *retinal hazard region*. Wavelengths between 400 nm and 550 nm are particularly hazardous for long-term retinal exposures or exposures lasting for several minutes or hours. This is sometimes referred to as the *blue light hazard*.

Far infrared (IR-C) radiation with wavelengths of 3 μm to 1 mm is absorbed in the front surface of the eye. However, some middle infrared (IR-B) radiation between 1.4 μm and 3 μm penetrates deeper and may contribute to "glass-blower's cataract". Extensive exposure to near infrared (IR-A) radiation may also contribute to such cataracts.

The localization of injury is always the result of strong absorption in the specific tissue for the particular wavelength.

5.4 Skin Hazards

From a safety standpoint, skin effects have been usually considered of secondary importance. However, with the more widespread use of lasers emitting in the ultraviolet spectral region as well as higher power lasers, skin effects have assumed greater importance.

Erythema[1], *skin cancer*, and accelerated *skin aging* are possible in the 230 nm to 380 nm wavelength range (actinic ultraviolet). The most severe effects occur in the UV-B (280–315 nm). Increased *pigmentation* can result following chronic exposures in the 280 nm to 480 nm wavelength range. At high irradiances, these wavelengths also produce "long-wave" erythema of the skin. In addition, photosensitive reactions are possible in the 310 nm to 400 nm (near ultraviolet) and 400 nm to 600 nm (visible) wavelength regions. The most significant effects in the 700 nm to 1000 nm range (infrared) will be skin burns and excessive dry skin.

---

[1] Sunburn.

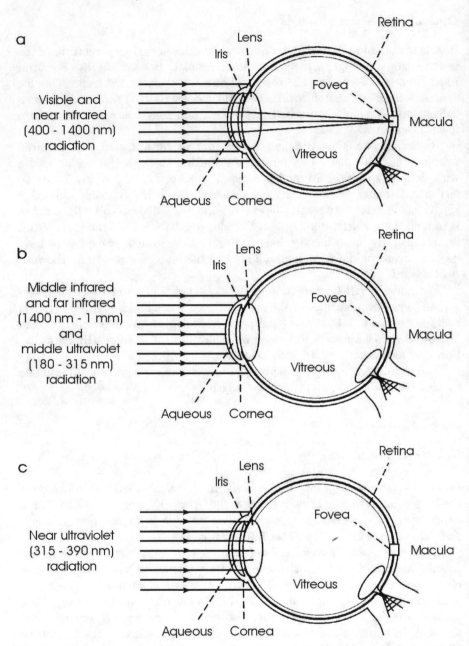

Fig. 5.1. (a) Absorption sites of visible and near infrared radiation. **(b)** Absorption sites of middle infrared, far infrared, and middle ultraviolet radiation. **(c)** Absorption sites of near ultraviolet radiation

5.5 Associated Hazards from High Power Lasers

Some applications of high-power lasers, especially in materials processing, can give rise to respiratory hazards. Laser welding, cutting, and drilling procedures can create potentially hazardous fumes and vapors. Fortunately, the same localized and general ventilation procedures developed for similar conventional operations apply to this type of laser application.

The most lethal hazards associated with the laser involves electricity. There have been several fatal accidents associated with lasers due to electrocution. These occurred when commonly accepted safety procedures were not followed when individuals were working with dangerous, high-voltage components of a laser system. Proper electrical hazards controls should be used at all times when working with laser systems.

Fire hazards may exist with some high-power laser devices, normally those with continuous wave (CW) lasers having an ouput power above 0.5 W. Another hazard sometimes associated with high-power laser systems involves the use of cryogenic coolants used in the laser system. Skin contact can cause burns, improper plumbing can cause explosions, and insufficient ventilation can result in the displacement of oxygen in the air by liquefied gas vaporizing (most commonly nitrogen). Cryogenic hazards are normally, but not exclusively, limited to research laboratories. Noise hazards are rarely present in laser operations.

5.6 Laser Safety Standards and Hazard Classification

The basic approach of virtually all laser safety standards has been to classify lasers by their hazard potential which is based upon their optical emission. The next step is to specify control measures which are commensurate with the relative hazard classification. In other words, the laser is classified based upon the hazard it presents, and for each classification a standard set of control measures applies. In this manner, unnecessary restrictions are not placed on the use of many lasers which are engineered to assure safety.

This philosophy has given rise to a number of specific classification schemes such as the one employed in the *American National Standards Institute's (ANSI) Z136.1 Safe Use of Lasers (1993)* standard. This standard was developed by the accredited standards committee Z136, and the Laser Institute of America is the secretariat. The standard has been used as a source by many organizations including the Occupational Health and Safety Agency (OSHA) and the American Conference of Governmental Industrial Hygienists (ACGIH) in developing their laser safety guidelines[2].

---

[2] Meanwhile, major parts of the ANSI classification scheme have been adapted by most European safety organizations, as well.

The ANSI scheme has four hazard classifications. The classification is based upon the beam output power or energy from the laser (emission) if it is used by itself. If the laser is a component within a laser system where the raw beam does not leave the enclosure, but instead a modified beam is emitted, the modified beam is normally used for classification. Basically, the classification scheme is used to describe the capability of the laser or laser system to produce injury to personnel. The higher the classification number, the greater is the potential hazard. Brief descriptions of each laser class are given as follows:

— *Class 1* denotes lasers or laser systems that do not, under normal operating conditions, pose a hazard.

— *Class 2* denotes low-power visible lasers or visible laser systems which, because of the normal human aversion response (i.e. blinking, eye movement, etc.), do not normally present a hazard, but may present some potential for hazard if viewed directly for extended periods of time (like many conventional light sources). Safety glasses are required for prolonged viewing only.

— *Class 3a* denotes the lowest class of lasers or laser systems that always require protective eyewear. These lasers would not injure the eye if viewed for only momentary periods (e.g. within the aversion response period of approximately 0.25 s) with the unaided eye, but may present a greater hazard if viewed using collecting optics or if viewed without the possibility of an aversion response (as for UV or IR radiation).

— *Class 3b* denotes lasers or laser systems that can produce a hazard if viewed directly. This includes intrabeam viewing of specular reflections. Normally, Class 3b lasers will not produce a hazardous diffuse reflection. Protective eyewear is always required.

— *Class 4* denotes lasers and laser systems that produce a hazard not only from direct or specular reflections, but may also produce hazardous diffuse reflections. Such lasers may produce significant skin hazards as well as fire hazards. Protective eyewear is always required.

Although the process of laser hazard evaluation does not rely entirely on the laser classification, the laser classification must be known. If the laser classification has not been provided by the manufacturer (as usually required by law), the class can be determined by measurement and/or calculation. A list of typical laser classifications is found in Table 5.1. Since the relative hazard of a laser may also vary depending upon use and/or environmental effects, measurements and/or calculations may be necessary to determine the degree of hazard in such cases.

Table 5.1. Typical classifications for selected CW and pulsed lasers, assuming that both skin and eye may be exposed (beam diameter: 1.0 mm)

Laser (CW)	Wavelength (nm)	Class 1 (W)	Class 2 (W)	Class 3b (W)	Class 4 (W)
Nd:YAG (4ω)	266	$\leq 9.6 \times 10^{-9}$	–	≤ 0.5	> 0.5
Argon ion	488/514	$\leq 0.4 \times 10^{-6}$	$\leq 10^{-3}$	≤ 0.5	> 0.5
Krypton ion	530	$\leq 0.4 \times 10^{-6}$	$\leq 10^{-3}$	≤ 0.5	> 0.5
Nd:YAG (2ω)	532	$\leq 0.4 \times 10^{-6}$	$\leq 10^{-3}$	≤ 0.5	> 0.5
Dye	400–550	$\leq 0.4 \times 10^{-6}$	$\leq 10^{-3}$	≤ 0.5	> 0.5
He-Ne	632	$\leq 7 \times 10^{-6}$	$\leq 10^{-3}$	≤ 0.5	> 0.5
Krypton ion	647	$\leq 11 \times 10^{-6}$	$\leq 10^{-3}$	≤ 0.5	> 0.5
Diode	670	$\leq 24 \times 10^{-6}$	$\leq 10^{-3}$	≤ 0.5	> 0.5
Diode	780	$\leq 0.18 \times 10^{-3}$	–	≤ 0.5	> 0.5
Diode	850	$\leq 0.25 \times 10^{-3}$	–	≤ 0.5	> 0.5
Diode	905	$\leq 0.32 \times 10^{-3}$	–	≤ 0.5	> 0.5
Nd:YAG	1064	$\leq 0.64 \times 10^{-3}$	–	≤ 0.5	> 0.5
Ho:YAG	2120	$\leq 9.6 \times 10^{-3}$	–	≤ 0.5	> 0.5
Er:YAG	2940	$\leq 9.6 \times 10^{-3}$	–	≤ 0.5	> 0.5
CO_2	10600	$\leq 9.6 \times 10^{-3}$	–	≤ 0.5	> 0.5

Laser (pulsed)	Wavelength (nm)	Duration (s)	Class 1 (W)	Class 3b (W)	Class 4 (W)
ArF	193	20×10^{-9}	$\leq 23.7 \times 10^{-6}$	≤ 0.125	> 0.125
KrF	248	20×10^{-9}	$\leq 23.7 \times 10^{-6}$	≤ 0.125	> 0.125
Nd:YAG (4ω)	266	20×10^{-9}	$\leq 23.7 \times 10^{-6}$	≤ 0.125	> 0.125
XeCl	308	20×10^{-9}	$\leq 52.6 \times 10^{-6}$	≤ 0.125	> 0.125
XeF	351	20×10^{-9}	$\leq 52.6 \times 10^{-6}$	≤ 0.125	> 0.125
Dye	450–650	10^{-6}	$\leq 0.2 \times 10^{-6}$	≤ 0.03	> 0.03
Nd:YAG	532	20×10^{-9}	$\leq 0.2 \times 10^{-6}$	≤ 0.03	> 0.03
Ruby	694	10^{-3}	$\leq 4 \times 10^{-6}$	≤ 0.03	> 0.03
Ti:Sapphire	700–1000	6×10^{-6}	$\leq 0.19 \times 10^{-6}$	≤ 0.03	> 0.03
Alexandrite	720–800	0.1×10^{-3}	$\leq 0.76 \times 10^{-6}$	≤ 0.03	> 0.03
Nd:YAG	1064	20×10^{-9}	$\leq 2 \times 10^{-6}$	≤ 0.15	> 0.15
Ho:YAG	2120	0.25×10^{-3}	$\leq 9.7 \times 10^{-3}$	≤ 0.125	> 0.125
Er:YAG	2940	0.25×10^{-3}	$\leq 6.8 \times 10^{-3}$	≤ 0.125	> 0.125
CO_2	10600	0.1×10^{-6}	$\leq 0.97 \times 10^{-3}$	≤ 0.125	> 0.125
CO_2	10600	10^{-3}	$\leq 9.6 \times 10^{-3}$	≤ 0.125	> 0.125

The term *limiting aperture* is often used when discussing laser classification. Limiting aperture is defined as the maximum circular area over which irradiance and radiant exposure can be averaged. It is a function of wavelength region and use.

In the ANSI classification system, the user or the Laser Safety Officer uses his judgement to establish the longest reasonable possible exposure duration for a CW or repetitively pulsed laser. This is called the classification duration t_{max} which cannot exceed an eight hour day equal to 3×10^4 seconds.

Very important is the so-called *MPE value* which denotes maximum permissible exposure. The MPE value depends on both exposure time and wavelength. In Fig. 5.2, some typical MPE values for maximum ocular exposure are graphically presented. The respective values for skin exposure are usually higher, since skin is not as sensitive as the retina. A comparison of ocular and skin exposure limits is provided in Table 5.2. For pulse durations shorter than 1 ns, the damage threshold of the energy density scales approximately with the square root of the pulse duration as discussed in Sect. 3.4. For instance, when evaluating an appropriate exposure limit for laser pulses with a duration of 10 ps, the energy densities listed for 1 ns pulses should be multiplied by a factor of $1/\sqrt{100} = 1/10$.

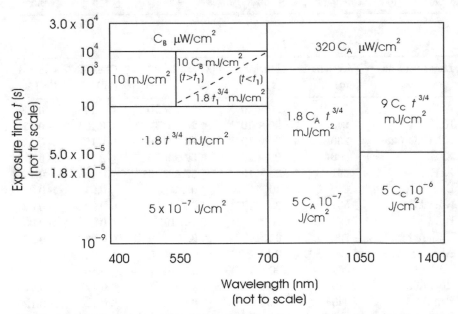

Fig. 5.2. Visible and near-IR MPE values for direct ocular exposure. Note that the correction factors (C) vary by wavelength. $C_A = 10^{2(\lambda-0.700)}$ for 0.700–1.050 μm. $C_A = 5$ for 1.050–1.400 μm. $C_B = 1$ for 0.400–0.550 μm. $C_B = 10^{15(\lambda-0.550)}$ for 0.550–0.700 μm. $t_1 = 10 \times 10^{20(\lambda-0.550)}$ for 0.550–0.700 μm. $C_C = 1$ for 1.050–1.150 μm. $C_C = 10^{18(\lambda-1.150)}$ for 1.150–1.200 μm. $C_C = 8$ for 1.200–1.400 μm

Table 5.2. Ocular and skin exposure limits of some representative lasers. Repetitive pulses at rates less than one pulse per second were assumed for any repetitive exposures. Higher repetition rates require more adjustment of the exposure limits. τ: pulse duration

Laser type	Wavelength (nm)	Ocular exposure limit[a] (MPE value)
Argon ion	488/514	$0.5\,\mu\mathrm{J/cm^2}$ for 1 ns to 18 µs
		$1.8\,\tau^{3/4}\,\mathrm{mJ/cm^2}$ for 18 µs to 10 s
		$10\,\mathrm{mJ/cm^2}$ for 10 s to 10 000 s
		$1\,\mu\mathrm{W/cm^2}$ for greater durations
He-Ne	632.8	$0.5\,\mu\mathrm{J/cm^2}$ for 1 ns to 18 µs
		$1.8\,\tau^{3/4}\,\mathrm{mJ/cm^2}$ for 18 µs to 430 s
		$170\,\mathrm{mJ/cm^2}$ for 430 s to 10 000 s
		$17\,\mu\mathrm{W/cm^2}$ for greater durations
Nd:YAG	1064	$5\,\mu\mathrm{J/cm^2}$ for 1 ns to 50 µs
		$9\,\tau^{3/4}\,\mathrm{mJ/cm^2}$ for 50 µs to 1 000 s
		$1.6\,\mathrm{mW/cm^2}$ for greater durations
Diode	910	$1.3\,\mu\mathrm{J/cm^2}$ for 1 ns to 18 µs
		$4.5\,\tau^{3/4}\,\mathrm{mJ/cm^2}$ for 18 µs to 1 000 s
		$0.8\,\mathrm{mW/cm^2}$ for greater durations
CO_2	10600	$10\,\mathrm{mJ/cm^2}$ for 1 ns to 100 ns
		$0.56\,\tau^{1/4}\,\mathrm{J/cm^2}$ for 100 ns to 10 s
		$0.1\,\mathrm{W/cm^2}$ for greater durations

Laser type	Wavelength (nm)	Skin exposure limit[b] (MPE value)
Argon ion	488/514	$0.02\,\mathrm{J/cm^2}$ for 1 ns to 100 ns
		$1.1\,\tau^{1/4}\,\mathrm{J/cm^2}$ for 100 ns to 10 s
		$0.2\,\mathrm{W/cm^2}$ for greater durations
He-Ne	632.8	$0.02\,\mathrm{J/cm^2}$ for 1 ns to 100 ns
		$1.1\,\tau^{1/4}\,\mathrm{J/cm^2}$ for 100 ns to 10 s
		$0.2\,\mathrm{W/cm^2}$ for greater durations
Nd:YAG	1064	$0.1\,\mathrm{J/cm^2}$ for 1 ns to 100 ns
		$5.5\,\tau^{1/4}\,\mathrm{J/cm^2}$ for 100 ns to 10 s
		$1.0\,\mathrm{W/cm^2}$ for greater durations
Diode	910	$0.05\,\mathrm{J/cm^2}$ for 1 ns to 100 ns
		$2.8\,\tau^{1/4}\,\mathrm{J/cm^2}$ for 100 ns to 10 s
		$0.5\,\mathrm{W/cm^2}$ for greater durations
CO_2	10600	$10\,\mathrm{mJ/cm^2}$ for 1 ns to 100 ns
		$0.56\,\tau^{1/4}\,\mathrm{J/cm^2}$ for 100 ns to 10 s
		$0.1\,\mathrm{W/cm^2}$ for greater durations

[a] The exposure limit is averaged over a 7 mm aperture for wavelengths between 400 nm and 1400 nm, and over 1.0 mm for the CO_2 laser wavelength at 10.6 µm.

[b] The exposure limit is defined for a 3.5 mm measuring aperture.

Any completely enclosed laser is classified as a Class 1 laser if emissions from the enclosure cannot exceed the MPE values under any conditions inherent in the laser design. During service procedures, however, the appropriate control measures are temporarily required for the class of laser contained within the enclosure. In the United States, a Federal government safety standard for laser products, which regulates laser manufacturers, was developed by the Center for Devices and Radiological Health (CDRH).

5.7 Viewing Laser Radiation

From a safety point of view, the laser can be considered as a highly collimated source of extremely intense monochromatic electromagnetic radiation. Due to these unique beam properties, most laser devices can be considered as a point source of great brightness. Conventional light sources or a diffuse reflection of a Class 2 or Class 3 laser beam are extended sources of very low brightness because the light radiates in all directions. This is of considerable consequence from a hazard point of view, since the eye will focus the rays (400–1400 nm) from a point source to a small spot on the retina while the rays from an extended source will be imaged, in general, over a much larger area. Only when one is relatively far away from a diffuse reflection (far enough that the eye can no longer resolve the image) will the diffuse reflection approximate a "point source". Diffuse reflections are only of importance with extremely high-power Class 4 laser devices emitting visible and IR-A radiation between 400 nm and 1400 nm.

Different geometries of ocular exposure are demonstrated in Figs. 5.3–5.6. Intrabeam viewing of the direct (primary) laser beam is shown in Fig. 5.3. This type of viewing is most hazardous. Intrabeam viewing of a specularly reflected (secondary) beam from a flat surface is illustrated in Fig. 5.4. Specular reflections are most hazardous when the reflecting surface is either flat or concave. On the other hand, intrabeam viewing of a specularly reflected (secondary) beam from a convex surface is less hazardous, since the divergence of the beam has increased after reflection (see Fig. 5.5). Finally, Fig. 5.6 illustrates extended source viewing of a diffuse reflection. Usually, diffuse reflections are not hazardous except with very high power Class 4 lasers.

Fig. 5.3. Intrabeam viewing of a direct beam

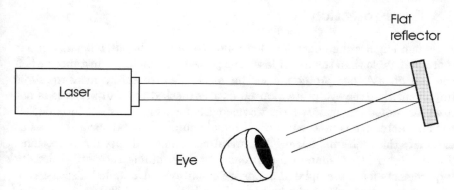

Fig. 5.4. Intrabeam viewing of a specularly reflected beam from a flat surface

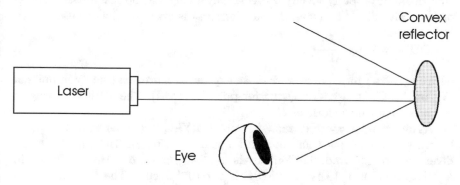

Fig. 5.5. Intrabeam viewing of a specularly reflected beam from a curved surface

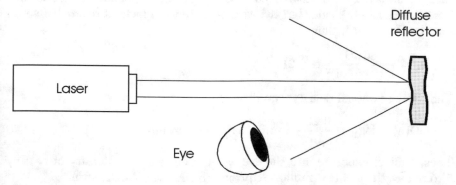

Fig. 5.6. Extended source viewing of a diffuse reflection

5.8 Eye Protection

Although engineering controls, such as enclosure of the laser beam, are far more preferable than the use of laser eye protection, there are instances when the use of laser eye protection is the most effective safety measure. It is important that the eye protection be clearly marked to insure that it is not used for protection against laser wavelengths for which it is not intended. If the eye protection is used outdoors it should employ curved-surface lenses to eliminate the additional hazard of generating collimated specular reflections. Table 5.3 provides a simplified approach for determining the optical density requirements for contemplated exposure conditions. A detailed discussion of safe eye protection was given by Sliney and Wolbarsht (1980).

Specifying Protective Filters

How does one properly specify protective eyewear? The optical density $OD(\lambda)$ of protective filters at a specific wavelength λ is given by the equation

$$OD(\lambda) = \log_{10} \frac{H_0}{MPE} ,$$

where H_0 is the anticipated worst case exposure (expressed usually in units of W/cm^2 for CW sources or J/cm^2 for pulsed sources). The MPE is expressed in units identical to those of H_0.

As an example, consider a single-pulse Nd:YAG laser operating at a wavelength of 1064 nm with an emergent beam diameter of 2 mm and a beam divergence of 1.0 mrad. The output is a TEM_{00} pulse of 80 mJ delivered in a pulse duration of 15 ns. The MPE is $5 \times 10^{-6} \, J/cm^2$. The most likely exposure is intrabeam viewing of the raw beam, i.e. $H_{raw} \simeq 2.55 \, J/cm^2$ (the fluence of the 80 mJ pulse in the 2 mm beam diameter). Since ANSI values of the MPEs are determined using a limiting aperture of 7 mm (maximum pupil size), the value of H_0 for beams smaller than 7 mm should be calculated as though the beam were distributed over the limiting aperture (which is the worst case exposure, since larger beam diameters are focused to smaller spot sizes on the retina), thus

$$H_0 = \frac{80 \, mJ}{\pi \, (0.35 \, cm)^2} \simeq 0.21 \, J/cm^2 .$$

The required optical density would be

$$OD(\lambda) = \log_{10} \frac{0.21}{5 \times 10^{-6}} \simeq 4.6 \quad at \quad \lambda = 1064 \, nm .$$

Thus, a direct exposure into the eye would require a filter density of nearly 5 to reduce the incident radiant exposure to the "safe" MPE level.

Table 5.3. Simplified method for selecting laser eye protection intrabeam viewing for wavelengths between 400 nm and 1400 nm. Data from American National Standards Institute's (ANSI) Z136.1 (1993)

Q-switched lasers (1 ns to 0.1 ms)		Non-Q-switched lasers (0.4 ms to 10 ms)		Attenuation	
Maximum output energy (J)	Maximum radiant exposure (J/cm^2)	Maximum output energy (J)	Maximum radiant exposure (J/cm^2)	Attenuation factor	OD
10	20	100	200	10^8	8
1	2	10	20	10^7	7
10^{-1}	2×10^{-1}	1	2	10^6	6
10^{-2}	2×10^{-2}	10^{-1}	2×10^{-1}	10^5	5
10^{-3}	2×10^{-3}	10^{-2}	2×10^{-2}	10^4	4
10^{-4}	2×10^{-4}	10^{-3}	2×10^{-3}	10^3	3
10^{-5}	2×10^{-5}	10^{-4}	2×10^{-4}	10^2	2
10^{-6}	2×10^{-6}	10^{-5}	2×10^{-5}	10^1	1

CW lasers momentary (0.25 s to 10 s)		CW lasers long-term staring (greater than 3 hours)		Attenuation	
Maximum output power (W)	Maximum irradiance (W/cm^2)	Maximum output power (W)	Maximum irradiance (W/cm^2)	Attenuation factor	OD
NR	NR	NR	NR	NR	NR
NR	NR	NR	NR	NR	NR
NR	NR	1	2	10^6	6
NR	NR	10^{-1}	2×10^{-1}	10^5	5
10	20	10^{-2}	2×10^{-2}	10^4	4
1	2	10^{-3}	2×10^{-3}	10^3	3
10^{-1}	2×10^{-1}	10^{-4}	2×10^{-4}	10^2	2
10^{-2}	2×10^{-2}	10^{-5}	2×10^{-5}	10^1	1

OD: optical density.
NR: not recommended.

5.9 Laser Beam Calculations

Only two laser beam calculations are provided in this section, since these are the most commonly encountered.

The emergent beam diameter is an important parameter in classifying a laser and in evaluating a hazard to the eye or skin. Most laser manufacturers specify output beam power or energy and generally provide the emergent beam diameter. It is often left to the user to calculate the output irradiance (intensity) in units of W/cm^2 and radiant exposure (fluence) in units of J/cm^2 for comparison with protection standards like the MPE values defined by ANSI.

The beam diameter specified by many laser manufacturers is the diameter of an aperture that will just accept 90 % of the output energy of a pulsed laser or 87 % of the output power of a CW laser having a Gaussian beam (also known as $1/e^2$ points). Regrettably for safety purposes it is necessary to calculate or measure *peak* radiant exposure or irradiance for a large beam diameter, or the fraction of power or energy passing through a specified "limiting aperture".

For ocular exposure limits in the visible and near infrared (400–1400 nm) spectral region, the limiting aperture is 7 mm corresponding to the dilated pupil. For ultraviolet and most of the infrared spectral regions, this aperture is normally 1 mm or 3.5 mm. Only for the extreme infrared region (0.1–1.0 mm), a limiting aperture of 1 cm is assumed.

Using a laser power meter at two distances z_1 and z_2 with a fixed aperture, one can determine the change in beam radius w to calculate the beam divergence Φ by means of

$$\Phi = \frac{2w(z_2) - 2w(z_1)}{z_2 - z_1} = \frac{2\Delta w}{\Delta z} \ ,$$

where w and z are in units of cm, and Φ is in units of radians[3].

The other beam calculation concerns the central-beam values of intensity and fluence, since they are usually the highest and thus determine potential hazards. Often, these values are not provided in the specification of a laser. However, if the laser has a Gaussian-shaped beam profile and emits in the fundamental TEM_{00} mode, the corresponding central-beam values can be obtained from the beam radius $w_{1/e}$ specified at the $1/e$ points. This beam radius is defined as the radius of an aperture that will just accept 63 % of the incident power, i.e. $1/e$ of the incident power is blocked.

---

[3] More accurately, the beam radius of a Gaussian beam changes with distance according to a hyperbolic function rather than linearly. When the beam waist occurs at or near the exit port of the laser, the correct equation for the beam radius is $w(z) = \sqrt{w_0^2 + (\Phi/2)^2 \, z^2}$, where w_0 is the smallest beam waist. Only at distances far apart from the smallest beam waist may w_0 be omitted without loss of accuracy.

The relations for the central-beam intensity I_0 and fluence E_0 can then be expressed by

$$I_0 = \frac{P}{\pi\, w_{1/e}^2}\;,$$

$$E_0 = \frac{Q}{\pi\, w_{1/e}^2}\;,$$

where P is the radiant power in units of W, and Q is the radiant energy in units of J.

Often, the beam radius is specified at the $1/e^2$ points rather than at the $1/e$ points, so that the above relations would provide lower values than the central-beam values. In this case, the radius specified at the $1/e^2$ points should be divided by $\sqrt{2}$ to obtain the corresponding $1/e$ value, hence

$$w_{1/e} = \frac{w_{1/e^2}}{\sqrt{2}} \simeq 0.707\; w_{1/e^2}\;.$$

5.10 Questions to Chapter 5

Q5.1. Which is the lowest laser class that always requires the use of safety glasses?
A: Class 2.   B: Class 3a.   C: Class 3b.

Q5.2. Which radiation is able to burn the fovea?
A: UV-B.   B: VIS.   C: IR-C.

Q5.3. Which radiation is able to cause skin cancer?
A: UV-B.   B: VIS.   C: IR-C.

Q5.4. Typical CW lasers in Class 4 have an output power
A: $< 0.5\,\mathrm{W}$.   B: $> 0.5\,\mathrm{W}$.   C: $> 1\,\mathrm{kW}$.

Q5.5. When comparing the MPE values for eyes and skin, then
A: $\mathrm{MPE_{eye}} < \mathrm{MPE_{skin}}$.   B: $\mathrm{MPE_{eye}} = \mathrm{MPE_{skin}}$.   C: $\mathrm{MPE_{eye}} > \mathrm{MPE_{skin}}$.

Q5.6. Which are the three basic hazards from laser equipment?

Q5.7. Why is a collinear, visible or near infrared laser beam most dangerous to our vision?

Q5.8. Why is there no ultraviolet or infrared Class 2 laser?

Q5.9. What laser class is a high power 100 W Nd:YAG laser at a wavelength of 1064 nm that is completely enclosed?

Q5.10. An ophthalmologist uses a 200 mW argon ion laser at a wavelength of 514 nm to perform a retinal coagulation. The procedure takes ten seconds. What is the optical density needed in his eyewear to protect his eyes from accidental damage (consider Fig. 5.2)?

A. Appendix

A.1 Medical Neodymium Laser System

Several of the tissue samples shown throughout this book were exposed to a picosecond Nd:YLF laser. Actually, Nd:YLF as well as Nd:YAG lasers are very versatile medical laser systems, since they can be used for inducing almost any type of laser–tissue interaction described in Chap. 3. Plasma-induced ablation and photodisruption are evoked by mode locking of these lasers. Photoablation occurs with their fourth harmonic, i.e. after two times frequency-doubling. Thermal effects are induced under either CW or Q-switched operation. Photochemical effects were also investigated as listed in Table 3.2. Moreover, both Nd:YLF and Nd:YAG lasers are solid-state lasers with several additional advantages for clinical applications, e.g.

- non-toxic laser media,
- low maintenance,
- compactness,
- operation in fundamental laser mode.

For the reasons just stated, a detailed description of the picosecond Nd:YLF laser system is given in this appendix. The other medical lasers mentioned in this book are described elsewhere.

The Nd:YLF laser system discussed below combines short pulse durations with moderate pulse energies and high repetition rates. It is designed as a two stage oscillator/regenerative amplifier combination to provide laser pulses with durations down to 30 ps and energies up to 1 mJ at a wavelength of 1053 nm. This arrangement was developed by Bado et al. (1987). First medical applications were reported by Niemz et al. (1991). A scheme of the complete laser system is shown in Fig. A.1.

The laser crystal in the oscillator is pumped by a temperature tuned 1 watt diode laser (DL) using beam shaping and collimating optics (CO). An acoustooptic mode locker (AOM) is placed near the flat 10 % output coupler (OC). For active amplitude modulation, an amplified 80 MHz RF signal is applied to this device generating a train of short laser pulses with typical durations of 25 ps each as illustrated in Fig. A.2. A real-time autocorrelation system allows for continuous supervision of the pulse width. To select the 1053 nm transition, a Brewster plate polarizer (BP) is added to the cavity.

© Springer Nature Switzerland AG 2019
M. H. Niemz, *Laser-Tissue Interactions*,
https://doi.org/10.1007/978-3-030-11917-1

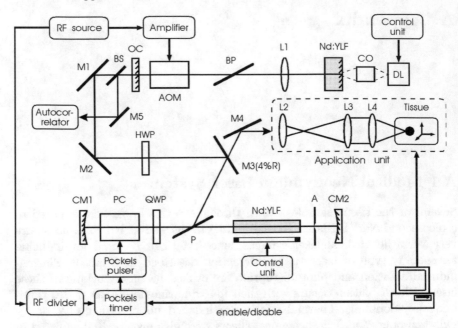

Fig. A.1. Complete picosecond Nd:YLF laser system consisting of laser oscillator *(top part)*, regenerative amplifier *(bottom part)*, and application unit. Layout according to Niemz (1994a) and Dr. Loesel (Heidelberg)

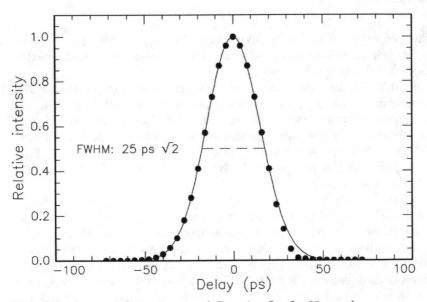

Fig. A.2. Autocorrelation trace and Gaussian fit of a 25 ps pulse

At the half-wave plate (HWP), the 160 MHz pulse train consisting of 0.2 nJ pulses experiences a 90° rotation of the polarization vector. Using a 4 % reflecting mirror (M3) and a polarizing beamsplitter (P), the oscillator pulses are then injected into the regenerative amplifier unit. A 76 mm amplifier Nd:YLF rod is pumped by a single xenon lamp controlled by a standard power supply. The cavity employs two highly reflecting mirrors with a radius of curvature of 1 m each. Applying a 2 kV voltage signal at a repetition rate of up to 1 kHz to a LiNbO$_3$ Pockels cell (PC) provides half-wave retardation per round-trip. In combination with the double-pass half-wave retardation of the intracavity quarter-wave plate (QWP), a selected oscillator pulse is seeded and trapped in the amplifier unit. The driving of the Pockels cell is synchronized to the mode locking process by feeding the 80 MHz RF signal into a special divider and timer logic. After about 100 roundtrips in the cavity, the seeded pulse reaches its saturation limit. At maximum gain, the Pockels cell driver switches back to 0 V causing no retardation. The polarization vector is now rotated by 90° as the pulse double-passes the quarter-wave plate and the Pockels cell in the left part of the cavity. Consequently, the amplified pulse is then reflected at the polarizing beamsplitter and dumped out of the regenerative amplifier as shown in Fig. A.3. By this means, the pulse energy can be boosted up to 1 mJ corresponding to an amplification of 10^6 of the oscillator output energy. Installation of an aperture (A) restricts the laser operation to the fundamental TEM$_{00}$ mode.

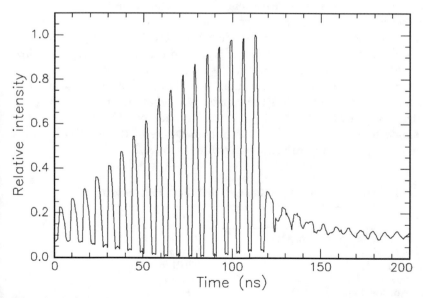

Fig. A.3. Amplification of a seeded pulse inside the regenerative amplifier unit. When reaching the limit of energy saturation, the pulse is dumped out of the cavity

After dumping the pulse out of the regenerative amplifier, mirror M3 is now transmitting 96 % of the amplified pulse energy. Using mirror M4, the pulse train is finally injected into a specially designed application unit as shown in Fig. A.1. Autocorrelation of the amplified pulses yields that their pulse duration slightly increases to about 30 ps due to dispersion inside the amplifier cavity. In Table A.1, the parameters of both laser oscillator and regenerative amplifier are listed.

Table A.1. Parameters of the picosecond Nd:YLF laser system. Data according to Niemz (1991)

Parameter	Laser oscillator	Regenerative amplifier
Wavelength	1053 nm	1053 nm
Pulse energy	0.2 nJ	1 mJ
Pulse duration	25 ps	30 ps
Repetition rate	160 MHz	1 kHz
Mean output power	32 mW	1 W

The application unit basically consists of delivering optics, focusing optics, and a computer controlled three-dimensional translation stage. After expanding the laser beam four times with the lenses L2 and L3, it is tightly focused on the tissue sample with lens L4. The diameter of the focus spot can be measured using the knife-edge method. For a dumped beam diameter of 2 mm and a focal length of 50 mm of lens L4, a focal spot of approximately 30 μm is obtained. Stepping motors connected to the translation stage allow precise spot-to-spot movements of the tissue sample at an accuracy of 1 μm. A software package gives the user a choice of different ablation patterns. The motor control software also has the capability of enabling and disabling the dumping mechanism of the Nd:YLF amplifier, thus providing the ability of exact single-shot operation. The optical parameters of the application unit and the focus parameters are summarized in Table A.2.

Table A.2. Parameters of application unit and focus parameters

Parameter	Value
Beam diameter (amplifier output)	2 mm
Focal length of L2	50 mm
Focal length of L3	200 mm
Focal length of L4	50 mm
Diameter of focal spot	30 μm
Peak energy density	140 J/cm^2
Peak power density	4.7×10^{12} W/cm^2

The laser system shown in Fig. A.1 can also be operated using Nd:YAG crystals. Actually, Nd:YLF is a solid-state optical gain medium closely related to Nd:YAG, but without some of its undesirable features. Unlike Nd:YAG, it is only weakly affected by thermal birefringence as reported by Murray (1983). Hence, a simpler cavity design and a higher ratio of TEM_{00} to multi-mode average power can be achieved according to Pollak et al. (1982). Nd:YLF also has a longer upper-level fluorescence lifetime compared to Nd:YAG which enables greater energy storage and thus higher peak pulse powers under Q-switched or mode locked operation. The physical parameters of both Nd:YLF and Nd:YAG are compared in Table A.3.

Table A.3. Comparison of Nd:YAG and Nd:YLF parameters. Data according to Koechner (1992), Murray (1983), and Dallas (1994)

Parameter	Nd:YAG	Nd:YLF
Chemical formula	$Nd:Y_3Al_5O_{12}$	$Nd:LiYF_4$
Wavelength	1064 nm	1053/1047 nm
Bandwidth	140 GHz	420 GHz
Pulse duration (theoretical limit)	7.1 ps	2.4 ps
Fluorescence lifetime	230 μs	480 μs
Cross-section (stimulated emission)	$2.8 \times 10^{-19}\,cm^2$	$1.2-1.8 \times 10^{-19}\,cm^2$
Thermal conductivity (at 300 K)	$13\,W\,m^{-1}\,K^{-1}$	$6\,W\,m^{-1}\,K^{-1}$
Thermal lensing[a]	0.12 diopters	< 0.006 diopters

[a] Measured at 633 nm for 10 mm rods pumped at 700 W average power.

A.2 Physical Constants and Parameters

In Table A.4, physical constants are listed as they appear in the book.

Table A.4. Physical constants

Quantity	Symbol	Value
Speed of light in vacuum	c	$2.9979 \times 10^8\,m\,s^{-1}$
Planck's constant	h	$6.6256 \times 10^{-34}\,J\,s^{-1}$
Boltzmann's constant	k	$1.38 \times 10^{-23}\,J\,K^{-1}$
Electron mass	m_e	$9.1091 \times 10^{-31}\,kg$
Gas constant	R	$8.3143\,J\,K^{-1}\,mol^{-1}$
Dielectric constant	ε_0	$8.854 \times 10^{-12}\,As\,V^{-1}\,m^{-1}$
Permeability constant	μ_0	$4\pi\,10^{-7}\,Vs\,A^{-1}\,m^{-1}$

In the literature, the use of radiometric parameters is not always uniform. For the purpose of selecting the appropriate term, a list of all significant parameters and their units[1] is given in Table A.5. The other physical parameters used throughout this book are listed in Table A.6.

Table A.5. Radiometric parameters and units

Parameter	Symbol	Unit
Exposure time	τ	s
Beam radius	w	m
Beam divergence	Φ	–
Wavelength	λ	m
Electromagnetic frequency	ω	Hz
Propagation vector	$\boldsymbol{k}$	m^{-1}
Radiant power	P	W
Power density, intensity, irradiance	I	$\mathrm{W\ cm}^{-2}$
Radiance	J	$\mathrm{W\ cm}^{-2}\ \mathrm{sr}^{-1}$
Vector flux	$\boldsymbol{F}$	$\mathrm{W\ cm}^{-2}$
Heat source	S	$\mathrm{W\ cm}^{-3}$
Radiant energy	Q	J
Energy density, fluence, radiant exposure	E	$\mathrm{J\ cm}^{-2}$
Energy dose	q	$\mathrm{J\ cm}^{-3}$
Maximum permissible exposure	MPE	$\mathrm{J\ cm}^{-2}$ or $\mathrm{W\ cm}^{-2}$
Damage threshold (power density)	I_{th}	$\mathrm{W\ cm}^{-2}$
Damage threshold (energy density)	E_{th}	$\mathrm{J\ cm}^{-2}$
Reflectance, transmittance	$R_{\mathrm{i}}, T_{\mathrm{i}}$	–
Kubelka–Munk coefficients	$A_{\mathrm{KM}}, S_{\mathrm{KM}}$	cm^{-1}
Absorption length	L	cm
Absorption coefficient of tissue	α	cm^{-1}
Absorption coefficient of plasma	α_{pl}	cm^{-1}
Scattering coefficient of tissue	α_{s}	cm^{-1}
Attenuation coefficient of tissue	α_{t}	cm^{-1}
Index of absorption	$\tilde{\alpha}$	–
Albedo	a	–
Optical depth	d	–
Coefficient of anisotropy	g	–
Index of refraction	n	–
Scattering phase function	p	–

---

[1] While the meter is the preferred unit of length throughout the MKS system, the centimeter is the most commonly used unit of length for power densities, energy densities, and absorption coefficients when dealing with cm-sized tissues.

Table A.6. Non-radiometric parameters and units

Parameter	Symbol	Unit
Arrhenius' constant	A	s^{-1}
Transition probability	A_i	s^{-1}
Magnetic induction	$\boldsymbol{B}$	Vs m^{-2}
Specific heat capacity	c	$\text{J kg}^{-1}\,\text{K}^{-1}$
Ablation depth	d	m
Dielectric induction	$\boldsymbol{D}$	As m^{-2}
Electric field strength	$\boldsymbol{E}$	V m^{-1}
Energy level	E_i	eV
Statistical weight	g_i	–
Magnetic field strength	$\boldsymbol{H}$	A m^{-1}
Electric current	$\boldsymbol{j}$	A m^{-2}
Density of free electric currents	$\boldsymbol{j}_f$	A m^{-2}
Heat flow	$\boldsymbol{j}_Q$	W m^{-2}
Heat conductivity	k	$\text{W m}^{-1}\,\text{K}^{-1}$
Mass	m	kg
Density of free electrons	N	m^{-3}
Pressure	p_i	N m^{-2}
Heat content	Q	J
Direction vector	$\boldsymbol{s}$	–
Time	t	s
Temperature	T	K
Particle speed	u_p	m s^{-1}
Shock front speed	u_s	m s^{-1}
Speed of light in medium	v	m s^{-1}
Coordinates	x, y, z, r	m
Thermal penetration depth	z_{therm}	m
Rate parameter for avalanche ionization	β	s^{-1}
Rate parameter for inelastic collision	γ	$\text{cm}^3\,\text{s}^{-1}$
Rate parameter for electron diffusion	δ	s^{-1}
Dielectric factor	ϵ	–
Complex dielectric factor	ϵ'	–
Temperature conductivity	κ	$\text{m}^2\,\text{s}^{-1}$
Relative permeability	μ	–
Collision frequency (electron–ion)	ν_{ei}	Hz
Mass density	ρ	kg m^{-3}
Particle density	ρ	m^{-3}
Density of free electric charges	ρ_f	As m^{-3}
Electric conductivity	σ	$\text{A V}^{-1}\,\text{m}^{-1}$
Time constant of inelastic collision	τ_c	s
Time constant of diffusion	τ_d	s
Thermal relaxation time	τ_{therm}	s
Plasma frequency	ω_{pl}	Hz

B. Solutions

The solutions given here are not arranged by chapter. B2.1. is the solution to question Q2.1.

B2.1. A
B3.1. C
B4.1. C
B5.1. B
B2.2. C
B3.2. B
B4.2. C
B5.2. B
B2.3. C
B3.3. A
B4.3. B
B5.3. A
B2.4. A
B3.4. C
B4.4. B
B5.4. B
B2.5. A
B3.5. C
B4.5. C
B5.5. A

B2.6. $R_\mathrm{p} \simeq R_\mathrm{s} \simeq \left(\frac{1.5-1}{1.5+1}\right)^2 \simeq 4\,\%$. Since the laser beam is reflected at both glass surfaces, the total loss in intensity is approximately $8\,\%$.

B3.6. An appropriate time gap is needed for the photosensitizer to be mostly cleared from healthy tissue, while it is still present in tumor cells at a high concentration.

B4.6. Since the corneal surface after a LASIK treatment is the same natural surface as before the treatment, irritations due to scattering effects are significantly reduced.

B5.6. Laser radiation hazards, chemical hazards, and electrical hazards.

B2.7. $I(1\,\mathrm{mm}) = 5\,\mathrm{mW} \times \exp(-10\,\mathrm{cm}^{-1} \times 0.1\,\mathrm{cm}) \simeq 1.8\,\mathrm{mW}$.

© Springer Nature Switzerland AG 2019
M. H. Niemz, *Laser-Tissue Interactions*,
https://doi.org/10.1007/978-3-030-11917-1

B3.7. The 1 ms pulse will locally coagulate the tissue, while the 1 ps pulse will not induce a significant effect.

B4.7. Because (a) there are no vibrations and (b) the pulse duration is so short that heat cannot diffuse into the tooth.

B5.7. Because only visible or near infrared wavelengths are transmitted to the retina, and because a collinear beam is focused to a tiny spot on the retina.

B2.8. $I_s(532\,\text{nm}) \simeq 16\ I_s(1064\,\text{nm})$.

B3.8. Because only UV photons provide an energy sufficient for the photodissociation of molecular bonds, which is the basic mechanism of photoablation.

B4.8. The CO_2 laser.

B5.8. Visibility is a mandatory requirement of a Class 2 laser beam, because its definition implies that the normal human aversion response (blinking of the eye) does not present a hazard.

B2.9. $a = \frac{310\,\text{cm}^{-1}}{2.3\,\text{cm}^{-1}+310\,\text{cm}^{-1}} \simeq 0.9926$.

B3.9. Avalanche ionization, inelastic collisions, and electron diffusion.

B4.9. By either holmium laser ablation of the prostate (HoLAP) or holmium laser enucleation of the prostate (HoLEP).

B5.9. Class 1.

B2.10. At 60°C, the coagulation of egg white strongly increases its scattering coefficient, thus giving it a white appearance.

B3.10. Both types of interaction are based on the formation of a plasma. The laser pulse energy has to be increased to switch from plasma-induced ablation to photodisruption.

B4.10. A stereotactic ring defines a coordinate system serving as a valuable means of orientation during surgery.

B5.10. From Fig. 5.2 we find that MPE $= 10\,\text{mJ/cm}^2$. Hence:

$H_0 = \frac{200\,\text{mW}\ 10\,\text{s}}{\pi\ (0.35\,\text{cm})^2} \simeq 5200\,\text{mJ/cm}^2$ and $\text{OD} = \log_{10} \frac{5200\,\text{mJ/cm}^2}{10\,\text{mJ/cm}^2} \simeq 2.7$.

Therefore, a filter with an optical density of 3 is required.

References

Abela, G., Norman, S., Cohen, R., Feldman, R., Geiger, F., Conti, C.R. (1982): Effects of carbon dioxide, Nd-YAG and argon laser radiation on coronary atheromatous plaque. Am. J. Cardiol. **50**, 1129–1205

Abergel, R.P., Meeker, C.A., Dwyer, R.M., Lesavoy, M.A. (1984): Nonthermal effects of Nd:YAG laser on biological functions of human skin fibroblast in culture. Lasers Surg. Med. **3**, 279–284

Agostinis, P., Berg, K., Cengel, K.A., Foster, T.H., Girotti, A.W., Gollnick, S.O., Hahn, S.M., Hamblin, M.R., Juzeniene, A., Kessel, D., Korbelik, M., Moan, J., Mroz, P., Nowis, D., Piette, J., Wilson, B.C., Golab, J. (2011): Photodynamic therapy of cancer: an update. CA Cancer J. Clin. **61**, 250–281

Alora, M.B., Anderson, R.R. (2000): Recent developments in cutaneous lasers. Lasers Surg. Med. **26**, 108–118

Anderson, R.R., Parrish, J.A. (1983): Selective photothermolysis: precise microsurgery by selective absorption of pulsed radiation. Science **220**, 524–527

Andreeva, L.I., Vodop'yanov, K.L., Kaidalov, S.A., Kalinin, Y.M., Karasev, M.E., Kulevskii, L.A., Lukashev, A.V. (1986): Picosecond erbium-doped ErYAG laser ($\lambda = 2.94\mu$) with active mode locking. Sov. J. Qu. Electron. **16**, 326–333

Andrew, J.E., Dyer, P.E., Forster, D., Key, P.H. (1983): Direct etching of polymeric materials using a XeCl laser. Appl. Phys. Lett. **43**, 717–719

Ang, M., Tan, D., Mehta, J.S. (2012): Small incision lenticule extraction (SMILE) versus laser in-situ keratomileusis (LASIK): study protocol for a randomized, non-inferiority trial. Trials **13**, 75

Aoki, A., Mizutani, K., Schwarz, F., Sculean, A., Yukna, R.A., Takasaki, A.A., Romanos, G.E., Taniguchi, Y., Sasaki, K.M., Zeredo, J.L., Koshy, G., Coluzzi, D.J., White, J.M., Abiko, Y., Ishikawa, I., Izumi, Y. (2015): Periodontal and peri-implant wound healing following laser therapy. Periodontol. 2000 **68**, 217–269

Apfelberg, D.B., Maser, M.R., Lash, H. (1978): Argon laser treatment of cutaneous vascular abnormalities: progress report. Ann. Plast. Surg. **1**, 14–18

Apfelberg, D.B., Maser, M.R., Lash, H. (1979a): Extended clinical use of the argon laser for cutaneous lesions. Arch. Dermatol. **115**, 719–721

Apfelberg, D.B., Maser, M.R., Lash, H. (1979b): Argon laser treatment of decorative tattoos. Br. J. Plast. Surg. **32**, 141–144

Aron-Rosa, D., Aron, J., Griesemann, J., Thyzel, R. (1980): Use of the neodym-YAG laser to open the posterior capsule after lens implant surgery: a preliminary report. J. Am. Intraocul. Implant Soc. **6**, 352–354

Aron-Rosa, D.S., Boerner, C.F., Bath, P., Carre, F., Gross, M., Timsit, J.C., True, L., Hufnagel, T. (1987): Corneal wound healing after laser keratotomy in a human eye. Am. J. Ophthalmol. **103**, 454–464

Ascher, P.W. (1979): Newest ultrastructural findings after the use of CO_2 laser on CNS tissue. Acta Neurochir. **28**, 572–581

© Springer Nature Switzerland AG 2019
M. H. Niemz, *Laser-Tissue Interactions*,
https://doi.org/10.1007/978-3-030-11917-1

Ascher, P.W., Heppner, F. (1984): CO_2-Laser in Neurosurgery. Neurosurg. Rev. **7**, 123–133

Ascher, P.W., Justich, E., Schröttner, O. (1991): A new surgical but less invasive treatment of central brain tumours. Preliminary report. Acta Neurochirur. Suppl. **52**, 78–80

Auler, H., Banzer, G. (1942): Untersuchungen über die Rolle der Porphyrine bei geschwulstkranken Menschen und Tieren. Z. Krebsforsch. **53**, 65–72

Austin, G.E., Ratliff, N.B., Hollman, J., Tabei, S., Phillips, D.F. (1985): Intimal proliferation of smooth muscle cells as an explanation for recurrent coronary artery stenosis after percutaneous transluminal coronary angioplasty. J. Am. Coll. Cardiol. **6**, 369–375

Azma, E., Safavi, N. (2013): Diode laser application in soft tissue oral surgery. J. Lasers Med. Sci. **4**, 206–211

Bach, G., Neckel, C., Mall, C., Krekeler, G. (2000): Conventional versus laser-assisted therapy of periimplantitis: a five-year comparative study. Implant Dent. **9**, 247–251

Bader, M., Dittler, H.J., Ultsch, B., Ries, G., Siewert, J.R. (1986): Palliative treatment of malignant stenoses of the upper gastrointestinal tract using a combination of laser and afterloading therapy. Endoscopy **18**, 27–31

Bader, M.J., Stepp, H., Beyer, W., Pongratz, T., Sroka, R., Kriegmair, M., Zaak, D., Welschof, M., Tilki, D., Stief, C.G., Waidelich, R. (2013): Photodynamic therapy of bladder cancer – a phase I study using hexaminolevulinate (HAL). Urol. Oncol. **31**, 1178–1183

Bado, P., Bouvier, M., Coe, J.S. (1987): Nd:YLF mode-locked oscillator and regenerative amplifier. Opt. Lett. **12**, 319–321

Baggish, M.S. (1980): Carbon dioxide laser treatment for condyloma accuminata veneral infections. Obstet. Gynecol. **55**, 711–715

Baggish, M.S., Dorsey, J. (1981): CO_2-laser for the treatment of vulvar carcinoma in situ. Obstet. Gynecol. **57**, 371–375

Baggish, M.S., Dorsey, J.H., Adelson, M. (1989): A ten-year experience treating cervical intraepithelial neoplasia with the carbon dioxide laser. Am. J. Obstet. Gynecol. **161**, 60–68

Baggish, M.S., Barash, F., Noel, Y., Brooks, M. (1992): Comparison of thermal injury zones in loop electrical and laser cervical excisional conization. Am. J. Obstet. Gynecol. **166**, 545–548

Bailer, P. (1983): Tubensterilisation durch Laser-Koagulation. Fortschr. Med. **43**, 1977

Bailes, J.E., Cozzens, J.W., Hudson, A.R., Kline, D.G., Ciric, I., Gianaris, P., Bernstein, L.P., Hunter, D. (1989): Laser-assisted nerve repair in primates. J. Neurosurg. **71**, 266–272

Barkana, Y., Belkin, M. (2007): Selective laser trabeculoplasty. Surv. Ophthalmol. **52**, 634–654

Barr, H., Bown, S.G., Krasner, N., Boulos, P.B. (1989): Photodynamic therapy for colorectal disease. Int. J. Colorectal Dis. **4**, 15–19

Barr, H., Krasner, N., Boulos, P.B., Chatlani, P., Bown, S.G. (1990): Photodynamic therapy for colorectal cancer: a quantitative pilot study. Br. J. Surg. **77**, 93–96

Barton, T.G., Christ, M., Foth, H.-J., Hörmann, K., Stasche, N. (1995): Ablation of hard tissue with the holmium laser investigated by a surface profile measurement system and a confocal laserscanning microscope. Proc. SPIE **2323**, 185–195

Beck, O.J. (1980): The use of the Nd-YAG and the CO_2 laser in neurosurgery. Neurosurg. Rev. **3**, 261–266

Beckman, H., Rota, A., Barraco, R., Sugar, H.S., Gaynes, E. (1971): Limbectomies, keratectomies and keratostomies performed with a rapid-pulsed carbon dioxide laser. Am. J. Ophthalmol. **72**, 1277–1283

Beckman, H., Sugar, H.S. (1973): Laser iridectomy therapy of glaucoma. Arch. Ophthalmol. **90**, 453–455

Beisland, H.O., Stranden, E. (1984): Rectal temperature monitoring during neodymion-YAG laser irradiation for prostatic carcinoma. Urol. Res. **12**, 257–259

Bell, C.E., Landt, J.A. (1967): Laser-induced high-pressure shock waves in water. Appl. Phys. Lett. **10**, 46–48

Benson, R.C. (1985): Treatment of diffuse carcinoma in situ by whole bladder hematoporphyrin derivative photodynamic therapy. J. Urol. **134**, 675–678

van Benthem, H. (1992): Laseranwendung in der zahnärztlichen Prothetik und der dentalen Technologie. In: Laser in der Zahnmedizin (Eds.: Vahl, J., van Benthem, H.). Quintessenz-Verlag, Berlin

Berns, M.W., Dahlman, A., Johnson, F., Burns, R., Sperling, D., Guiltinan, M., Siemans, A., Walter, R., Wright, R., Hammer-Wilson, M., Wile, A. (1982): In vitro cellular effects of hematoporphyrin derivative. Cancer Res. **42**, 2325–2329

Berry, S.J., Coffey, D.S., Walsh, P.C., Ewing, L.L. (1984): The development of human benign prostatic hyperplasia with age. J. Urol. **132**, 474–479

Bille, J.F., Dreher, A.W., Zinser, G. (1990): Scanning laser tomography of the living human eye. In: Noninvasive diagnostic techniques in ophthalmology (Ed.: Masters, D.). Springer-Verlag, Berlin, Heidelberg, New York

Bille, J.F., Schlegel, W., Sturm, V. (1993): Stereotaktische Laser-Neurochirurgie. Physik in unserer Zeit **24**, 280–286

Bird, A.C. (1974): Recent advances in the treatment of senile disciform macular degeneration by photocoagulation. Br. J. Ophthalmol. **58**, 367–376

Bjordal, J.M., Lopes-Martins, R.A.B., Iversen, V.V. (2006): A randomised, placebo controlled trial of low level laser therapy for activated Achilles tendinitis with microdialysis measurement of peritendinous prostaglandin E_2 concentrations. Br. J. Sports Med. **40**, 76–80

Bloembergen, N. (1974): Laser-induced electric breakdown in solids. IEEE J. Qu. Electron. **QE-10**, 375–386

Boenninghaus, H.-G. (1980): Hals-Nasen-Ohrenheilkunde. Springer-Verlag, Berlin, Heidelberg, New York

Böwering, R., Hofstetter, A., Weinberg, W., Kronester, A., Keiditsch, E., Frank, F. (1979): Irradiation of prostatic carcinoma by neodymium-YAG-laser. In: Proc. 3rd Int. Congr. Laser Surgery, Graz (Eds.: Kaplan, I., Ascher, P.W.)

Boggan, J.E., Edwards, M.S., Davis, R.L., Bolger, C.A., Martin, N. (1982): Comparison of the brain tissue response in rats to injury by argon and carbon dioxide lasers. Neurosurgery **11**, 609–616

Bonney, W.W., Fallon, B., Gerber, W.L., Hawtrey, C.E., Loening, S.A., Narayana, A.S., Platz, C.E., Rose, E.F., Sall, J.C., Schmidt, J.D., Culp, D.A. (1982): Cryosurgery in prostatic cancer: survival. Urology **19**, 37–42

Boulnois, J.-L. (1986): Photophysical processes in recent medical laser developments: a review. Lasers Med. Sci. **1**, 47–66

Bradrick, J.P., Eckhauser, M.L., Indresano, A.T. (1989): Morphologic and histologic changes in canine temporomandibular joint tissues following arthroscopic guided neodymium:YAG laser exposure. J. Oral Maxillofac. Surg. **47**, 1177–1181

Brannon, J.H., Lamkard, J.R., Baise, A.I., Burns, F., Kaufman, J. (1985): Excimer laser etching of polyimide. J. Appl. Phys. **58**, 2036–2043

Brunner, R., Landthaler, M., Haina, D., Waidelich, W., Braun-Falco, O. (1984): Experimentelle Untersuchungen zum Einfluß von Laserlicht niedriger Leistungsdichte auf die Epidermisregeneration. In: Optoelektronik in der Medizin (Ed.: Waidelich, W.). Springer-Verlag, Berlin, Heidelberg, New York

Campbell, C.J., Rittler, M.C., Koester, C.J. (1963): The optical maser as a retinal coagulator: an evaluation. Am. Acad. Ophthalmol. Otolaryngol. **67**, 58

Camps, J.L., Powers, S.K., Beckman, W.C., Brown, J.T., Weissmann, R.M. (1985): Photodynamic therapy of prostate cancer: an in vitro study. J. Urol. **134**, 1222–1226

Carome, E.F., Moeller, C.E., Clark, N.A. (1966): Intense ruby-laser-induced acoustic impulse in liquids. J. Acoust. Soc. Am. **40**, 1462–1466

Carrillo, J.S., Calatayud, J., Manso, F.J., Barberia, E., Martinez, J.M., Donado, M. (1990): A randomized double-blind clinical trial on effectiveness of helium-neon laser in the prevention of pain, swelling and trismus after removal of impacted third molars. Int. Dent. J. **40**, 31–36

Castro, D.J., Abergel, R.P., Meeker, C., Dwyer, R.M., Lesavoy, M.A., Uitto, J. (1983): Effects of the Nd:YAG laser on DNA synthesis and collagen production in human skin fibroblast cultures. Ann. Plast. Surg. **11**, 214–222

Charlton, A., Dickinson, M.R., King, T.A., Freemont, A.J. (1990): Erbium:YAG and holmium:YAG laser ablations of bones. Lasers Med. Sci. **5**, 365–373

Cheong, W.-F., Motamedi, M., Welch, A.J. (1987): Optical modeling of laser photocoagulation of bladder tissue. Lasers Surg. Med. **7**, 72

Cheong, W.-F., Prahl, S.A., Welch, A.J. (1990): A review of the optical properties of biological tissues. IEEE J. Qu. Electron. **QE-26**, 2166–2185

Cho, B.Y., Cho, J.O. (1986): Experimental study of the effect of the laser irradiation in treating oral soft tissue damage. J. Dent. Res. **65**, 600 (A34)

Choy, D.S.J., Stertzer, S.H., Rotterdam, H.Z., Bruno, M.S. (1982): Laser coronary angioplasty; experience with 9 cadaver hearts. Am. J. Cardiol. **50**, 1209–1211

Choy, D.S.J., Stertzer, S.H., Myler, R.K., Marco, J., Fournial, G. (1984): Human coronary laser recanalization. Clin. Cardiol. **7**, 377–381

Chung, H., Dai, T., Sharma, S.K., Huang, Y.-Y., Carroll, J.D., Hamblin, M.R. (2012): The nuts and bolts of low-level laser (light) therapy. Ann. Biomed. Eng. **40**, 516–533

Christ, M., Barton, T.G., Hörmann, K., Foth, H.-J., Stasche, N. (1995): A new approach to determining laser effects on bone. Proc. SPIE **2327**, 394–401

Cilesiz, I.F., Welch, A.J. (1993): Light dosimetry: effects of dehydration and thermal damage on the optical properties of the human aorta. Appl. Opt. **32**, 477–487

Clayman, L., Fuller, T., Beckman, H. (1978): Healing of continuous-wave and rapid superpulsed carbon dioxide laser-induced bone defects. J. Oral Maxillofac. Surg. **36**, 932–937

Coker, N.J., Ator, G.A., Jenklins, H.A., Neblett, C.R. (1986): Carbon dioxide laser stapedotomy: a histology study. Am. J. Otolaryngol. **7**, 253–257

Colin, P., Nevoux, P., Marqa, M., Auger, F., Leroy, X., Villers, A., Puech, P., Mordon, S., Betrouni, N. (2011): Focal laser interstitial thermotherapy (LITT) at 980 nm for prostate cancer: treatment feasibility in Dunning R3327–AT2 rat prostate tumour. BJU Internat. **109**, 452–458

Coluzzi, D.J. (2008): Fundamentals of lasers in dentistry: basic science, tissue interaction, and instrumentation. J. Laser Dent. **16**, 4–10

Cooley, D.A., Frazier, O.H., Kadipasaoglu, K.A., Lindenmeir, M.H., Pehlivanoglu, S., Kolff, J.W., Wilansky, S., Moore, W.H. (1996): Transmyocardial laser revascularization: clinical experience with twelve-month follow-up. J. Thorac. Cardiovasc. Surg. **111**, 791–799

Cotlair, A.M., Schubert, H.D., Mandek, E.R., Trokel, S.L. (1985): Excimer laser radial keratotomy. Ophthalmology **92**, 206–208

Cubeddu, R., Sozzi, C., Taroni, P., Valentini, G., Bottiroli, G., Croce, A.C. (1994): Ablation of brain by Erbium laser: study of dynamic behavior and tissue damage. Proc. SPIE **2077**, 13–20

Cumberland, D.C., Sanborn, T.A., Taylor, D.I., Moore, D.J., Welsh, C.L., Greenfield, A.J., Guben, J.K., Ryan, T.J. (1986): Percutaneous laser thermal angioplasty. Lancet **1**, 1457–1459

Daehlin, L., Frugard, J. (2007): Interstitial laser coagulation in the management of lower urinary tract symptoms suggestive of bladder outlet obstruction from benign prostatic hyperplasia: long-term follow-up. BJU Int. **100**, 89–93

Dallas, J.L. (1994): Frequency-modulation mode-locking performance for four Nd^{3+}-doped laser crystals. Appl. Opt. **33**, 6373–6376

de Lia, J.E., Kuhlmann, R.S., Harstad, T.W., Cruikshank, D.P. (1995): Fetoscopic laser ablation of placental vessels in severe previable twin-twin transfusion syndrome. Am. J. Obstet. Gynecol. **172**, 1202–1211

Derbyshire, G.J., Bogen, D.K., Unger, M. (1990): Thermally induced optical property changes in myocardium at 1.06 µm. Lasers Surg. Med. **10**, 28–34

Desmet, K.D., Paz, D.A., Corry, J.J., EElls, J.T., Wong-Riley, M.T., Henry, M.M., Buchmann, E.V., Connelly, M.P., Dovi, J.V., Liang, H.L., Henshel, D.S., Yeager, R.L., Millsap, D.S., Lim, J., Gould, L.J., Das, R., Jett, M., Hodgson, B.D., Margolis, D., Whelan, H.T. (2006): Clinical and experimental applications of NIR-LED photobiomodulation. Photomed. Laser Surg. **24**, 121–128

Deutsch, T.F., Geis, M.W. (1983): Self-developing UV photoresist using excimer laser exposure. J. Appl. Phys. **54**, 7201–7204

Diethrich, E.B., Timbadia, E., Bahadir, I., Coburn, K., Zenzen, S. (1988): Argon laser-assisted peripheral angioplasty. Vasc. Surg. **22**, 77–87

Dimaras, H., Kimani, K., Dimba, E.A.O., Gronsdahl, P., White, A., Chan, H.S.L., Gallie, B.L. (2012): Retinoblastoma. Lancet **379**, 1436–1446

Dixon, J.A., Gilbertson, J.J. (1986): Argon and neodymium YAG laser therapy of dark nodular port wine stains in older patients. Lasers Surg. Med. **6**, 5–11

Dobberstein, H., Dobberstein, H., Zuhrt, R., Thierfelder, C., Ertl, T. (1991): Laserbearbeitung von Dentalkeramik und Dentallegierungen. In: Angewandte Lasermedizin (Eds.: Berlien, H.-P., Müller, G.). Ecomed-Verlag, Landsberg

Docchio, F., Sacchi, C.A., Marshall, J. (1986): Experimental investigation of optical breakdown thresholds in ocular media under single pulse irradiation with different pulse durations. Lasers Ophthalmol. **1**, 83–93

Docchio, F., Regondi, P., Capon, M.R.C., Mellerio, J. (1988a): Study of the temporal and spatial dynamics of plasmas induced in liquids by nanosecond Nd:YAG laser pulses. 1: Analysis of the plasma starting times. Appl. Opt. **27**, 3661–3668

Docchio, F., Regondi, P., Capon, M.R.C., Mellerio, J. (1988b): Study of the temporal and spatial dynamics of plasmas induced in liquids by nanosecond Nd:YAG laser pulses. 2: Plasma luminescence and shielding. Appl. Opt. **27**, 3669–3674

Dolphin, D. (1979): The porphyrins I–VII. Academic Press, New York

Dong, Z., Zhou, X., Wu, J., Zhang, Z., Li, T., Zhou, Z. (2014): Small incision lenticule extraction (SMILE) and femtosecond laser LASIK: comparison of corneal wound healing and inflammation. Br. J. Ophthalmol. **98**, 263–269

Dörschel, K., Berlien, H.-P., Brodzinski, T., Helfmann, J., Müller, G.J., Scholz, C. (1988): Primary results in the laser lithotripsy using a frequency doubled Q-switched Nd:YAG laser. In: Laser lithotripsy – clinical use and technical aspects (Ed.: Steiner, R.). Springer-Verlag, Berlin, Heidelberg, New York

Dorsey, J.H., Diggs, E.S. (1979): Microsurgical conization of the cervix by carbon dioxide laser. Obstet. Gynecol. **54**, 565–570

Dotter, C.T., Judkins, M.P. (1964): Transluminal treatment of arteriosclerotic obstruction: description of a new technique and a preliminary report of its application. Circulation **30**, 654–670

Doukas, A.G., Zweig, A.D., Frisoli, J.K., Birngruber, R., Deutsch, T.F. (1991): Noninvasive determination of shock wave pressure generated by optical breakdown. Appl. Phys. **B 53**, 237–245

Du, D., Liu, X., Korn, G., Squier, J., Mourou, G. (1994): Laser-induced breakdown by impact ionization in SiO_2 with pulse widths from 7 ns to 150 fs. Appl. Phys. Lett. **64**, 3071–3073

Dumon, J.F., Reboud, E., Garbe, L., Aucomte, F., Meric, B. (1982): Treatment of tracheobronchial lesions by laser photoresection. Chest **81**, 278–284

Duncavage, J.A., Ossoff, R.H., Toohill, R.J. (1985): Carbon dioxide laser management of laryngeal stenosis. Ann. Otol. Rhinol. Laryngol. **94**, 565–569

Dyson, M., Young, S. (1986): Effect of laser therapy on wound contraction and cellularity in mice. Lasers Med. Sci. **1**, 125–130

Ebeling, K.J. (1978): Zum Verhalten kugelförmiger, lasererzeugter Kavitationsblasen in Wasser. Acustica **40**, 229–239

Eells, J.T., Henry, M.M., Summerfelt, P., Wong-Riley, M.T., Buchmann, E.V., Kane, M., Whelan, N.T., Whelan, H.T. (2003): Therapeutic photobiomodulation for methanol-induced retinal toxicity. Proc. Natl. Acad. Sci. **100**, 3439–3444

Eichler, J., Seiler, T. (1991): Lasertechnik in der Medizin. Springer-Verlag, Berlin, Heidelberg, New York

Eichler, H.J., Albertz, J., Below, F., Kummrow, A., Leitert, T., Kaminskii, A.A., Jakab, L. (1992): Acousto-optic mode locking of 3–μm Er lasers. Appl. Opt. **31**, 4909–4911

Epifanov, A.S. (1981): Theory of electron-avalanche ionization induced in solids by electromagnetic waves. IEEE J. Qu. Electron. **QE-17**, 2018–2022

L'Esperance, F. (1983): Laser trabeculosclerostomy in ophthalmic lasers: photocoagulation, photoradiation, and surgery. C.V.Mosby Co., St. Louis

L'Esperance, F.A., Warner, J.W., Telfair, W.B., Yoder, P.R., Martin, C.A. (1989): Excimer laser instrumentation and technique for human corneal surgery. Arch. Ophthalmol. **107**, 131–139

Eyrich, G.K., Bruder, E., Hilfiker, P., Dubno, B., Quick, H.H., Patak, M.A., Gratz, K.W., Sailer, H.F. (2000): Temperature mapping of magnetic resonance-guided laser interstitial thermal therapy (LITT) in lymphangiomas of the head and neck. Lasers Surg. Med. **26**, 467–476

Fankhauser, F., Roussel, P., Steffen, J., van der Zypen, E., Chrenkova, A. (1981): Clinical studies on the efficiency of high power laser radiation upon some structures of the anterior segment of the eye. Int. Ophthalmol. **3**, 129–139

Fantes, F.E., Waring, G.O. (1989): Effect of excimer laser radiant exposure on uniformity of ablated corneal surface. Lasers Surg. Med. **9**, 533–542

Farah, S.G., Azar, D.T., Gurdal, C., Wong, J. (1998): Laser in situ keratomileusis: literature review of a developing technique. J. Cataract Refract. Surg. **24**, 989–1006

Fasano, V.A. (1981): Treatment of vascular malformations of the brain with laser source. Lasers Surg. Med. **1**, 347–356

Fasano, V.A., Urciuoli, R., Ponzio, R.M. (1982): Photocoagulation of cerebral arteriovenous malformations and arterial aneurysms with Nd:YAG and argon laser. Neurosurgery **11**, 754–760

Felix, M.P., Ellis, A.T. (1971): Laser-induced liquid breakdown – a step-by-step account. Appl. Phys. Lett. **19**, 484–486

Fischer, J.P., Dams, J., Götz, M.H., Kerker, E., Loesel, F.H., Messer, C.J., Niemz, M.H., Suhm, N., Bille, J.F. (1994): Plasma-mediated ablation of brain tissue with picosecond laser pulses. Appl. Phys. B 58, 493–499

Fjodorov, S.N., Durnev, V.V. (1979): Operation of dosaged dissection of corneal circular ligament in cases of myopia of a mild degree. Ann. Ophthalmol. 11, 1885–1890

Fleischer, D., Kessler, F., Haye, O. (1982): Endoscopic Nd:YAG laser therapy for carcinoma of the esophagus: a new palliative approach. Am. J. Surg. 143, 280–283

Fleischer, D., Sivak, M.V. (1985): Endoscopic Nd:YAG laser therapy as palliation for esophagastric cancer. Gastroenterology 89, 827–831

Foote, C.S. (1968): Mechanisms of photosensitized oxidation. Science 162, 963–970

Forrer, M., Frenz, M., Romano, V., Altermatt, H.J., Weber, H.P., Silenok, A., Istomyn, M., Konov, V.I. (1993): Bone-ablation mechanism using CO_2 lasers of different pulse duration and wavelength. Appl. Phys. B 56, 104–112

Fradin, D.W., Bloembergen, N., Letellier, J.P. (1973a): Dependence of laser-induced breakdown field strength on pulse duration. Appl. Phys. Lett. 22, 635–637

Fradin, D.W., Yablonovitch, E., Bass, M. (1973b): Confirmation of an electron avalanche causing laser-induced bulk damage at 1.06 µm. Appl. Opt. 12, 700–709

Frame, J.W., Das Gupta, A.R., Dalton, G.A., Rhys Evans, P.H. (1984): Use of the carbon dioxide laser in the management of premalignant lesions of the oral mucosa. J. Laryngol. Otol. 98, 1251–1260

Frame, J.W. (1985): Carbon dioxide laser surgery for benign oral lesions. Br. Dent. J. 158, 125–128

Frank, F., Keiditsch, E., Hofstetter, A., Pensel, J., Rothenberger, K. (1982): Various effects of the CO_2-, the neodymium-YAG-, and the argon-laser irradiation on bladder tissue. Lasers Surg. Med. 2, 89–96

Frentzen, M., Koort, H.-J., Kermani, O., Dardenne, M.U. (1989): Bearbeitung von Zahnhartgeweben mit einem Excimer-Laser – eine in-vitro Studie. Dtsch. Zahnärztl. Z. 44, 431–435

Frentzen, M., Koort, H.-J. (1990): Lasers in dentistry. Int. Dent. J. 40, 323–332

Frentzen, M., Koort, H.-J. (1991): Lasertechnik in der Zahnheilkunde. Dtsch. Zahnärztl. Z. 46, 443–454

Frentzen, M., Koort, H.-J. (1992): Excimer Laser – Grundlagen und mögliche Anwendungen in der Zahnheilkunde. In: Laser in der Zahnmedizin (Eds.: Vahl, J., van Benthem, H.). Quintessenz-Verlag, Berlin, Chicago, London

Frentzen, M., Winkelsträter, C., van Benthem, H., Koort, H.-J. (1994): Bearbeitung der Schmelzoberflächen mit gepulster Laserstrahlung. Dtsch. Zahnärztl. Z. 49, 166–168

Frenz, M., Romano, V., Pratisto, H., Weber, H.P., Altermatt, H.J., Delix, D., Grossenbacher, R. (1994): Stapedotomie: neue experimentelle Resultate mit dem Erbium-Laser. In: Aktuelle Probleme der Otorhinolaryngologie ORL 17 (Ed.: Sopko, J.). Verlag Huber AG, Bern

Fujikawa, S., Akamatsu, T. (1980): Effects of the non-equilibrium condensation of vapour on the pressure wave produced by the collapse of a bubble in a liquid. J. Fluid Mech. 97, 481–512

Fung, Y.C. (1981): Biomechanics: mechanical properties of living tissue. Springer-Verlag, Berlin, Heidelberg, New York

Gantz, B.J., Jenkins, H.A., Kishimoto, S., Fish, U. (1982): Argon laser stapedotomy. Ann. Otol. 91, 25–26

Gao, H., Ding, X., Wei, D., Cheng, P., Su, X., Liu, H., Zhang, T. (2011): Endoscopic management of benign tracheobronchial tumors. J. Thorac. Dis. 3, 255–261

284   References

Garden, J.M., Polla, L.L., Tan, O.T. (1988): The treatment of port-wine stains by the pulsed dye laser. Arch. Dermatol. **124**, 889–896

Garrison, B.J., Srinivasan, R. (1985): Laser ablation of organic polymers: microscopic models for photochemical and thermal processes. J. Appl. Phys. **57**, 2909–2914

van Gemert, M.J.C., Welch, A.J., Star, W.M., Motamedi, M., Cheong, W.F. (1987): Tissue optics for a slab geometry in the diffusion approximation. Lasers Med. Sci. **2**, 295–302

van Gemert, M.J.C., Welch, A.J. (1989): Time constants in thermal laser medicine. Lasers Surg. Med. **9**, 405–421

van Gemert, M.J.C., Prahl, S.A., Welch, A.S. (1990): Lichtausbreitung und Streuung in trüben Medien. In: Angewandte Lasermedizin (Eds.: Berlien, H.-P., Müller, G.). Ecomed-Verlag, Landsberg

van Gemert, M.J.C., Welch, A.J., Pickering, J.W., Tan, O.T., Gijsbers, G.H. (1995): Wavelengths for laser treatment of port wine stains and telangiectasia. Lasers Surg. Med. **16**, 147–155

Genovese, W.J., dos Santos, M.T., Faloppa, F., de Souza Merli, L.A. (2010): The use of surgical diode laser in oral hemangioma: a case report. Photomed. Laser Surg. **28**, 147–151

Gertzbein, S.D., de Demeter, D., Cruickshank, B., Kapasouri, A. (1981): The effect of laser-osteotomy on bone healing. Lasers Surg. Med. **1**, 361–373

Gilling, P.J., Cass, C.B., Malcolm, A.R., Fraundorfer, M.R. (1995): Combination holmium and Nd:YAG laser ablation of the prostate: initial clinical experience. J. Endourol. **9**, 151–153

Gilling, P.J., Cass, C.B., Cresswell, M.D., Malcolm, A.R., Fraundorfer, M.R. (1995): The use of the holmium laser in the treatment of benign prostatic hyperplasia. J. Endourol. **10**, 459–461

Gilling, P.J., Fraundorfer, M.R. (1998): Holmium laser prostatectomy: a technique in evolution. Curr. Opin. Urol. **8**, 11–15

Ginsburg, R., Wexler, L., Mitchell, R.S., Profitt, D. (1985): Percutaneous transluminal laser angioplasty for treatment of peripheral vascular disease. Radiology **156**, 619–624

Glick, J. (1981): YAG-laser meniscectomy. Proc. Triannual Meeting Int. Arthrosc. Assoc., Rio de Janeiro.

Goebel, D.G. (1967): Generalized integrating-sphere theory. Appl. Opt. **6**, 125–128

Goldman, L., Hornby, P., Mayer, R., Goldman, B. (1964): Impact of the laser on dental caries. Nature **203**, 417

Gonzales, C., van de Merwe, W.P., Smith, M., Reinisch, L. (1990): Comparison of the erbium-yttrium aluminium garnet and carbon dioxide lasers for in vitro bone and cartilage ablation. Laryngoscope **100**, 14–17

Gossner, L., Ell, C. (1993): Photodynamische Therapie in der Gastroenterologie. In: Angewandte Lasermedizin (Eds.: Berlien, H.-P., Müller, G.). Ecomed-Verlag, Landsberg

Gossner, L., Borrmann, J., Ernst, H., Sroka, R., Hahn, E.G., Ell, C. (1994): Photodynamische Therapie. Lasermedizin **10**, 183–189

Graaff, R., Koelink, M.H., de Mul, F.F.M., Zijlstra, W.G., Dassel, A.C.M., Aarnoudse, J.G. (1993a): Condensed Monte Carlo simulations for the description of light transport. Appl. Opt. **32**, 426–434

Graaff, R., Dassel, A.C.M., Koelink, M.H., de Mul, F.F.M., Aarnoudse, J.G., Zijlstra, W.G. (1993b): Optical properties of human dermis in vitro and in vivo. Appl. Opt. **32**, 435–447

Gravas, S., Bachmann, A., Reich, O., Roehrborn, C.G., Gilling, P.J., de la Rosette, J. (2011): Critical review of lasers in benign prostatic hyperplasia (BPH). BJU Int. **107**, 1030–1043

Green, H., Boll, J., Parrish, J.A., Kochevar, I.E., Oseroff, A.R. (1987): Cytotoxicity and mutagenicity of low intensity, 248 and 193 nm excimer laser radiation in mammalian cells. Canc. Res. **47**, 410–413

Griem, H.R. (1964): Plasma spectroscopy. McGraw-Hill, New York

Groenhuis, R.A.J., Ferwerda, H.A., Ten Bosch, J.J. (1983): Scattering and absorption of turbid materials determined from reflection measurements. Appl. Opt. **22**, 2456–2462

Grundfest, W.S., Litvack, F., Forrester, J.S., Goldenberg, T., Swan, H.J.C., Morgenstern, L., Fishbein, M., McDermid, S., Rider, D.M., Pacala, T.J., Laudenslager, J.B. (1985): Laser ablation of human atherosclerotic plaque without adjacent tissue injury. J. Am. Coll. Cardiol. **5**, 929–933

Grüntzig, A. (1978): Transluminal dilatation of coronary-artery stenosis. Lancet **1**, 263

Grüntzig, A.R., Senning, A., Siegenthaler, W.E. (1979): Nonoperative dilatation of coronary stenosis. N. Engl. J. Med. **301**, 61–68

Grüntzig, A.R., Meier, B. (1983): Percutaneous transluminal coronary angioplasty. The first five years and the future. Int. J. Cardiol. **2**, 319–323

Haina, D., Seipp, W., Landthaler, M., Waidelich, W. (1988): Improvement of therapy results in treatment of port wine stains with the argonlaser. Proc. LASER'87. Springer-Verlag, Berlin, Heidelberg, New York

Hale, G.M., Querry, M.R. (1973): Optical constants of water in the 200-nm to 200-μm wavelength region. Appl. Opt. **12**, 555–563

Halldorsson, T., Langerholc, J. (1978): Thermodynamic analysis of laser irradiation of biological tissue. Appl. Opt. **17**, 3948–3958

Hamblin, M.R., Ferraresi, C., Huang, Y.Y., de Freitas, L.F., Carroll, J.D. (2018): Low-level light therapy: photobiomodulation. SPIE Press Book, Bellingham

Hanke, H., Haase, K.K., Hanke, S., Oberhoff, M., Hassenstein, S., Betz, E., Karsch, K.R. (1991): Morphological changes and smooth muscle cell proliferation after experimental excimer laser treatment. Circulation **83**, 1380–1389

Hanna, K.D., Chastang, J.C., Pouliquen, Y., Renard, G., Asfar, L., Waring, G.O. (1988): Excimer laser keratectomy for myopia with a rotating-slit delivery system. Arch. Ophthalmol. **106**, 245–250

Hanna, K.D., Chastang, J.C., Asfar, L., Samson, J., Pouliquen, Y., Waring, G.O. (1989): Scanning slit delivery system. J. Cataract Refract. Surg. **15**, 390–396

Harty, J.I., Amin, M., Wieman, T.J., Tseng, M.T., Ackerman, D., Broghamer, W. (1989): Complications of whole bladder dihematoporphyrin ether photodynamic therapy. J. Urol. **141**, 1341–1346

Hassan, M.T., Luu, T.T., Moulet, A., Raskazovskaya, O., Zhokov, P., Garg, M., Karpowicz, N., Zheltikov, A.M., Pervak, V., Krausz, F., Goulielmakis, E. (2016): Optical attosecond pulses and tracking the nonlinear response of bound electrons. Nature **530**, 66–70

Hassenstein, S., Hanke, H., Kamenz, J., Oberhoff, M., Hanke, S., Rießen, R., Haase, K.K., Betz, E., Karsch, K.R. (1992): Vascular injury and time course of smooth muscle cell proliferation after experimental holmium laser angioplasty. Circulation **86**, 1575–1583

Hayes, J.R., Wolbarsht, M.L. (1968): Thermal model for retinal damage induced by pulsed lasers. Aerospace Med. **39**, 474–480

Heckmann, U. (1992): CO_2-laser excisional conization: five years experience treating cervical intraepithelial neoplasia. In: Lasers in gynecology (Eds.: Bastert, G., Wallwiener, D.). Springer-Verlag, Berlin, Heidelberg, New York

Heitz-Mayfield, L.J. (2008): Peri-implant diseases: diagnosis and risk indicators. J. Clin. Periodontol. **35**, 292–304

Henriques, F.C. (1947): Studies of thermal injury. V: The predictability and the significance of thermally induced rate processes leading to irreversible epidermal injury. Am. J. Pathol. **23**, 489–502

Henyey, L, Greenstein, J. (1941): Diffuse radiation in the galaxy. Astrophys. J. **93**, 70–83

Hering, P. (1987): Limits of optical fiber systems for pulsed lasers. In: Laser lithotripsy – clinical use and technical aspects (Ed.: Steiner, R.). Springer-Verlag, Berlin, Heidelberg, New York

Hessel, S., Frank, F. (1990): Technical prerequisites for the interstitial thermo-therapy using the Nd:YAG laser. Proc. SPIE **1201**, 233–238

Hetzel, M.R., Nixon, C., Edmondstone, W.M., Mitchell, D.M., Millard, F.J., Nanson, E.M., Woodcock, A.A., Bridges, C.E., Humberstone, A.M. (1985): Laser therapy in 100 tracheobronchial tumours. Thorax **40**, 341–345

Hibst, R., Keller, U. (1989): Experimental studies of the application of the Er:YAG laser on dental hard substances: I. Measurement of the ablation rate. Lasers Surg. Med. **9**, 338–344

Hibst, R., Keller, U. (1991): Removal of dental filling materials by Er:YAG laser radiation. Proc. SPIE **1200**, 120–126

Hoffman, R.M., Macdonald, R., Slaton, J.W., Wilt, T.J. (2003): Laser prostatectomy versus transurethral resection for treating benign prostatic obstruction: a systematic review. J. Urol. **169**, 210–215

Hofmann, R., Hartung, R. (1987): Clinical experience with laser-induced shock wave lithotripsy. In: Laser lithotripsy – clinical use and technical aspects (Ed.: Steiner, R.). Springer-Verlag, Berlin, Heidelberg, New York

Hofstetter, A., Böwering, R., Frank, F., Keiditsch, E., Pensel, J., Rothenberger, K.H., Staehler, G. (1980): Laserbehandlung von Blasentumoren. Dtsch. Med. Wochenschr. **105**, 1442–1444

Hofstetter, A., Pensel, J. (1993): Der Laser in der Urologie. In: Angewandte Laser-medizin (Eds.: Berlien, H.-P., Müller, G.). Ecomed-Verlag, Landsberg

Höh, H. (1990): Neodymium:YAG laser keratotomy. A new method of refractive surgery. Ann. Inst. Barraquer **21**, 23–27

Hohenberger, W., Altendorf, A., Hermanek, P., Gall, F.P. (1986): The laser in gastroenterology: malignant tumors in the lower gastrointestinal tract – therapeutic alternatives. Endoscopy **18**, 47–52

Holinger, L.D. (1982): Treatment of severe subglottic stenosis without tracheotomy: a preliminary report. Ann. Otol. Rhinol. Laryngol. **91**, 407–412

Hombach, V., Waltenberger, J., Voisard, R., Höher, M. (1995): Rezidivstenose nach Koronarangioplastie. Z. Kardiol. **84**, 5–21

Hopkins, J.T., McLoda, T.A., Seegmiller, J.G., Baxter, G.D. (2004): Low-level laser therapy facilitates superficial wound healing in humans: a triple-blind, sham-controlled study. J. Athl. Train. **39**, 223–229

Horch, H.-H., Gerlach, K.L. (1982): CO_2 laser treatment of oral dysplastic precancerous lesions: a preliminary report. Lasers Surg. Med. **2**, 179–185

Horch, H.-H. (1992): Laser in der Mund-Kiefer-Gesichts-Chirurgie. In: Laser in der Zahnmedizin (Eds.: Vahl, J., van Benthem, H.). Quintessenz-Verlag, Berlin, Chicago, London

Horvath, K.A., Smith, W.J., Laurence, R.G., Schoen, F.J., Appleyard, R.F., Cohn, L.H. (1995): Recovery and viability of an acute myocardial infarct after transmyocardial laser revascularization. J. Am. Coll. Cardiol. **25**, 258–263

Horvath, K.A. (2008): Transmyocardial laser revascularization. J. Card. Surg. **23**, 266–276

Hövener, G. (1980): Photokoagulation bei Retinopathia diabetica proliferans. Klin. Mbl. Augenheilk. **176**, 938–949

Huber, J., Hosmann, J., Vytiska-Binstorfer, E. (1989): Laparoscopic surgery for tubal pregnancies utilizing laser. Int. J. Gynecol. Obstet. **29**, 153–157

van de Hulst, H.C. (1957): Light scattering by small particles. Wiley & Sons, New York

van de Hulst, H.C. (1962): A new look at multiple scattering. Tech. Report, NASA Institute for Space Studies, New York

Hunter, J., Leonard, L., Wilson, R., Snider, G., Dixon, J. (1984): Effects of low energy laser on wound healing in a porcine model. Lasers Surg. Med. **3**, 285–290

Hussein, H. (1986): A novel fiberoptic laserprobe for treatment of occlusive vessel disease. Proc. SPIE **605**, 59–66

Inoue, K. (2017): 5-aminolevulinic acid-mediated photodynamic therapy for bladder cancer. Int. J. Urol. **24**, 97–101

Ishimaru, A. (1978): Wave propagation and scattering in random media. Academic Press, New York

Ishimaru, A., Yeh, C. (1984): Matrix representations of the vector-radiative transfer theory for randomly distributed non-spherical particles. J. Opt. Soc. Am. **A 1**, 359–364

Ishimaru, A. (1989): Diffusion of light in turbid media. Appl. Opt. **28**, 2210–2215

Ith, M., Pratisto, H., Altermatt, H.J., Frenz, M., Weber, H.P. (1994): Dynamics of laser-induced channel formation in water and influence of pulse duration on the ablation of biotissue under water with pulsed erbium-laser radiation. Appl. Phys. **B 59**, 621–629

Izatt, J.A., Sankey, N.D., Partovi, F., Fitzmaurice, M., Rava, R.P., Itzkan, I., Feld, M.S. (1990): Ablation of calcified biological tissue using pulsed hydrogen fluoride laser radiation. IEEE J. Qu. Electron. **QE-26**, 2261–2270

Jacques, S.L., Prahl, S.A. (1987a): Modeling optical and thermal distributions in tissue during laser irradiation. Lasers Surg. Med. **6**, 494–503

Jacques, S.L., Alter, C.A., Prahl, S.A. (1987b): Angular dependence of HeNe laser light scattering by human dermis. Lasers Life Sci. **1**, 309–333

Jacquez, J.A., Kuppenheim, H.F. (1955): Theory of the integrating sphere. J. Opt. Soc. Am. **45**, 460–470

Jain, K.K. (1980): Sutureless microvascular anastomosis using a neodymium-yag laser. J. Microsurg. **1**, 436–439

Jain, K.K. (1983): Handbook of laser neurosurgery. Charles Thomas Publications, Springfield

Jain, K.K. (1984a): Sutureless extra-intracranial anastomosis by laser. Lancet **2**, 816–817

Jain, K.K. (1984b): Current status of laser applications in neurosurgery. IEEE J. Qu. Electron. **QE-20**, 1401–1406

Jako, G.J. (1972): Laser surgery of the vocal cords. An experimental study with carbon dioxide lasers on dogs. Laryngoscope **82**, 2204–2216

Jocham, D., Stepp, H., Waidelich, R. (2008): Photodynamic diagnosis in urology: state-of-the-art. Eur. Urol. **53**, 1138–1148

Johnson, F.H., Eyring, H., Stover, B.J. (1974): The theory of rate processes in biology and medicine. Wiley & Sons, New York

Johnson, D.E., Price, R.E., Cromeens, D.M. (1992): Pathologic changes occurring in the prostate following transurethral laser prostatectomy. Lasers Surg. Med. **12**, 254–263

Jongsma, F., Bogaard, A., van Gemert, M., Hennig, J. (1983): Is closure of open skin wounds in rats accelerated by argon laser exposure? Lasers Surg. Med. **3**, 75–80

Joseph, J.H., Wiscombe, W.J., Weinman, J.A. (1976): The delta-Eddington approximation for radiative flux transfer. J. Atmos. Sci. **33**, 2452–2459

Jue, B., Maurice, D.M. (1989): The mechanical properties of the rabbit and human cornea. J. Biomech. **19**, 847–853

Kaplan, I., Giler, S. (1984): CO_2 laser surgery. Springer-Verlag, Berlin, Heidelberg, New York

Karanov, S., Shopova, M., Getov, H. (1991): Photodynamic therapy in gastrointestinal tumors. Lasers Surg. Med. **11**, 395–398

Karsch, K.R., Haase, K.K., Mauser, M., Ickrath, O., Voelker, W., Duda, S., Seipel, L. (1989): Percutaneous coronary excimer laser angioplasty: initial clinical results. Lancet **2**, 647–650

Karsch, K.R., Haase, K.K., Wehrmann, M., Hassenstein, S., Hanke, H. (1991): Smooth muscle cell proliferation and restenosis after stand alone coronary excimer laser angioplasty. J. Am. Coll. Cardiol. **17**, 991–994

Katsuyuki, A., Waseda, T., Ota, H., Azuma, T., Nagasawa, A. (1983): A preliminary study on clinical application of Nd:YAG laser to the face and head. Lasers Surg. Med. **3**, 231–239

Kayano, T., Ochiai, S., Kiyono, K., Yamamoto, H., Nakajima, S., Mochizuki, T. (1989): Effects of Er:YAG laser irradiation on human extracted teeth. J. Stomat. Soc. Jap. **56**, 381–392

Keates, R.H., Pedrotti, L.S., Weichel, H., Possel, W.H. (1981): Carbon dioxide laser beam control for corneal surgery. Ophthalmic Surg. **12**, 117–122

Keijzer, M., Richards-Kortum, R.R., Jacques, S.L., Feld, M.S. (1989): Fluorescence spectroscopy of turbid media: autofluorescence of human aorta. Appl. Opt. **28**, 4286–4292

Keller, U., Hibst, R. (1989): Experimental studies of the application of the Er:YAG laser on dental hard substances: II. Light microscopic and SEM investigations. Lasers Surg. Med. **9**, 345–351

Kelly, J.F., Snell, M.E. (1976): Hematoporphyrin derivative: a possible aid in the diagnosis and therapy of carcinoma of the bladder. J. Urol. **115**, 150–151

Kelly, P.J., Alker, G.J., Goerss, S. (1982): Computer assisted stereotactic laser microsurgery for the treatment of intracranial neoplasma. Neurosurgery **10**, 324–331

Kelman, C.D. (1967): Phaco-emulsification and aspiration: a new technique of cataract extraction. Am. J. Ophthalmol. **64**, 23–35

Kessel, D., Dougherty, T.J. (1983): Porphyrin photosensitization. Plenum Press, New York

Kiefhaber, P., Nath, G., Moritz, K. (1977): Endoscopical control of massive gastrointestinal haemorrhage by irradiation with a high-power neodymium-YAG laser. Prog. Surg. **15**, 140–155

Kiefhaber, P., Huber, F., Kiefhaber, K. (1987): Palliative and preoperative endoscopic neodymium-YAG laser treatment of colorectal carcinoma. Endoscopy **19**, 43–46

Kimel, S., Svaasand, L.O., Milner, T.E., Hammer-Wilson, M., Schell, M.J., Nelson, J.S., Berns, M.W. (1994): Laser photothermolysis of single blood vessels in the chick chorioallantoic membrane (CAM). Proc. SPIE **2077**, 216–226

Kimura, Y., Wilder-Smith, P., Matsumoto, K. (2000): Lasers in endodontics: a review. Int. Endod. J. **33**, 173–185

Kinoshita, S. (1988) Fluorescence of hematoporphyrin in living cells and in solution. J. Photochem. Photobiol. **B 2**, 195–208

Klingenberg, M., Bohris, C., Niemz, M.H., Bille, J.F., Kurek, R., Wallwiener, D. (2000): Multifibre application in laser-induced interstitial thermotherapy under on-line MR control. Lasers Med. Sci. **15**, 6–14

Knorz, M.C., Liermann, A., Seiberth, V., Steiner, H., Wiesinger, B. (1996): Laser in situ keratomileusis to correct myopia of -6.00 to -29.00 diopters. J. Refract. Surg. **12**, 575–584

Kochevar, I.E. (1989): Cytotoxicity and mutagenicity of excimer laser radiation. Lasers Surg. Med. **9**, 440–445

Koechner, W. (1992): Solid-state laser engineering. Springer-Verlag, Berlin, Heidelberg, New York

Koizumi, N., Harada, Y., Minamikawa, T., Tanaka, H., Otsuji, E., Takamatsu, T. (2016): Recent advances in photodynamic diagnosis of gastric cancer using 5-aminolevulinic acid. World J. Gastroenterol. **22**, 1289–1296

Kottler, F. (1960): Turbid media with plane-parallel surfaces. J. Opt. Soc. Am. **50**, 483–490

Kovacs, I.B., Mester, E., Görög, P. (1974): Stimulation of wound healing with laser beam in the rat. Experientia **30**, 1275–1276

Krasnov, M.M. (1973): Laserpuncture of anterior chamber angle in glaucoma. Am. J. Ophthalmol. **75**, 674–678

Kronenberg, P., Somani, B. (2018): Advances in lasers for the treatment of stones—a systematic review. Curr. Urol. Rep. **19**, 45

Kubasova, T., Kovacs, L., Somosy, Z., Unk, P., Kokai, A. (1984): Biological effect of He-Ne laser: investigations on functional and micromorphological alterations of cell membranes, in vitro. Lasers Surg. Med. **4**, 381–388

Kubelka, P., Munk, F. (1931): Ein Beitrag zur Optik der Farbanstriche. Z. Techn. Phys. **12**, 593–601

Kubelka, P. (1948): New contributions to the optics of intensely light-scattering materials. J. Opt. Soc. Am. **38**, 448–457

Kucera, A., Blake, J.R. (1988): Computational modelling of cavitation bubbles near boundaries. In: Computational techniques and applications (Eds.: Noyce, J., Fletcher, C.). North-Holland, Amsterdam

Kuo, R.L., Paterson, R.F., Kim, S.C., Siqueira Jr., T.M., Elhilali, M.M., Lingeman, J.E. (2003): Holmium laser enucleation of the prostate (HoLEP): a technical update. World J. Surg. Oncol. **1**, 6

Kurtz, R.M., Horvath, C., Liu, H.H., Krueger, R.R., Juhasz, T. (1998): Lamellar refractive surgery with scanned intrastromal picosecond and femtosecond laser pulses in animal eyes. J. Refract. Surg. **14**, 541–548

Laatikainen, L., Kohner, E.M., Khourey, D., Blach, R.K. (1977): Panretinal photocoagulation in central vein occlusion: a randomised controlled clinical study. Br. J. Ophthalmol. **61**, 741–753

Labs, J.D., White, R.I., Anderson, J.H., Williams, G.M. (1987): Thermodynamic correlates of hot tip laser angioplasty. Invest. Radiol. **22**, 954–959

Laguna, M.P., Alivizatos, G., de la Rosette, J.J. (2003): Interstitial laser coagulation treatment of benign prostatic hyperplasia: is it to be recommended? J. Endourol. **17**, 595–600

Lam, S. (1994): Photodynamic therapy of lung cancer. Semin. Oncol. **21**, 15–19

Landau, C., Lange, R.A., Hillis, L.D. (1994): Percutaneous transluminal coronary angioplasty. N. Engl. J. Med. **330**, 981–993

Lang, G.K., Schroder, E., Koch, J.W., Yanoff, M., Naumann, G.O.H. (1989): Excimer laser keratoplasty. Part I: basic concepts. Ophthalmic Surg. **20**, 262–265

Lange, R.A., Willard, J.E., Hillis, L.D. (1993): Restenosis: the Achilles heel of coronary angioplasty. Am. J. Med. Sci. **306**, 265–275

Lanzafame, R.J., Rogers, D.W., Naim, J.O., de France, C.A., Ochej, H., Hinshaw, J.R. (1986): Reduction of local tumor recurrence by excision with the CO_2 laser. Lasers Surg. Med. **6**, 439–441

Latina, M., Goode, S., de Kater, A.W., Long, F.H., Deutsch, T.F., Epstein, D.L. (1988): Experimental ab interno sclerostomies using a pulsed-dye laser. Lasers Surg. Med. **8**, 233–240

Latina, M.A., Dobrogowski, M., March, W.F., Birngruber, R. (1990): Laser sclerostomy by pulsed-dye laser and goniolens. Lasers Surg. Med. **10**, 1745–1750

Latina, M.A., de Leon, J.M. (2005): Selective laser trabeculoplasty. Ophthalmol. Clin. North Am. **18**, 409–419

Lauterborn, W. (1972): High-speed photography of laser-induced breakdown in liquids. Appl. Phys. Lett. **21**, 27–29

Lauterborn, W. (1974): Kavitation durch Laserlicht. Acustica **31**, 51–78

Lauterborn, W., Bolle, H. (1975): Experimental investigations of cavitation-bubble collapse in the neighbourhood of a solid boundary. J. Fluid Mech. **72**, 391–399

Lee, G., Ikeda, R.M., Kozina, J., Mason, D.T. (1981): Laser dissolution of atherosclerotic obstruction. Am. Heart J. **102**, 1074–1075

Lee, M.T., Bird, P.S., Walsh, L.J. (2004): Photo-activated disinfection of the root canal: a new role for lasers in endodontics. Aust. Endod. J. **30**, 93–98

Le Grand, Y., El Hage, S.G. (1980): Physiological optics. Springer-Verlag, Berlin, Heidelberg, New York

Lenz, H., Eichler, J. (1984): Endonasale Chirurgie mit dem Argonlaser. Laryng. Rhinol. Otol. **63**, 534–540

Lesinski, S.G., Palmer, A. (1989): Laser for otosclerosis: CO_2 vs. argon and KT-532. Laryngoscope **99**, 1–8

Liesenhoff, T., Bende, T., Lenz, H., Seiler, T. (1989): Abtragen von Zahnhartsubstanzen mit Excimer-Laserstrahlen. Dtsch. Zahnärztl. Z. **44**, 426–430

Limbird, T.J. (1990): Application of laser Doppler technology to meniscal injuries. Clin. Orthop. **252**, 88–91

Lin, T., Chan, C. (1984): Effect of laser microbeam irradiation of the nucleus on the cleavage of mouse eggs in culture. Rad. Res. **98**, 549–560

Lin, F., Xu, Y., Yang, Y. (2014): Comparison of the visual results after SMILE and femtosecond laser-assisted LASIK for myopia. J. Refract. Surg. **30**, 248–254

Lipshitz, I., Man, O., Shemesh, G., Lazar, M., Loewenstein, A. (2001): Laser in situ keratomileusis to correct hyperopic shift after radial keratotomy. J. Cataract Refract. Surg. **27**, 273–276

Lipson, R., Baldes, E. (1961): Hematoporphyrin derivative: a new aid for endoscopic detection of malignant disease. J. Thorac. Cardiovasc. Surg. **42**, 623–629

Liu, Y., Niemz, M.H. (2007): Ablation of femural bone with femtosecond laser pulses – a feasibility study. Lasers Med. Sci. **22**, 171–174

Lochte-Holtgreven, W. (1968): Plasma diagnostics. North-Holland, Amsterdam

Loertscher, H., Mandelbaum, S., Parel, J.M., Parrish, R.K. (1987): Noncontact trephination of the cornea using a pulsed hydrogen fluoride laser. Am. J. Ophthalmol. **104**, 471–475

Lundeberg, T., Hode, L., Zhou, J. (1987): A comparative study of the pain-relieving effect of laser treatment and acupuncture. Acta Phsiol. Scand. **131**, 161–162

Lyons, R.F., Abergel, R.P., White, R.A., Dwyer, R.M., Castel, J.C., Uitto, J. (1987): Biostimulation of wound healing in vivo by a helium-neon laser. Ann. Plast. Surg. **18**, 47–50

Macha, H.-N., Koch, K., Stadler, M., Schumacher, W., Krumhaar, D. (1987): New technique for treating occlusive and stenosing tumours of the trachea and main bronchi: endobronchial irradiation by high dose iridium 192 combined with laser canalisation. Thorax **42**, 511–515

Macha, H.-N. (1991): Die Möglichkeiten der Lasertherapie und ihre Kombination mit der endobronchialen Kleinraumbestrahlung zur Behandlung tracheobronchialer Tumoren. In: Angewandte Lasermedizin (Eds.: Berlien, H.-P., Müller, G.). Ecomed-Verlag, Landsberg

Machemer, R., Laqua, H. (1978): A logical approach to the treatment of massive periretinal proliferation. Ophthalmology **85**, 584–593

Mackey, M.J., Chilton, C.P., Gilling, P.J., Fraundorfer, M., Cresswell, M.D. (1998): The results of holmium laser resection of the prostate. Br. J. Urol. **81**, 518–519

Macleod, I.A., Mills, P.R., MacKenzie, J.F., Joffe, S.N., Russell, R.I., Carter, D.C. (1983): Neodymium yttrium aluminium garnet laser photocoagulation for major haemorrhage from peptic ulcers and single vessels: a single blind controlled study. Br. Med. J. Clin. Res. **286**, 345–348

Macruz, R., Martins, J.R.M., Tupinambas, A.S., Lopes, E.A., Vargas, H., Pena, A.F., Carvalho, V.B., Armelin, E., Delcourt, L.V. (1980): Possibilidades terapeuticas do raio laser em ateromas. Arg. Bras. Cardiol. **34**, 9

Maiman, T. (1960): Optical and microwave-optical experiments in ruby. Phys. Rev. Lett. **4**, 564–566

Maitland, D.J., Walsh, J.T., Prystowsky, J.B. (1993): Optical properties of human gallbladder tissue and bile. Appl. Opt. **32**, 586–590

Malik, Z., Hanania, J., Nitzan, Y. (1990): Bactericidal effects of photoactivated porphyrins – an alternative approach to antimicrobial drugs. J. Photochem. Photobiol. **5**, 281–293

March, W.F., Gherezghiher, T., Koss, M.C., Nordquist, R.E. (1984): Experimental YAG laser sclerostomy. Arch. Ophthalmol. **102**, 1834–1836

March, W.F., Gherezghiher, T., Koss, M.C., Shaver, R.P., Heath, W.D., Nordquist, R.D. (1985): Histologic study of a neodymium-YAG laser sclerostomy. Arch. Ophthalmol. **103**, 860–863

Marchesini, R., Bertoni, A., Andreola, S., Melloni, E., Sichirollo, A.E. (1989): Extinction and absorption coefficients and scattering phase functions of human tissues in vitro. Appl. Opt. **28**, 2318–2324

Marshall, J., Bird, A.C. (1979): A comparative histopathological study of argon and krypton laser irradiations of the human retina. Br. J. Ophthalmol. **63**, 657–668

Marshall, J., Trokel, S., Rothery, S., Schubert, H. (1985): An ultrastructural study of corneal incisions induced by an excimer laser at 193 nm. Ophthalmology **92**, 749–758

Marshall, J., Trokel, S., Rothery, S., Krueger, R.R. (1986): Photoablative reprofiling of the cornea using an excimer laser: photorefractive keratotomy. Lasers Ophthalmol. **1**, 21–48

Marynissen, J.P.A., Jansen, H., Star, W.M. (1989): Treatment system for whole bladder wall photodynamic therapy with in vivo monitoring and control of light dose rate and dose. J. Urol. **142**, 1351–1355

Mathews-Roth, M.M. (1982): Beta-carotenotherapy for erythropoietic protoporphyria and other photosensitivity diseases. In: Science of photomedicine (Eds.: Regan, J.D., Parrish, J.A.). Plenum Press, New York

Mathus-Vliegen, E.M.H., Tytgat, G.N.J. (1990): Analysis of failures and complications of neodymium:YAG laser photocoagulation in gastrointestinal tract tumors. Endoscopy **22**, 17–23

McCaughan, J.S., Bethel, B.H., Johnston, T., Janssen, W. (1985): Effect of low-dose argonlaser irradiation on rate of wound closure. Lasers Surg. Med. **5**, 607–614

McCullough, D.L., Roth, R.A., Babayan, R.K., Gordon, J.O., Reese, J.H., Crawford, E.D., Fuselier, H.A., Smith, J.A., Murchison, R.J., Kaye, K.W. (1993): Transurethral ultrasound-guided laser-induced prostatectomy: national human cooperative study results. J. Urol. **150**, 1607–1611

McDonald, M.B., Frantz, J.M., Klyce, S.D., Beuerman, R.W., Varnell, R., Munnerlyn, C.R., Clapham, T.N., Salmeron, B., Kaufman, H.E. (1990): Central photorefractive keratectomy for myopia. The blind eye study. Arch. Ophthalmol. **108**, 799–808

McHugh, J.D.A., Marshall, J., Capon, M., Tothery, S., Raven, A., Naylor, R.P. (1988): Transpupillary retinal photocoagulation in the eyes of rabbit and human using a diode laser. Lasers Light Ophthalmol. **2**, 125–143

McNicholas, T.A., Carter, S.S.C., Wickham, J.E.A., O'Donoghue, E.P.N. (1988): YAG laser treatment of early carcinoma of the prostate. Br. J. Urol. **61**, 239–243

McNicholas, T.A., Steger, A.C., Bown, S.G., O'Donoghue, N. (1991): Interstitial laser coagulation of the prostate: experimental studies. Proc. SPIE **1421**, 30–35

Meador, W.E., Weaver, W.R. (1979): Diffusion approximation for large absorption in radiative transfer. Appl. Opt. **18**, 1204–1208

Mebust, W.K. (1993): Benign prostatic hypertrophy: standards and guidelines. In: Alternate methods in the treatment of benign prostatic hyperplasia (Eds.: Romas, N.A., Vaughan, E.D.). Springer-Verlag, Berlin, Heidelberg, New York

Meier, R.R., Lee, J.S., Anderson, D.E. (1978): Atmospheric scattering of middle UV radiation from an internal source. Appl. Opt. **17**, 3216–3225

Melcer, J., Chaumette, M.T., Melcer, F. (1987): Dental pulp exposed to CO_2 laser beam. Lasers Surg. Med. **7**, 347–352

Mester, E., Szende, B., Gärtner, P. (1968): Die Wirkung der Laserstrahlen auf den Haarwuchs der Maus. Radiobiol. Radiother. **9**, 621–626

Mester, E., Gyenes, G., Tota, J.G. (1969): Experimentelle Untersuchungen über die Wirkung der Laserstrahlen auf die Wundheilung. Z. Exp. Chirur. **2**, 94–101

Mester, E., Spiry, T., Szende, B., Tota, J.G. (1971): Effect of laser rays on wound healing. Am. J. Surg. **122**, 532–535

Metropolis, N., Ulam, S. (1949): The Monte Carlo method. J. Am. Stat. Assoc. **44**, 335–350

Meuleman, C., d'Hoore, A., van Cleynenbreugel, B., Beks, N., d'Hooghe, T. (2009): Outcome after multidisciplinary CO_2 laser laparoscopic excision of deep infiltrating colorectal endometriosis. Reprod. Biomed. Online **18**, 282–289

Meyer-Schwickerath, G. (1949): Koagulation der Netzhaut mit Sonnenlicht. Ber. Dtsch. Ophthalmol. Ges. **55**, 256

Meyer-Schwickerath, G. (1956): Erfahrungen mit der Lichtkoagulation der Netzhaut und der Iris. Doc. Ophthalmol. **10**, 91–131

Michalak, J., Tzou, D., Funk, J. (2015): HoLEP: the gold standard for the surgical management of BPH in the 21st century. Am. J. Clin. Exp. Urol. **3**, 36–42

Miller, G.E., Sant, A.J. (1958): Incomplete integrating sphere. J. Opt. Soc. Am. **48**, 828–831

Mirohseini, M., Muckerheide, M., Cayton, M.M. (1982): Transventricular revascularization by laser. Lasers Surg. Med. **2**, 187–198

Mirohseini, M., Cayton, M.M., Shelgikar, S., Fisher, J.C. (1986): Clinical report: laser myocardial revascularization. Lasers Surg. Med. **6**, 459–461

Moan, J., Christensen, T. (1981): Photodynamic effects on human cells exposed to light in the presence of hematoporphyrin. Localisation of the active dye. Cancer Lett. **11**, 209–214

Molchanov, A.G. (1970): Avalanche ionization in transparent dielectrics induced by intense light pulses. Sov. Phys. Solid State **12**, 749

Moore, J.H. (1973): Laser energy in orthopedic surgery. Orthoped. Surg. Traumat. Excerpta Medica, Amsterdam

Morelli, J.G., Tan, O.T., Garden, J., Margolis, R., Seki, Y., Boll, J., Carney, J.M., Anderson, R.R., Furumoto, H., Parrish, J.A. (1986): Tunable dye laser (577 nm) treatment of port wine stains. Lasers Surg. Med. **6**, 94–99

Moritz, A., Gutknecht, N., Schoop, U., Goharkhay, K., Doertbudak, O., Sperr, W. (1997a): Irradiation of infected root canals with a diode laser in vivo: results of microbiological examinations. Lasers Surg. Med. **21**, 221–226

Moritz, A., Gutknecht, N., Doertbudak, O., Goharkhay, K., Schoop, U., Schauer, P., Sperr, W. (1997b): Bacterial reduction in periodontal pockets through irradiation with a diode laser: a pilot study. J. Clin. Laser Med. Surg. **15**, 33–37

Mottet, N., Anidjar, M., Bourdon, O., Louis, J.F., Teillac, P., Costa, P., le Duc, A. (1999): Randomized comparison of transurethral electroresection and holmium:YAG laser vaporization for symptomatic benign prostatic hyperplasia. J. Endourol. **13**, 127–130

Mudgett, P.S., Richards, L.W. (1971): Multiple scattering calculations for technology. Appl. Opt. **10**, 1485–1502

Mulser, P., Sigel, R., Witkowski, S. (1973): Plasma production by laser. Phys. Lett. **6 C**, 187–239

Mulvaney, W.P., Beck, C.W. (1968): The laser beam in urology. J. Urol. **99**, 112

Murray, J.E. (1983): Pulsed gain and thermal lensing of Nd:LiYF$_4$. IEEE J. Qu. Electron. **QE-19**, 488–491

Muschter, R., Hofstetter, A., Hessel, S., Keiditsch, E., Rothenberger, K.-H., Scheede, P., Frank, F. (1992): Hi-tech of the prostate: interstitial laser coagulation of benign prostatic hypertrophy. Proc. SPIE **1643**, 25–34

Muschter, R., Hessel, S., Hofstetter, A., Keiditsch, E., Rothenberger, K.-H., Schneede, P., Frank, F. (1993): Die interstitielle Laserkoagulation der benignen Prostatahyperplasie. Urologe **A 32**, 273–281

Nath, G., Gorisch, W., Kiefhaber, P. (1973): First laser endoscopy via a fiber optic transmission system. Endoscopy **5**, 208

Neblett, C.R., Morris, J.R., Thomson, S. (1986): Laser-assisted microsurgical anastomosis. Neurosurgery **19**, 914–934

Nelson, J.S., Orenstein, A., Liaw, L.-H.L., Berns, M.W. (1989a): Mid-infrared erbium:YAG laser ablation of bone: the effect of laser osteotomy on bone healing. Lasers Surg. Med. **9**, 362–374

Nelson, J.S., Orenstein, A., Liaw, L.-H.L., Zavar, R.B., Gianchandani, S., Berns, M.W. (1989b): Ultraviolet 308-nm excimer laser ablation of bone: an acute and chronic study. Appl. Opt. **28**, 2350–2357

Nicolai, H., Semmelmann, M., Rößler, W., Wieland, W.F. (1995): Langzeitergebnisse in der Behandlung von Strikturen des unteren Harntraktes mit dem Holmium-YAG-Laser. Proc. LASERmed'95. Springer-Verlag, Berlin, Heidelberg, New York

Niemz, M.H., Klancnik, E.G., Bille, J.F. (1991): Plasma-mediated ablation of corneal tissue at 1053 nm using a Nd:YLF oscillator/regenerative amplifier laser. Lasers Surg. Med. **11**, 426–431

Niemz, M.H., Hoppeler, T.P., Juhasz, T., Bille, J.F. (1993a): Intrastromal ablations for refractive corneal surgery using picosecond infrared laser pulses. Lasers Light Ophthalmol. **5**, 149–155

Niemz, M.H., Eisenmann, L., Pioch, T. (1993b): Vergleich von drei Lasersystemen zur Abtragung von Zahnschmelz. Schweiz. Monatsschr. Zahnmed. **103**, 1252–1256

Niemz, M.H. (1994a): Investigation and spectral analysis of the plasma-induced ablation mechanism of dental hydroxyapatite. Appl. Phys. **B 58**, 273–281

Niemz, M.H., Loesel, F.H., Fischer, M., Lappe, C., Bille, J.F. (1994b): Surface ablation of corneal tissue using UV, green and IR picosecond laser pulses. Proc. SPIE **2079**, 131–139

Niemz, M.H. (1995a): Threshold dependence of laser-induced optical breakdown on pulse duration. Appl. Phys. Lett. **66**, 1181–1183

Niemz, M.H. (1995b): Cavity preparation with the Nd:YLF picosecond laser. J. Dent. Res. **74**, 1194–1199

Niemz, M.H. (1995c): Evaluation of physical parameters during the plasma-induced ablation of teeth. Proc. SPIE **2323**, 170–178

Niemz, M.H., Herschel, A., Willms, L. (1995d): Application of a picosecond Nd:YLF laser in dentistry. Proc. LASERmed'95. Springer-Verlag, Berlin, Heidelberg, New York

Niemz, M.H. (1998): Ultrashort laser pulses in dentistry – advantages and limitations. Proc. SPIE **3255**, 84–91

Niemz, M.H., Kasenbacher, A., Strassl, M., Bäcker, A., Beyertt, A., Nickel, D., Giesen, A. (2004): Tooth ablation using a CPA-free thin disk femtosecond laser system. Appl. Phys. **B 93**, 269–271

Niemz, M.H. (2018): How schience can help us live in peace – Darwin, Einstein, Whitehead. Universal Publishers, Irvine, 77

Nseyo, U.O., Dougherty, T.J., Boyle, D.G., Potter, W.R., Wolf, R., Huben, R., Pontes, J.E. (1985): Whole bladder photodynamic therapy for transitional cell carcinoma of bladder. Urology **26**, 274–280

Nuss, R.C., Fabian, R.L., Sarkar, R., Puliafito, C.A. (1988): Infrared laser bone ablation. Lasers Surg. Med. **8**, 381–392

O'Brien, S.J., Miller, D.V. (1990): The contact neodymium-yttrium aluminium garnet laser. A new approach to arthroscopic laser surgery. Clin. Orthop. **252**, 310

Overholt, B., Panjehpour, M., Tefftellar, E., Rose, M. (1993): Photodynamic therapy for treatment of early adenocarcinoma in Barrett's esophagus. Gastrointest. Endosc. **39**, 73–76

Pallikaris, I.G., Papatzanaki, M.E., Stathi, E.Z., Frenschock, O., Georgiadis, A. (1990): Laser in situ keratomileusis. Lasers Surg. Med. **10**, 463–468

Pallikaris, I.G., Papatzanaki, M.E., Siganos, D.S., Tsilimbaris, M.K. (1991): A corneal flap technique for laser in situ keratomileusis: human studies. Arch. Ophthalmol. **109**, 1699–1702

Pallikaris, I.G., Siganos, D.S. (1994): Excimer laser in situ keratomileusis and photorefractive keratectomy for correction of high myopia. J Refract. Corneal Surg. **10**, 498–510

Pao-Chang, M., Xiou-Qui, X., Hui, Z., Zheng, L., Rui-Peng, Z. (1981): Preliminary report on the application of CO_2 laser scalpel for operations on the maxillo-facial bones. Lasers Surg. Med. **1**, 375–384

Parkin, J., Dixon, J. (1985): Argon laser treatment of head and neck vascular lesions. Otolaryngol. Head Neck Surg. **93**, 211–216

Parrish, J., Anderson, R. (1983): Considerations of selectivity in laser therapy. In: Cutaneous laser therapy (Eds.: Arndt, K., Noc, J., Rosen, S.). Wiley & Sons, New York

Parrish, J.A., Deutsch, T.F. (1984): Laser photomedicine. IEEE J. Qu. Electron. **QE-20**, 1386–1396

Parsa, P., Jacques, S.L., Nishioka, N.S. (1989): Optical properties of rat liver between 350 and 2200 nm. Appl. Opt. **28**, 2325–2330

Patel, N.P., Clinch, T.E., Weis, J.R., Ahn, C., Lundergan, M.K., Heidenreich, K. (2000): Comparison of visual results in initial and re-treatment laser in situ keratomileusis procedures for myopia and astigmatism. Am. J. Ophthalmol. **130**, 1–11

Pauling, L. (1962): Die Natur der chemischen Bindung. Verlag Chemie, Weinheim

Pedersen, G.D., McCormick, N.J., Reynolds, L.O. (1976): Transport calculations for light scattering in blood. Biophys. J. **16**, 199–207

Pelz, B., Schott, M.K., Niemz, M.H. (1994): Electro-optic mode locking of an erbium:YAG laser with a rf resonance transformer. Appl. Opt. **33**, 364–367

Pensel, J. (1986): Dosimetry of the neodymium-YAG-laser in urological applications. Eur. Urol. **12**, 17–20

Perkins, R.C. (1980): Laser stapedotomy for otosclerosis. Laryngoscope **90**, 228–240

Pfander, F. (1975): Das Knalltrauma. Springer-Verlag, Berlin, Heidelberg, New York

Philipp, C., Poetke, M., Berlien, H.-P. (1992): Klinik und Technik der Laserbehandlung angeborener Gefäßerkrankungen. In: Angewandte Lasermedizin (Eds.: Berlien, H.-P., Müller, G.). Ecomed-Verlag, Landsberg

Pickering, J.W., Moes, C.J., Sterenborg, H.J.C.M., Prahl, S.A., van Gemert, M.J.C. (1992): Two integrating spheres with an intervening scattering sample. J. Opt. Soc. Am. **A 9**, 621–631

Pickering, J.W., Prahl, S.A., van Wieringen, N., Beek, J.F., Sterenborg, H.J.C.M., van Gemert, M.J.C. (1993): Double-integrating-sphere system for measuring the optical properties of tissue. Appl. Opt. **32**, 399–410

Pioch, T., Niemz, M., Mindermann, A., Staehle, H.J. (1994): Schmelzablationen durch Laserimpulse im Pikosekundenbereich. Dtsch. Zahnärztl. Z. **49**, 163–165

Plass, G.N., Kattawar, G.W., Catchings, F.E. (1973): Matrix operator theory of radiative transfer. 1: Rayleigh scattering. Appl. Opt. **12**, 314–329

Pohl, R.W. (1976): Optik und Atomphysik. Springer-Verlag, Berlin, Heidelberg, New York

Pollack, I.P., Patz, A. (1976): Argon laser iridotomy: an experimental and clinical study. Ophthalmic. Surg. **7**, 22–30

Pollak, T.M., Wing, W.F., Gasso, R.J., Chicklis, E.P., Jenssen, J.P. (1982): Laser operation of Nd:YLF. IEEE J. Qu. Electron. **QE-18**, 159–162

Prahl, S.A., van Gemert, M.J.C., Welch, A.J. (1993): Determining the optical properties of turbid media by using the adding-doubling method. Appl. Opt. **32**, 559–568

Pratisto, H., Frenz, M., Ith, M., Romano, V., Felix, D., Grossenbacher, R., Altermatt, H.J., Weber, H.P. (1996): Temperature and pressure effects during erbium laser stapedotomy. Lasers Surg. Med. **18**, 100–108

Prauser, R., Eitner, K., Donnerhacke, K.-H., Fritze, C., Schneider, A. (1991): Endoskopische Lasertherapie villöser and tubulovillöser Adenome des Rektosigmoids. In: Angewandte Lasermedizin (Eds.: Berlien, H.-P., Müller, G.). Ecomed-Verlag, Landsberg

Prendeville, W., Davis, R., Berry, P.G. (1986): A low voltage diathermy loop for taking cervical biopsies: a qualitative comparison with punch biopsy forceps. Br. J. Obstet. Gynaecol. **93**, 773–776

Puliafito, C.A., Steinert, R.F. (1984): Short-pulsed Nd:YAG laser microsurgery of the eye: biophysical considerations. IEEE J. Qu. Electron. **QE-20**, 1442–1448

Puliafito, C.A., Steinert, R.F., Deutsch, T.F., Hillenkamp, F., Dehm, E.J., Adler, C.M. (1985): Excimer laser ablation of the cornea and lens: experimental studies. Ophthalmology **92**, 741–748

Qiu, J., Teichman, J.M., Wang, T., Neev, J., Glickman, R.D., Chan, K.F., Milner, T.E. (2010): Femtosecond laser lithotripsy: feasibility and ablation mechanism. J. Biomed. Opt. **15**, 028001

Quickenden, T.I., Daniels, L.L. (1993): Attempted biostimulation of division in saccharomyces cerevisiae using red coherent light. Proc. ICEMS, 79–82

Rasmussen, R.E., Hammer-Wilson, M., Berns, M.W. (1989): Mutation and sister chromatid exchange induction in chinese hamster ovary (CHO) cells by pulsed excimer laser radiation at 193 nm and 308 nm and continuous UV radiation at 254 nm. Photochem. Photobiol. **49**, 413–418

Ratkay-Traub, I., Juhasz, T., Horvath, C., Suarez, C., Kiss, K., Ferincz, I., Kurtz, R. (2001): Ultra-short pulse (femtosecond) laser surgery: initial use in LASIK flap creation. Ophthalmol. Clin. North Am. **14**, 347–355

Ratkay-Traub, I., Ferincz, I.E., Juhasz, T., Kurtz, R.M., Krueger, R.R. (2003): First clinical results with the femtosecond neodynium-glass laser in refractive surgery. J. Refract. Surg. **19**, 94–103

Rayleigh, Lord (1917): On the pressure developed in a liquid during the collapse of a spherical cavity. Phil. Mag. **34**, 94–98

Ready, J.F. (1971): Effects of high-power laser radiation. Academic Press, New York, 279–283

Rechmann, P., Hennig, T., von den Hoff, U., Kaufmann, R. (1993): Caries selective ablation: wavelength 377 nm versus 2.9 µm. Proc. SPIE **1880**, 235–239

Reid, R., Muller, S. (1980): Tattoo removal by CO_2 laser dermabrasion. Plast. Reconstr. Surg. **65**, 717–728

Reinstein, D.Z., Archer, T.J., Gobbe, M. (2014): Small incision lenticule extraction (SMILE) history, fundamentals of a new refractive surgery technique and clinical outcomes. Eye Vis. **1**, 3

Reynolds, L.O., Johnson, C.C., Ishimaru, A. (1976): Diffuse reflectance from a finite blood medium: applications to the modeling of fiber optic catheters. Appl. Opt. **15**, 2059–2067

Reynolds, L.O., McCormick, N.J. (1980): Approximate two-parameter phase function for light scattering. J. Opt. Soc. Am. **70**, 1206–1212

Rice, M.H., Walsh, J.M. (1957): Equation of state of water to 250 kilobars. J. Chem. Phys. **26**, 824–830

Roggan, A., Müller, G. (1993): Computer simulations for the irradiation planning of LITT. Med. Tech. **4**, 18–24

Roggan, A., Handke, A., Miller, K., Müller, G. (1994): Laser induced interstitial thermotherapy of benign prostatic hyperplasia. Min. Invas. Med. **5**, 55–63

Roggan, A., Albrecht, H., Dörschel, K., Minet, O., Müller, G. (1995a): Experimental set-up and Monte-Carlo model for the determination of optical tissue properties in the wavelength range 330–1100 nm. Proc. SPIE **2323**, 21–36

Roggan, A., Müller, G. (1995b): 2D-computer simulations for real-time irradiation planning of laserinduced interstitial thermotherapy (LITT). Proc. SPIE **2327**, 242–252

Roggan, A., Albrecht, D., Berlien, H.-P., Beuthan, J., Germer, C., Koch, H., Wodrich, W., Müller, G. (1995c): Development of an application-set for intraoperative and percutaneous laserinduced interstitial thermotherapy (LITT). Proc. SPIE **2327**, 253–260

Rol, P., Nieder, P., Dürr, U., Henchoz, P.-D., Fankhauser, F. (1990): Experimental investigations on the light scattering properties of the human sclera. Lasers Light Ophthalmol. **3**, 201–212

Romano, V., Rodriguez, R., Altermatt, H.J., Frenz, M., Weber, H.P. (1994): Bone microsurgery with IR-lasers: a comparative study of the thermal action at different wavelengths. Proc. SPIE **2077**, 87–97

Romanos, G.E., Everts, H., Nentwig, G.H. (2000): Effects of diode and Nd:YAG laser irradiation on titanium discs: a scanning electron microscope examination. J. Periodontol. **71**, 810–815

Romanos, G.E., Gutknecht, N., Dieter, S., Schwarz, F., Crespi, R., Sculean, A. (2009): Laser wavelengths and oral implantology. Lasers Med. Sci. **24**, 961–970

Rosemberg, S. (1983): The use of CO_2-laser in urology. Lasers Surg. Med. **3**, 152

Rosen, E.S. (2001): What is LASIK? J. Cataract Refract. Surg. **27**, 339–340

Rosomoff, H.L., Caroll, F. (1966): Reaction of neoplasm and brain to laser. Arch. Neurol. **1**, 143–148

Roth, R.A., Aretz, H.T. (1991): Transurethral ultrasound-guided laser-induced prostatectomy (TULIP procedure): a canine prostate feasibility study. J. Urol. **146**, 1128–1135

Rothauge, C.F. (1980): Urethroscopic recanalization of urethra stenosis using argon laser. Urology **16**, 158–161

Roussel, R., Fankhauser, F. (1983): Contactglas for use with high powerlaser – geometrical and optical aspects. Int. Ophthalmol. **6**, 183–190

Roynesdal, A.K., Bjornland, T., Barkvoll, P., Haanaes, H.R. (1993): The effect of soft-laser application on postoperative pain and swelling. A double-blind crossover study. Int. J. Oral Maxillofac. Surg. **22**, 242–245

Rutgeerts, P., Vantrappen, G., Broeckaert, L., Janssens, J., Coremans, G., Geboes, K., Schurmans, P. (1982): Controlled trial of YAG laser treatment of upper digestive hemorrhage. Gastroenterology **83**, 410–416

Sacchi, C.A. (1991): Laser-induced electric breakdown in water. J. Opt. Soc. Am. **B 8**, 337–345

Sachse, H. (1974): Zur Behandlung der Harnröhrenstriktur: die transurethrale Schlitzung unter Sicht mit scharfem Messer. Fortschr. Med. **1992**, 12–15

Salomao, M.Q., Wilson, S.E. (2010): Femtosecond laser in laser in situ keratomileusis. J. Cataract Refract. Surg. **36**, 1024–1032

Sanborn, T.A., Faxon, D.P., Kellet, M.A., Ryan, T.J. (1986): Percutaneous coronary laser thermal angioplasty. J. Am. Coll. Cardiol. **8**, 1437–1440

Sander, S., Beisland, H.O., Fossberg, E. (1982): Neodymion YAG laser in the treatment of prostatic cancer. Urol. Res. **10**, 85–86

Sander, R., Poesl, H., Frank, F., Meister, P., Strobel, M., Spuhler, A. (1988): A Nd:YAG laser with a water-guided laser beam – a new transmission system. Gastrointest. Endosc. **34**, 336–338

Sander, R., Poesl, H., Zuern, W., Spuhler, A., Braida, M. (1989): The water jet-guided Nd:YAG laser in the treatment of gastroduodenal ulcer with a visible vessel – a randomized, controlled and prospective study. Endoscopy **21**, 217–220

Sander, R., Poesl, H., Spuhler, A. (1990): Lasertherapie von Blutungen und nicht neoplastischen Stenosen. In: Angewandte Lasermedizin (Eds.: Berlien, H.-P., Müller, G.). Ecomed-Verlag, Landsberg

Scheibner, A., Kenny, G., White, W., Wheeland, R.G. (1990): A superior method of tattoo removal using the Q-switched ruby laser. J. Dermatol. Surg. Oncol. **16**, 1091–1098

Scholz, C., Grothves-Spork, M. (1992): Die Bearbeitung von Knochen mit dem Laser. In: Angewandte Lasermedizin (Eds.: Berlien, H.-P., Müller, G.). Ecomed-Verlag, Landsberg

Schulenburg, W.E., Hamilton, A,M., Black, R.K. (1979): A comparative study of argon laser and krypton laser in the treatment of diabetic optic disc neovascularization. Br. J. Ophthalmol. **63**, 412–417

Schwartz, L.W., Spaeth, G.L. (1980): Argon laser iridotomy in primary angle-closure or pupillary block glaucoma. Lasers Surg. Med. **1**, 153–164

Sea, J., Jonat, L.M., Chew, B.H., Qiu, J., Wang, B., Hoopman, J., Milner, T., Teichman, J.M. (2012): Optimal power settings for Holmium:YAG lithotripsy. J. Urol. **187**, 914–919

Seiler, T., Bende, T., Wollensak, J., Trokel, S. (1988): Excimer laser keratectomy for correction of astigmatism. Am. J. Ophthalmol. **105**, 117–124

Seiler, T., Genth, U. (1994): Zum gegenwärtigen Stand der chirurgischen Myopie-Korrektur. Dtsch. Ärztebl. **91**, 3344–3350

Seipp, W., Haina, D., Seipp, V. (1989): Laser in der Dermatologie. In: Angewandte Lasermedizin (Eds.: Berlien, II.-P., Müllcr, G.). Ecomed-Verlag, Landsberg

Seitz, F. (1949): On the theory of electron multiplication in crystals. Phys. Rev. **76**, 1376–1393

Selzer, P.M., Murphy-Chutorian, D., Ginsburg, R., Wexler, L. (1985): Optimizing strategies for laser angioplasty. Invest. Radiol. **20**, 860–866

Semler, P. (1989): Tumortherapie im Gastrointestinaltrakt mit einer Kombination von Laserstrahl und Bestrahlung in "afterloading"-Technik. In: Angewandte Lasermedizin (Eds.: Berlien, H.-P., Müller, G.). Ecomed-Verlag, Landsberg

Senior, K. (2005): Photodynamic therapy for bladder cancer. Lancet Oncol. **6**, 546

Serruys, P.W., Luijten, H.E., Beat, K.J., Geuskens, R., de Feyter, P.J., van den Brand, M., Reiber, J.H.C., ten Katen, H.J., van Es, G.A., Hugenholtz, P.G. (1988): Incidence of restenosis after successful coronary angioplasty: a time-related phenomenon. Circulation **77**, 361–371

Shafirstein, G., Battoo, A., Harris, K., Baumann, H., Gollnick, S.O., Lindenmann, J., Nwogu, C.E. (2016): Photodynamic therapy of non-small cell lung cancer. Narrative review and future directions. Ann. Am. Thorac. Soc. **13**, 265–275

Shortt, A.J., Bunce, C., Allan, B.D.S. (2006): Evidence for superior efficacy and safety of LASIK over photorefractive keratectomy for correction of myopia. Ophthalmology **113**, 1897–1908

Shumaker, B.P., Hetzel, F.W. (1987): Clinical laser photodynamic therapy in the treatment of bladder carcinoma. Photochem. Photobiol. **46**, 899–901

Siegel, Y.I., Zaidel, L., Hammel, I., Korczak, D., Lindner, A. (1991): Histopathology of benign prostatic hyperplasia after failure of hyperthermia treatment. Br. J. Urol. **68**, 383–386

Siegman, A.E. (1986): Lasers. University Science Books, Mill Valley, California

Sievers, M., Frentzen, M., Kosina, A., Koort, H.-J. (1993): Scaling of root surfaces with lasers – an in vitro study. Proc. SPIE **2080**, 82–87

Simone II, C.B., Friedberg, J.S., Glatstein, E., Stevenson, J.P., Sterman, D.H., Hahn, S.M., Cengel, K.A. (2012): Photodynamic therapy for the treatment of non-small cell lung cancer. J. Thorac. Dis. **4**, 63–75

Sliney, D.H., Wolbarsht, M.L. (1980): Safety with lasers and other optical sources. Plenum Pub. Corp., New York

Smith, D.C., Haught, A.F. (1966): Energy-loss processes in optical-frequency gas breakdown. Phys. Rev. Lett. **16**, 1085–1088

Smith, C., Johansen, E., Vangsness, T., Yamagudi, K., McEleney, E. (1987): Does success of arthroscopic laser surgery in the knee joint warrant its extension to "non-knee" joints? Proc. SPIE **712**, 214–217

Sohn, C., Wallwiener, D., Kurek, R., Hahn, U., Schiesser, M., Bastert, G. (1996): Treatment of the twin-twin transfusion syndrome: initial experience using laser-induced interstitial thermotherapy. Fetal Diagn. Ther. **11**, 390–397

Splinter, R., Cheong, W.-F., van Gemert, M.J.C., Welch, A.J. (1989): In vitro optical properties of human and canine brain and urinary bladder tissues at 633 nm. Lasers Surg. Med. **9**, 37–41

Splinter, R., Svenson, R.H., Littmann, L., Chuang, C.H., Tuntelder, J.R., Thompson, M., Tatsis, G.P., Keijzer, M. (1993): Computer simulated light distributions in myocardial tissues at the Nd-YAG wavelength of 1064 nm. Lasers Med. Sci. **8**, 15–21

Srinivasan, R., Mayne-Banton, V. (1982): Self-developing photoetching of poly (ethylene terephthalate) films by far-ultraviolet excimer laser radiation. Appl. Phys. Lett. **41**, 576–578

Srinivasan, R. (1986a): Ablation of polymers and biological tissue by ultraviolet lasers. Science **234**, 559–565

Srinivasan, R., Braren, B., Dreyfus, R.W., Hadel, L., Seeger, D.E. (1986b): Mechanism of the ultraviolet laser ablation of polymethyl methacrylate at 193 and 248 nm: laser-induced fluorescence analysis, chemical analysis, and doping studies. J. Opt. Soc. Am. **B 3**, 785–791

Stabholz, A., Sahar-Helft, S., Moshonov, J. (2004): Lasers in endodontics. Dent. Clin. North Am. **48**, 809–832

Staehler, G., Hofstetter, A., Gorisch, W., Keiditsch, E., Müssiggang, M. (1976): Endoscopy in experimental urology using an argon-laser beam. Endoscopy **8**, 1–4

Stafford, R.J., Fuentes, D., Elliott, A.A., Weinberg, J.S., Ahrar, K. (2010): Laser-induced thermal therapy for tumor ablation. Crit. Rev. Biomedical. Eng. **38**, 79–100

Stafl, A., Wilkinson, E.J., Mattingly, R.F. (1977): Laser treatment of cervical and vaginal neoplasia. Am. J. Obstet. Gynecol. **128**, 128–136

Steiger, E., Maurer, N., Geisel, G. (1993): The frequency-doubled alexandrite laser: an alternative dental device. Proc. SPIE **1880**, 149–152

Stein, E., Sedlacek, T., Fabian, R.L., Nishioka, N.S. (1990): Acute and chronic effects of bone ablation with a pulsed holmium laser. Lasers Surg. Med. **10**, 384–388

Steiner, W. (1989): Die endoskopische Lasertherapie im oberen Aero-Digestivtrakt. In: Angewandte Lasermedizin (Eds.: Berlien, H.-P., Müller, G.). Ecomed-Verlag, Landsberg

Steinert, R.F., Puliafito, C.A., Trokel, S. (1983): Plasma formation and shielding by three ophthalmic Nd-YAG lasers. Am. J. Ophthalmol. **90**, 973–991

Steinert, R.F., Puliafito, C.A. (1985): The Nd-YAG laser in ophthalmology. W.B. Saunders Co., Philadelphia

Stellar, S., Polanyi, T.G., Bredemeier, H.C. (1970): Experimental studies with the carbon dioxide laser as a neurosurgical instrument. Med. Biol. Eng. **8**, 549–558

Stellar, S. (1984): Present status of laser neurosurgery. Lasers Surg. Med. **4**, 215–220

Stern,, R.H., Sognnaes, R.F. (1964): Laser beam effect on dental hard tissues. J. Dent. Res. **43**, 873

Stern, R.H., Vahl, J., Sognnaes, R. (1972): Lased enamel: ultrastructural observations of pulsed carbon dioxide laser effects. J. Dent. Res. **51**, 455–460

Stern, R.H. (1974): Dentistry and the laser. In: Laser applications in medicine and biology (Ed.: Wolbarsht, M.L.). Plenum Press, New York

Stern, D., Schoenlein, R.W., Puliafito, C.A., Dobi, E.T., Birngruber, R., Fujimoto, J.G. (1989): Corneal ablation by nanosecond, picosecond and femtosecond lasers at 532 nm and 625 nm. Arch. Ophthalmol. **107**, 587–592

Strong, M.S., Jako, G.J., Vaughan, C.W., Healy, G.B., Polanyi, T. (1976): The use of CO_2 laser in otolaryngology: a progress report. Trans. Am. Acad. Ophthalmol. Otolaryngol. **82**, 595–602

Strong, M.S., Vaughan, C.W., Healy, G.B., Shapshay, S.M., Jako, G.J. (1979): Transoral management of localised carcinoma of the oral cavity using the CO_2 laser. Laryngoscope **89**, 897–905

Strube, D., Haina, D., Landthaler, M., Braun-Falco, O., Waidelich, W. (1988): Störeffekte bei Tierversuchen zur Stimulation der Wundheilung mit Laserlicht. Laser Med. Surg. **4**, 15–20

van Stryland, E.W., Soileau, M.J., Smirl, A.L., Williams, W.E. (1981): Pulse-width and focal-volume dependence of laser-induced breakdown. Phys. Rev. **B 23**, 2144–2151

Sutcliffe, E., Srinivasan, R. (1986): Dynamics of UV laser ablation of organic polymer surfaces. J. Appl. Phys. **60**, 3315–3322

Sutton, C., Hill, D. (1990): Laser laparoscopy in the treatment of endometriosis. A 5-year study. Br. J. Obstet. Gynaecol. **97**, 181–185

Svaasand, L.O., Boerslid, T., Oeveraasen, M. (1985): Thermal and optical properties of living tissues. Lasers Surg. Med. **5**, 589–602

Svaasand, L.O., Gomer, C.J., Profio, A.E. (1989): Laser-induced hyperthermia of ocular tumors. Appl. Opt. **28**, 2280–2287

Swain, C.P., Kirkham, J.S., Salmon, P.S., Bown, S.G., Northfield, T.C. (1986): Controlled trial of Nd:YAG laser photocoagulation in bleeding peptic ulcers. Lancet **1**, 1113–1117

Takata, A.N., Zaneveld, L., Richter, W. (1977): Laser-induced thermal damage in skin. Aerospace Med., Rep. SAM-TR-77-38

Tan, O.T., Sherwood, K., Gilchrest, B.A. (1989): Treatment of children with port-wine stains using the flashlamp-pulsed tunable dye laser. N. Engl. J. Med. **320**, 416–421

Tan, A.H., Gilling, P.J., Kennett, K.M., Fletcher, H., Fraundorfer, M.R. (2003): Long-term results of high-power holmium laser vaporization (ablation) of the prostate. BJU Int. **92**, 707–709

Tanigawa, M., Shinohara, T., Nishimura, K., Nagata, K., Ishizuka, M., Nagata, Y. (2010): Purification of helicobacter pylori NCTC 11637 cytochrome bc1 and respiration with D-proline as a substrate. J. Bacteriol. **192**, 1410–1415

von Tappeiner, H. (1900): Über die Wirkung fluorescierender Stoffe auf Infusorien nach Versuchen von O. Raab. Münch. Med. Wochenschr. **47**, 5–13

von Tappeiner, H., Jesionek, A. (1903): Therapeutische Versuche mit fluorescierenden Stoffen. Münch. Med. Wochenschr. **50**, 2042–2051

Taube, S., Piironen, J., Ylipaavalniemi, P. (1990): Helium-neon laser therapy in the prevention of post-operative swelling and pain after wisdom tooth extraction. Proc. Finn. Dent. Soc. **86**, 23–27

Taylor, R.S., Leopold, K.E., Mihailov, S. (1987): Damage measurements of fused silica fibres using long optical pulse XeCl lasers. Opt. Commun. **63**, 26–31

Taylor, D.M., L'Esperance, F.A., Del Pero, R.A., Roberts, A.D., Gigstad, J.E., Klintworth, G., Martin, C.A., Warner, J. (1989): Human excimer laser lamellar keratectomy. Ophthalmology **96**, 654–664

Taylor, C.R., Gange, R.W., Dover, J.S., Flotte, T.J., Gonzalez, E., Michaud, N., Anderson, R.R. (1990): Treatment of tattoos by Q-switched ruby laser. A dose-response study. Arch. Dermatol. **126**, 893–899

Teng, P., Nishioka, N.S., Anderson, R.R., Deutsch, T.F. (1987): Acoustic studies of the role of immersion in plasma-mediated laser ablation. IEEE J. Qu. Electron. **QE-23**, 1845–1852

Terry, A.C., Stark, W.J., Maumenee, H.E., Fagadau, W. (1983): Neodymium-Yag laser for posterior capsulotomy. Am. J. Ophthalmol. **96**, 716–720

Ticho, U., Zauberman, H. (1976): Argon laser application to the angle structures in the glaucoma. Arch. Ophthalmol. **94**, 61–64

Tierstein, P.S., Warth, D.C., Haq, N., Jenkins, M.S., McCowan, L.C., Aubanle-Reidel, P., Morris, N., Ginsburg, R. (1991): High speed rotational atherectomy for patients with diffuse coronary artery disease. J. Am. Coll. Cardiol. **18**, 1694–1701

Tomita, Y., Shima, A. (1986): Mechanism of impulsive pressure generation and damage pit fromation by bubble collapse. J. Fluid Mech. **169**, 535–564

Trauner, K., Nishioka, N., Patel, D. (1990): Pulsed holmium:yttrium-aluminum-garnet (Ho:YAG) laser ablation of fibrocartilage and articular cartilage. Am. J. Sports Med. **18**, 316–320

Trokel, S.L., Srinivasan, R., Braren, B.A. (1983): Excimer laser surgery of the cornea. Am. J. Ophthalmol. **96**, 710–715

Ulrich, F., Nicola, N., Bock, W.J. (1986): A micromanipulator to aid microsurgical removal of intracranial tumors with the Nd:YAG laser. Lasers Med. Sci. **1**, 131–133

Ulrich, F., Dürselen, F., Schober, R. (1988): Long-term investigations of laser-assisted microvascular anastomoses with the 1.318 μm Nd:YAG-laser. Lasers Surg. Med. **8**, 104–107

Vaidyanthan, A., Walker, T.W., Guenther, A.H. (1980): The relative roles of avalanche multiplication and multiphoton absorption in laser-induced damage of dielectrics. IEEE J. Qu. Electron. **QE-16**, 89–93

Vassar, G.J., Chan, K.F., Teichman, J.M., Glickman, R.D., Weintraub, S.T., Pfefer, T.J., Welch, A.J. (1999): Holmium:YAG lithotripsy: photothermal mechanism. J. Endourol. **13**, 181–190

Vassiliadis, A., Christian, H.C., Dedrick, K.G. (1971): Ocular laser threshold investigations. Aerospace Med., Rep. F41609-70-C-0002

Verschueren, R.C.J., Oldhoff, J. (1975): The carbon-dioxide laser, a new surgical tool. Arch. Chirur. Nederl. **27**, 199–207

Ville, Y., Hyett, J., Hecher, K., Nicolaides, K. (1995): Preliminary experience with endoscopic laser surgery for severe twin-twin transfusion syndrome. New Engl. J. Med. **332**, 224–227

Vogel, A., Lauterborn, W. (1988): Time resolved particle image velocimetry applied to the investigation of cavitation bubbles. Appl. Opt. **27**, 1869–1876

Vogel, A., Lauterborn, W., Timm, R. (1989): Optical and acoustic investigation of the dynamics of laser-produced cavitation bubbles near a solid boundary. J. Fluid Mech. **206**, 299–338

Vogel, A., Schweiger, P., Frieser, A., Asiyo, M.N., Birngruber, R. (1990): Intraocular Nd:YAG laser surgery: light-tissue interaction, damage range, and the reduction of collateral effects. IEEE J. Qu. Electron. **QE-26**, 2240–2260

Vogel, A., Busch, S., Jungnickel, K., Birngruber, R. (1994a): Mechanisms of intraocular photodisruption with picosecond and nanosecond laser pulses. Lasers Surg. Med. **15**, 32–43

Vogel, A., Capon, M.R.C., Asiyo-Vogel, M.N., Birngruber, R. (1994b): Intraocular photodisruption with picosecond and nanosecond laser pulses: tissue effects in cornea, lens, and retina. Invest. Ophthalmol. Vis. Sci. **35**, 3032–3044

Wacker, F., Reither, K., Ritz, J., Roggan, A., Germer, C., Wolf, K. (2001): MR-guided interstitial laser-induced thermotherapy of hepatic metastasis combined with arterial blood flow reduction: technique and first clinical results in an open MR system. J. Magn. Reson. Imaging **13**, 31–36

Waller, B.F. (1983): Early and late morphological changes in human coronary arteries after percutaneous transluminal coronary angioplasty. Clin. Cardiol. **6**, 363–372

Wallwiener, D., Kurek, R., Pollmann, D., Kaufmann, M., Schmid, H., Bastert, G., Frank, F. (1994): Palliative therapy of gynecological malignancies by laserinduced interstitial thermotherapy. Lasermedizin **10**, 44–51

Walsh, J.T., Deutsch, T.F. (1989): Er:YAG laser ablation of tissue: measurement of ablation rates. Lasers Surg. Med. **9**, 327–337

Wang, Y., Kong, W.-M., Wu, Y.-M., Wang, J.-D., Zhang, W.-Y. (2014): Therapeutic effect of laser vaporization for vaginal intraepithelial neoplasia following hysterectomy due to premalignant and malignant lesions. J. Obstet. Gynaecol. Res. **40**, 1740–1747

Waring, G.O. III, Lynn, M.J., McDonnell, P.J. (1994): Results of the prospective evaluation of radial keratotomy (PERK) study 10 years after surgery. Arch. Ophthalmol. **112**, 1298–1308

Watson, G.M., Wickham, J.E.A., Mills, T.N., Bown, S.G., Swain, P., Salmon, P.R. (1983): Laser fragmentation of renal calculi. Br. J. Urol. **55**, 613–616

Watson, G.M., Murray, S., Dretler, S.P., Parrish, J.A. (1987): The pulsed dye laser for fragmenting urinary calculi. J. Urol. **138**, 195–198

Weast, R.C. (1981): Handbook of chemistry and physics. CRC Press, Boca Raton

Weichmann, G., Johnson, J. (1971): Laser use in endodontics. A preliminary investigation. Oral Surg. **31**, 416–420

Weichmann, G., Johnson, J., Nitta, L. (1972): Laser use in endodontics. Part II. Oral Surg. **34**, 828–830

Weinberg, W.S., Birngruber, R., Lorenz, B. (1984): The change in light reflection of the retina during therapeutic laser-photocoagulation. IEEE J. Qu. Electron. **QE-20**, 1481–1489

Weishaupt, K.R., Gomer, C.J., Dougherty, T.J. (1986): Identification of singlet oxygen as the toxic agent in photoactivation of a murine tumor. Cancer Res. **36**, 2326–2331

Welch, A.J. (1984): The thermal response of laser irradiated tissue. IEEE J. Qu. Electron. **QE-20**, 1471–1481

Welch, A.J., van Gemert, M.J.C. (1995): Optical-thermal response of laser-irradiated tissue. Plenum Press, New York, London

Westenberg, A., Gilling, P., Kennett, K., Frampton, C., Fraundorfer, M. (2004): Holmium laser resection of the prostate versus transurethral resection of the prostate: results of a randomized trial with 4-year minimum long-term followup. J. Urol. **172**, 616–619

Wharen, R.E., Anderson, R.E. (1984a): The Nd:YAG laser in neurosurgery. Part 1: Laboratory investigations – dose related biological response of neural tissue. J. Neurosurg. **60**, 531–539

Wharen, R.E., Anderson, R.E. (1984b): The Nd:YAG laser in neurosurgery. Part 2: Clinical studies – an adjunctive measure for hemostasis in reactions of arteriovenous malformations. J. Neurosurg. **60**, 540–547

Whipple, T.L. (1981): Applications of the CO_2 laser to arthroscopic meniscectomy in a gas medium. Proc. Triannual Meeting Int. Arthrosc. Assoc., Rio de Janeiro

Whittaker, P., Kloner, R.A., Przyklenk, K. (1993): Laser-mediated transmural myocardial channels do not salvage acutely ischemic myocardium. J. Am. Coll. Cardiol. **22**, 302–309

Wieland, W.F., Nicolai, H., Rössler, W., Hofstädter, F. (1993): Erste Erfahrungen mit dem Holmium-YAG-Laser bei der Behandlung von Harnröhrenstrikturen. Proc. LASERmed'93. Springer-Verlag, Berlin, Heidelberg, New York

Wilder-Smith, P. (1988): The soft laser: therapeutic tool or popular placebo? Oral Surg. Oral Med. Oral Pathol. **66**, 654–658

Williams, J.A., Pearson, G.J., Colles, M.J. (2006): Antibacterial action of photoactivated disinfection (PAD) used on endodontic bacteria in planktonic suspension and in artificial and human root canals. J. Dent. **34**, 363–371

Wilson, B.C., Adam, G. (1983): A Monte Carlo model for the absorption and flux distributions of light in tissue. Med. Phys. **10**, 824–830

Wilson, B.C., Patterson, M.S. (1986): The physics of photodynamic therapy. Phys. Med. Biol. **31**, 327–360

Wilson, S.E. (1990): Excimer laser (193 nm) myopic keratomileusis in sighted and blind human eyes. Refract. Corneal Surg. **6**, 383–385

Wilson, M., Dobson, J., Harvey, W. (1993): Sensitization of streptococcus sanguis to killing by light from a helium/neon laser. Lasers Med. Sci. **8**, 69–73

Wise, J.B., Witter, S.L. (1979): Argon laser therapy for open-angle glaucoma. Arch. Ophthalmol. **97**, 319–322

Wise, J.B. (1987): Ten years results of laser trabeculoplasty – does the laser avoid glaucoma surgery or merely defer it? Eye **1**, 45–50

Wolbarsht, M.L. (1971): Laser applications in medicine and biology. Plenum Press, New York

Wolbarsht, M.L. (1984): Laser surgery: CO_2 or HF. IEEE J. Qu. Electron. **QE-20**, 1427–1432

Worthen, D.M., Wickham, M.G. (1974): Argon laser trabeculotomy. Trans. Am. Acad. Ophthalmol. Otolaryngol. **78**, 371–375

Wright, V.C., Davies, E., Riopelle, M.A. (1983): Laser surgery for cervical intraepithelial neoplasia: principles and results. Am. J. Obstet. Gynecol. **145**, 181–184

Yablonovitch, N., Bloembergen, N. (1972): Avalanche ionization and the limiting parameter of filaments induced by light pulses in transparent media. Phys. Rev. Lett. **29**, 907–910

Yamashita, M. (1984): Picosecond time resolved fluorescence spectroscopy of HpD. IEEE J. Qu. Electron. **QE-20**, 1363–1369

Yang, E., Roberts, C.J., Mehta, J.S. (2016): A review of corneal biomechanics after LASIK and SMILE and the current methods of corneal biomechanical analysis. J. Clin. Exp. Ophthalmol. **6**, 6

Yano, O.J., Bielefeld, M.R., Jeevanadam, V. (1993): Prevention of acute regional ischemia with endocardial laser channels. Ann. Thorac Surg. **56**, 46–53

Yoon, G., Welch, A.J., Motamedi, M., van Gemert, M.C.J. (1987): Development and application of three-dimensional light distribution model for laser irradiated tissue. IEEE J. Qu. Electron. **QE-23**, 1721–1733

Yow, L., Nelson, J.S., Berns, M.W. (1989): Ablation of bone and polymethylmethacrylate by an XeCl (308 nm) excimer laser. Lasers Surg. Med. **9**, 141–147

Zaret, M.M., Breinin, G.M., Schmidt, H., Ripps, H., Siegel, I.M., Solon, L.R. (1961): Ocular lesions produced by an optical maser (laser). Science **134**, 1525

Zhang, X., Shen, P., He, Q., Yin, X., Chen, Z., Gui, H., Shu, K., Tang, Q., Yang, Y., Pan, X., Wang, J., Chen, N., Zeng, H. (2016): Different lasers in the treatment of benign prostatic hyperplasia: a network meta-analysis. Sci. Rep. **6**, 23503

Zhou, J., Taft, G., Huang, C.-P., Murnane, M.M., Kapteyn, H.C., Christov, I.P. (1994): Pulse evaluation in a broad-bandwidth Ti:sapphire laser. Opt. Lett. **19**, 1149–1151

Zitzmann, N.U., Berglundh, T. (2008): Definition and prevalence of peri-implant diseases. J. Clin. Periodontol. **35**, 286–291

Zweig, A.D., Deutsch, T.F. (1992): Shock waves generated by XeCl excimer laser ablation of polyimide. Appl. Phys. **B 54**, 76–82

Zweng, H.C., Flocks, M., Kapany, N.S., Silbertrust, N., Peppers, N.A. (1964): Experimental laser photocoagulation. Am. J. Ophthalmol. **58**, 353–362

Zysset, B., Fujimoto, J.G., Deutsch, T.F. (1989): Time-resolved measurements of picosecond optical breakdown. Appl. Phys. **B 48**, 139–147

Index

Ablation
- pattern, 90, 270
- photoablation, *see* Decomposition
- plasma-induced, 105–127
- plasma-mediated, 107
- thermal, *see* Decomposition
Absorbance, 15, 37
Absorption
- coefficient, 15, 37, 67, 98
- index, 112
- length, 16, 67
- of dyes, 18
- of water, 67
- plasma, *see* Plasma
Adding–doubling, 35, 89
Adenosine triphosphate (ATP), 57
ALA, 55, 213
Albedo, 25, 28, 32, 36
Alexandrite laser, 3
Amalgam, 182, 199–200
Amido black, 18
Amplifier
- regenerative, 267–270
Anastomosis, 223
Angiography, 229
Anisotropy
- coefficient, 23, 37
ANSI classification, *see* Laser
Aorta, 16, 42
ArF laser, 3, 67, 94, 101–104
Argon ion laser, 3, 18, 56, 67, 94
Arndt–Schulz Law, 58
Arrhenius' constant, 81
Arrhenius' equation, 79, 81
Arteriosclerosis, 224
Arthroscopy, 234
Astigmatism, 166, 169, 172
Atherectomy, 225
Atherosclerosis, 224
Attenuation

- coefficient, 25, 27, 32, 37
Autocorrelation, 267
Avalanche, *see* Electron

Backscattering, 15
Balloon dilation, 224
Bandwidth, 4–5, 60
BBO, 95, 124
Beer's law, 15
Benign prostatic hyperplasia (BPH), 214–215
Biostimulation, 56
Bismarck brown, 18
Bladder, 42, 210–211, 213
Blood, 42
- coagulation, 83
- perfusion, 71, 85
Bone, 42, 230–232, 245–248
- ablation, 231, 248
Bowman's membrane, 165–166
Brain, 42, 51, 71, 216–223
- ablation, 217–220
- gray matter, 216
- white matter, 216
Brainstem, 216
Breakdown, *see* Optical breakdown
Breast, 42
Bremsstrahlung, 109
- inverse, 109
Brewster angle, 12
Brewster plate, 267
Brilliant black, 18
Brillouin scattering, *see* Scattering
Bypass surgery, 225

Cancer, 49, 104, 204, 206, 243, 249
Capsulotomy, 128
Carbonization, 41, 60–61, 64, 79–80, 82–83
Caries, 182–183
- diagnosis, 123–125

© Springer Nature Switzerland AG 2019
M. H. Niemz, *Laser-Tissue Interactions*,
https://doi.org/10.1007/978-3-030-11917-1

– therapy, 185–193
Carotenoid protection, 48, 52, 54
Cartilage, 232, 234
Catalyst, 48
Cataract, 159–161, 252–253
Catheter, 83–84
Cavitation, 128–133, 145–149
Cell
– critical temperature, 80
– immobility, 79–80
– necrosis, *see* Necrosis
– nucleus, 104
– oxidation, 48–49
– proliferation, 104, 227
Cerebellum, 216
Cerebrovascular occlusive disease, 223
Cerebrum, 216
Cervical intraepithelial neoplasia
   (CIN), 204, 206–208
Cervix, 204, 206–208
Chorioidea, 154, 156
Chromophore, 16, 47, 102
Clearance, 49
CO_2 laser, 3, 67, 82, 94
Coagulation, 18, 41, 60–62, 74, 79–84,
   156, 205, 217
Coherence, 60
Collagen, 16, 56, 128
– denaturation, *see* Denaturation
Collision
– inelastic, 66, 110, 114–119
Colon, 240
Color
– body, 15
– surface, 15
Computer simulation, 33, 85
Condylomata acuminata, 211
Confocal microscopy, 154, 197, 232
Conization, 208
Cornea, 110, 121, 131, 136, 154,
   164–180
– ablation, 62, 90, 96–99, 119–120, 130,
   168
– absorption, 16
– hazard, 252–253
– refractive surgery, 91, 166–180
Coulomb field, 107
Counterjet, 149–151
Cr:LiSAF laser, 5
Craniotomy, 223
Cryotherapy, 208, 214, 236, 242, 249
Cutis, 235
Cytochrome c oxidase, 57–58

Cytotoxicity, 102–104

Damage
– degree, 81
– irreversible, 80–82, 104
– mechanical, 135, 143, 149–151
– reversible, 82
– thermal, 74–82, 84
Decay
– fluorescence, 52
– non-radiative, 48, 66
– radiative, 48
– thermal, 66
– tooth, *see* Caries
Decomposition
– energy diagram, 93
– photoablative, 90–101
– thermal, 61, 74, 79–80, 100–101
Dehydration, 43
Denaturation, 79–80, 83
Dental alloy, 201
Dentin, 181–182, 193
Depth
– ablation, 90, 95, 98–99
– optical, 26, 28
Dermis, 235
Descemet's membrane, 165–166
DHE, 55
Diagnosis
– of caries, *see* Caries
– of tumor, *see* Tumor
Diathermy, 208
Diencephalon, 216
Diffusion
– approximation, 31–32, 35, 89
– electron, 110, 115–119
– heat, *see* Heat
– length, 32
Diode laser, 3, 56, 67, 82–83, 94, 267
Dispersion, 19, 111, 270
Dissociation energy, 94
DNA, 102, 104
Doppler angiography, 229
Doppler effect, 22
Dumping, 269
Dye laser, 3–4, 18, 141
Dye penetration test, 185–187

Ear, 248
Electrocoagulation, 249
Electron
– avalanche, 109, 115
– diffusion, *see* Diffusion

Electrophoresis, 18, 164
Enamel, 181–182, 193
Endodontics, 198
Endometriosis, 209
Endoscope, 229
Enzyme activity, 79–80
Eosin, 61
Epidermis, 235
Epiphysis, 216
Er:YAG laser, 3, 67–68, 76, 82, 94
Er:YLF laser, 68
Er:YSGG laser, 3, 68, 76
Erythrosine, 18
Esophagus, 240
Excimer laser, 46, 90, 94, 101–104
Exposure limit
– eye, 259
– skin, 259
Eye, 154
– protection, 13, 263

Fiber
– bare, 83
– frosted, 83–84
Fibroblast, 56, 182
Finite differences, 76, 85
Finite elements, 178–180
Fluorescence, 48–49, 52–54, 95, 104
Flux theory, 29–31
Fovea, 155
Franck–Condon principle, 93
Free electron laser, 3, 76
Frequency-doubling, 95, 124, 267
Fresnel's laws, 9, 12

Gallbladder, 42, 240
Gaussian beam, 69, 264
Gingiva, 181–182
Glaucoma, 154, 161–163
Glottis, 244
Greens' function, 73

Half-wave plate, 269
Hardness test, 187–191
Heat
– capacity, 70
– conduction, 70–77
–– equation, 72–73, 76, 85
– conductivity, 71
– convection, 70
– diffusion, 72–77
– effects, 68–69, 79–82
– generation, 68–70, 77
– radiation, 70

– source, 70–72
– transport, 68–77
Helium–cadmium laser, 56
Helium–neon laser, 3, 56, 67, 94, 141
Hematoporphyrin, see HpD
Hematoxylin, 61
Hemoglobin, 16, 66
Hemorrhage, 83, 158, 162, 217, 224, 230
HF* laser, 232
Hg lamp, 102–104
High-frequency rotational coronary
    angioplasty (HFRCA), 225
High-speed photography, 145, 149–150
Histology, 61, 191–192
Ho:YAG laser, 3, 67–68, 94
Holmium laser ablation of the prostate
    (HoLAP), 215
Holmium laser enucleation of the
    prostate (HoLEP), 215
Hot tip, 225–226
HpD, 51
– absorption, 52–53
– concentration, 52
– consistency, 49–51
– energy diagram, 53
– fluorescence, 52
Hydroxyapatite, 61, 124, 182, 230
Hyperopia, 166, 169, 172, 174
Hyperthermia, 79–82
Hypophysis, 216
Hypothalamus, 216

Ileum, 240
Implantology, 202–203
India ink, 18
Inflammation, 57
Integrating sphere, 37–38
– double, 39
Intersystem crossing, 48, 53
Intestine, 240
Intrastromal ablation, 178–180
Inverse Bremsstrahlung, see
    Bremsstrahlung
Ionization, 46, 107–109, 115, 130
Ionization probability, 115–116, 121
Iridotomy, 161–162
Iris, 154, 161–162

Jejunum, 240
Jet formation, 131–133, 149–151

Keratin, 235
Keratoconus, 166
Keratomileusis, 169–175

Kidney, 71, 210
KrF laser, 3, 67, 94, 101–102, 104
Krypton ion laser, 3, 18
Kubelka–Munk
– coefficients, 29, 40
– theory, 29–31, 35, 40–41, 89

Lambert's law, 15, 29, 32, 66, 98
Lambert–Beer law, 16
Larynx, 244
Laser
– bandwidth, see Bandwidth
– chopped pulse, 204–205
– classification scheme, 7, 255–260
– coherence, see Coherence
– continuous wave (CW), 3, 60,
  204–205, 255, 267
– maintenance, 267
– mode locking, 4–5, 76, 108, 136,
  267–269
– polarization, see Polarization
– Q-switch, 5, 108, 136, 267, 271
– superpulse, 204–205
Laser-assisted in situ keratomileusis
  (LASIK), 169, 175–178
– femto–LASIK, 175–176
– smile–LASIK, 176–178
Laser-induced interstitial thermo-
  therapy (LITT), 79, 83–89, 214–215
Laser-welding, 201
Lens, 110, 119, 121, 154, 159–161
Lens fragmentation, 161
Leukoplakia, 199
Lithotripsy, 128, 213–214
LITT, see Laser-induced interstitial
  thermotherapy
Liver, 42, 83, 85–87, 240
Low-intensity laser therapy (LILT), 56
Low-level light therapy (LLLT), 56
Lung, 42, 249

Macula, 154–155
– degeneration, 157–158
Magnetic resonance, 84, 221
Maxwell's equations, 27, 110–111
Medium
– opaque, 15
– transparent, 15
– turbid, 25, 28, 33, 43
Medulla oblongata, 216
Melanin, 16, 66, 235
Melting, 60–61, 65, 79–80, 82
Membrane permeability, 79–80

Meniscectomy, 234
Meniscus, 234
Mesencephalon, 216
Methylene blue, 18, 50
Minimally invasive surgery (MIS), 2,
  83, 128, 153, 223, 245
Mitochondria, 57
Mode locking, see Laser
Monomer, 91–92
Monte Carlo simulation, 33–35, 89
– condensed, 35
MPE value, 258–260, 262–264
mTHPC, 55
Müller matrix, 35
Multi-photon
– absorption, 100
– ionization, 108–109
Muscle, 42, 71
Mutagenesis, 102–104
Myocardial infarction, 224
Myocardium, 42
Myopia, 166, 169, 172–174

NaCl, 107, 119
Naevus flammeus, 236–238
Naphthalocyanin, see Phthalocyanin
Nd:Glass laser, 177
Nd:YAG laser, 3, 56, 67, 82–83, 94, 102,
  262, 267, 271
Nd:YLF laser, 3, 67, 94, 102, 267–271
Necrosis, 2, 52, 79–85, 104
Neoplasia, 204
Neovascularization, 158
Nigrosine, 18
Nitric oxide (NO), 57

OD value, 262–263
Odontoblast, 182
OPO, 76
Optical breakdown, 4, 18, 105–123, 128,
  132–136
Optical penetration depth, 16, 74
Osteotomy, 230
Ovarium, 204
Oxygen
– singlet, 48–49, 52
– triplet, 48–49, 52

Pain relief, 47, 56–60, 183, 223, 241,
  249
Palliative treatment, 241–242
Panretinal coagulation, 157–158
Papilla, 154–155
Pathline portrait, 150–151

Percutaneous transluminal coronary angioplasty (PTCA), 224–225
Perfusion, see Blood
Peri-implant hyperplasia, 202
Peri-implantitis, 202
Phase function, 23–28, 33, 37
– δ–Eddington, 23
– Henyey–Greenstein, 23–25, 34, 37
– Rayleigh–Gans, 23
– Reynolds, 23
Phase transition, 70, 79
Phorbide, 55
Phosphorescence, 48–49
Photoactivated disinfection (PAD), 47–49, 198
Photobiomodulation (PBM), 6, 47, 56–60
Photodecomposition, see Decomposition
Photodynamic diagnosis (PDD), 54
Photodynamic therapy (PDT), 47, 49–55, 213–214, 243
Photokeratitis, 253
Photorefractive keratectomy (PRK), see Keratomileusis
Photosensitizer, 48
Photosynthesis, 47
Photothermolysis, 238
Phthalocyanin, 55
Piezoelectric transducer, 142
Pigment, 16–18, 66, 110
Plaque
– blood vessel, 224–227
– tooth, 49, 182, 198
Plasma
– absorption, 99, 110–113, 136
– analysis, 123–127
– conductivity, 111
– dielectric factor, 111
– electron density, 107–110, 113–127, 133–136
– frequency, 113–114
– generation, 99–105, 113, 130–136
– ionization, 107–110, 130
– length, 135
– lifetime, 117, 122
– recombination, see Recombination
– shielding, 99, 113, 134
– spark, 105, 122–123
– spectrum, 123–127
– temperature, 109, 123–127, 133
– threshold, 99, 107, 114–123, 131
PMMA, 91–92, 101

Pockels effect, 5, 269
Polarization, 11, 35, 60
Polarized microscopy, 191–192
Polyimide, 91
Polymer, 91–92
Pons, 216
Porphyrin, 49
Port wine stain, see Naevus flammeus
Posterior capsulotomy, 159–160
Probe beam experiment, 141–143, 146–147
Prostate, 42, 83, 210
Protein, 16–18, 66, 104
– denaturation, see Denaturation
Pulp, 181–182, 196, 198
Pulse
– femtosecond, 4, 108, 123
– microsecond, 4, 75–77
– nanosecond, 75, 90, 108, 119, 128, 135, 141
– picosecond, 4, 75, 107–108, 117, 128, 135, 141, 267–270
Pupil, 161, 175
PVDF, 142–144
Pyrimidine, 102

Q-switch, see Laser
Quarter-wave plate, 269

Radial keratotomy, 166
Radial keratotomy (RK), 169
Radiation
– infrared (IR), 66, 100, 251–253
– ionizing, 251
– non-ionizing, 251
– ultraviolet (UV), 7, 66, 90–95, 100–104, 251–253
– visible (VIS), 66, 100, 251–253
Random walk, 33
Rayleigh scattering, see Scattering
Rayleigh's law, 19–21
Reactive black, 18
Reactive oxygen species (ROS), 57
Recanalization, 226–227
Recombination, 105
Rectum, 240
Reflectance, 11–14, 39
Reflection
– angle, 10–12
– diffuse, 10, 39, 260
– specular, 10, 39, 260
– total, 11
Reflectivity, 11

Refraction
- index, 11, 14
Refractive power, 165
Regenerative amplifier, see Amplifier
Relaxation time, see Thermal
  relaxation time
Repetition rate, 77
Restenosis
- arterial, 225, 227
- gastrointestinal, 241
- urethral, 211
Retina, 76, 83, 134, 154–158
- detachment, 157
- hazard, 252–253
- hole, 157
Retinoblastoma, 157–158
Retinopathy, 157–158
Root canal, 181–182, 198
Rotating disk mask, 174
Ruby laser, 2–3, 56, 67

Saha's equation, 126–127
Salpingectomy, 209
Salpingostomy, 209
Scanning slit, 174
Scattering
- angle, 21, 23, 34
- anisotropic, 23, 32
- backward, 23
- Brillouin, 22, 134
- coefficient, 20, 27, 37
- elastic, 19, 22
- first-order, 29, 89
- forward, 23
- inelastic, 19, 22
- isotropic, 23, 32, 83
- Mie, 22–23
- phase function, see Phase function
- Rayleigh, 19–23, 67
Schlemm's canal, 161, 164
Schrödinger function, 93
Sclera, 18, 154, 163–164
Sclerostomy, 18, 163–164
Seeding, 269
Shock wave, 128–145
Similarity relation, 32
Singlet state, 48, 53
SiO_2, 119
Sister chromatid, 102–104
Skin, 42, 71, 235–238
- absorption, 16
- cancer, 252
- erythema, 253

- hazard, 252–253
- wound healing, 56
Slit lamp, 156
Snell's law, 11
Soft laser, 238
Speed
- of light, 11, 111
- of particle, 137
- of shock wave, 137, 144–145
- of sound, 137, 143
Spike, 4
Spinal cord, 223
Stefan–Boltzmann law, 71
Stenosis
- arterial, 224, 227
- gastrointestinal, 241–242
- laryngeal, 244
- urethral, 211
Stereotactic neurosurgery, 221–223
Sterilization, 209, 217
Stokes vector, 35
Stomach, 240
Streptococcus sanguis, 49
Striae of Retzius, 61
Subcutis, 235

Tatrazine, 18
Tattoo, 238–239
Telegraph equation, 111
Temperature
- conductivity, 71
- tissue, see Tissue
Thalamus, 216
Therapeutic window, 16, 29, 52
Thermal birefringence, 271
Thermal penetration depth, 74
Thermal relaxation time, 74–75
Thermionic emission, 108–109
Thermomechanical interaction, 185
Thrombosis, 224
Thyroid gland, 71
Ti:Sapphire laser, 3–5, 194–195
Tissue
- ablation, see Decomposition
- cooling, 79
- damage, see Damage
- in vitro, 43
- in vivo, 43
- inhomogeneity, 10, 16, 43, 91, 95
- measurement of properties, 37–43
- necrosis, see Necrosis
- optical properties, 66, 86
- temperature, 79–81

– thermal properties, 66
Titanium disc, 203
Tooth, 121, 123–127, 143, 181–197
– ablation, 63–65, 78, 105–106,
  119–120, 193
– pain relief, 56
Trabeculoplasty, 162–163
Trabeculum, 154, 162–163
Transition probability, 127
Transmittance, 10, 39
Transmyocardial laser revascularization
  (TMLR), 229
Transport equation, 27, 89
– vector, 35
Transurethral resection (TUR),
  213–214
Transurethral resection of the prostate
  (TURP), 215
Transurethral ultrasound-guided laser-
  induced prostatectomy (TULIP),
  214
Triplet state, 48, 53
Tuba uterina, 204, 209
Tubal pregnancy, 209
Tumor
– diagnosis, 52–54
– necrosis, 2, 52, 83–85
– treatment, 49, 83–84, 158, 211, 213,
  216, 221, 241–244, 249
Twin-twin transfusion syndrome, 209
Type I reaction, 48–49, 52
Type II reaction, 48–49, 52

Ultrasound, 84, 229
Units, 272–273
Ureter, 210–212
Urethra, 210–212
Urethrotomy, 211
Urinary calculi, 128, 213–214
Urinary tract, 210
Uterus, 42, 62, 64, 83, 204

Vagina, 204, 206
Vaginal intraepithelial neoplasia
  (VAIN), 204, 206
Vaporization, 60–61, 63, 77, 79–80, 82,
  205
Vascularization, 83
Vasospasm, 226
Vein occlusion, 157–158
Vitreous body, 119, 121, 154, 158
Vulva, 204, 206
Vulvar intraepithelial neoplasia (VIN),
  204, 206

Water
– absorption, 66–68
– breakdown threshold, 119
– cavitation, 147
– content, 70
– heat conductivity, 71
– ionization probability, 121
– plasma absorption, 114
– reflection, 14
– refraction, 14
– shock wave, 137–145
– thermal penetration depth, 74
– vaporization heat, 77, 79
Wound healing, 47, 56–59, 199, 238

X-ray, 224, 229, 236, 251
XeCl laser, 3, 67, 94, 101–104
XeF laser, 3, 67, 94, 102
Xenon lamp, 1, 4, 269

Zirconium fluoride fiber, 234

About the Author

Prof. Dr. Markolf Niemz studied physics and bio-engineering at the Universities of Frankfurt, Heidelberg, and the University of California at San Diego (UCSD). He received the Master of Science in bioengineering 1989 from UCSD, the Diplom in physics 1989 from Heidelberg University, and the Ph.D. degree in physics 1992 also from Heidelberg University. In 1996, he joined the Wellman Labs of Photomedicine at Harvard Medical School. In 1999, he became head of the Optical Spectroscopy department at Fraunhofer IPM, Institute for Physical Measurement Techniques, in Freiburg.

Since 2000, he has been a director of MABEL, Mannheim Biomedical Engineering Laboratories, — a joint venture of Heidelberg University and the Mannheim University of Applied Sciences. His research is focused on physics and applications of laser–tissue interactions. For these studies, he was awarded the Karl–Freudenberg Prize by the Heidelberg Academy of Sciences.

© Springer Nature Switzerland AG 2019
M. H. Niemz, *Laser-Tissue Interactions*,
https://doi.org/10.1007/978-3-030-11917-1

Printed in the United States
By Bookmasters